Kruse · Stadler (Eds.)
Ambiguity in Mind and Nature

Springer Series in Synergetics

Editor: Hermann Haken

Synergetics, an interdisciplinary field of research, is concerned with the cooperation of individual parts of a system that produces macroscopic spatial, temporal or functional structures. It deals with deterministic as well as stochastic processes.

36 **Temporal Disorder in Human Oscillatory Systems** Editors: L. Rensing, U. an der Heiden, M. C. Mackey
37 **The Physics of Structure Formation** Theory and Simulation Editors: W. Güttinger, G. Dangelmayr
38 **Computational Systems – Natural and Artificial** Editor: H. Haken
39 **From Chemical to Biological Organization** Editors: M. Markus, S. C. Müller, G. Nicolis
40 **Information and Self-Organization** A Macroscopic Approach to Complex Systems By H. Haken
41 **Propagation in Systems Far from Equilibrium** Editors: J. E. Wesfreid, H. R. Brand, P. Manneville, G. Albinet, N. Boccara
42 **Neural and Synergetic Computers** Editor: H. Haken
43 **Cooperative Dynamics in Complex Physical Systems** Editor: H. Takayama
44 **Optimal Structures in Heterogeneous Reaction Systems** Editor: P. J. Plath
45 **Synergetics of Cognition** Editors: H. Haken, M. Stadler
46 **Theories of Immune Networks** Editors: H. Atlan, I. R. Cohen
47 **Relative Information** Theories and Applications By G. Jumarie
48 **Dissipative Structures in Transport Processes and Combustion** Editor: D. Meinköhn
49 **Neuronal Cooperativity** Editor: J. Krüger
50 **Synergetic Computers and Cognition** A Top-Down Approach to Neural Nets By H. Haken
51 **Foundations of Synergetics I** Distributed Active Systems 2nd Edition By A. S. Mikhailov
52 **Foundations of Synergetics II** Complex Patterns By A. S. Mikhailov, A.Yu. Loskutov
53 **Synergetic Economics** Time and Change in Nonlinear Economics By W.-B. Zhang
54 **Quantum Signatures of Chaos** By F. Haake
55 **Rhythms in Physiological Systems** Editors: H. Haken, H. P. Koepchen
56 **Quantum Noise** By C. W. Gardiner
57 **Nonlinear Nonequilibrium Thermodynamics I** Linear and Nonlinear Fluctuation-Dissipation Theorems By R. L. Stratonovich
58 **Self-Organization and Clinical Psychology** Empirical Approaches to Synergetics in Psychology Editors: W. Tschacher, G. Schiepek, E.J. Brunner
59 **Nonlinear Nonequilibrium Thermodynamics II** Advanced Theory By R. L. Stratonovich
60 **Limits of Predictability** Editor: Yu. A. Kravtsov
61 **On Self-Organization** An Interdisciplinary Search for a Unifying Principle Editors: R. K. Mishra, D. Maaß, E. Zwierlein
62 **Interdisciplinary Approaches to Nonlinear Complex Systems** Editors: H. Haken, A. Mikhailov
63 **Inside Versus Outside** Endo- and Exo-Concepts of Observation and Knowledge in Physics, Philosophy and Cognitive Science Editors: H. Atmanspacher, G. J. Dalenoort
64 **Ambiguity in Mind and Nature** Multistable Cognitive Phenomena Editors: P. Kruse, M. Stadler
65 **Modelling the Dynamics of Biological Systems** Nonlinear Phenomena and Pattern Formation Editors: E. Mosekilde, O. G. Mouritsen

Volumes 1–35 are listed at the end of the book

Peter Kruse   Michael Stadler (Eds.)

Ambiguity in Mind and Nature

Multistable Cognitive Phenomena

With 255 Figures

Springer

Dr. Peter Kruse
Professor Dr. Michael Stadler

Institut für Psychologie und Kognitionsforschung und Zentrum für Kognitionswissenschaften
Universität Bremen
Postfach 330440, D-28334 Bremen, Germany

Series Editor:

Professor Dr. Dr. h. c. Hermann Haken

Institut für Theoretische Physik und Synergetik der Universität Stuttgart
D-70550 Stuttgart, Germany and
Center for Complex Systems, Florida Atlantic University
Boca Raton, FL 33431, USA

ISBN-13: 978-3-642-78413-2   e-ISBN-13: 978-3-642-78411-8
DOI: 10.1007/978-3-642-78411-8

Library of Congress Cataloging-in-Publication Data. Ambiguity in mind and nature: multistable cognitive phenomena, with 255 figures / Peter Kruse, Michael Stadler, eds. p. cm. – (Springer series in synergetics; v. 64) Papers originally presented at the International Symposium on Perceptual Multistability and Semantic Ambiguity held at the University of Bremen, Germany, March 22-25, 1993. Includes bibliographical references and indexes. ISBN 978-3-642-78413-2 (Berlin: alk. paper) 1. Cognition–Congresses. 2. Perception–Congresses. 3. Schemas (Psychology)–Congresses. I. Kruse, Peter. II. Stadler, Michael, 1941- . III. International Symposium on Perceptual Multistability (1993: University of Bremen) IV. Title: Multistable cognitive phenomena. V. Series. BF311.A524   1995   153.7–dc20   94-37026

This work is subject to copyright. All rights are reserved, whether the whole or part of the material is concerned, specifically the rights of translation, reprinting, reuse of illustrations, recitation, broadcasting, reproduction on microfilm or in any other way, and storage in data banks. Duplication of this publication or parts thereof is permitted only under the provisions of the German Copyright Law of September 9, 1965, in its current version, and permission for use must always be obtained from Springer-Verlag. Violations are liable for prosecution under the German Copyright Law.

© Springer-Verlag Berlin Heidelberg 1995
Softcover reprint of the hardcover 1st edition 1995

The use of general descriptive names, registered names, trademarks, etc. in this publication does not imply, even in the absence of a specific statement, that such names are exempt from the relevant protective laws and regulations and therefore free for general use.

Typesetting: Camera-ready copy from the editors using a Springer T$_E$X macro package
SPIN: 10122058   55/3140 - 5 4 3 2 1 0 - Printed on acid-free paper

In memoriam
GAETANO KANIZSA
1913 – 1993

Preface

This book presents the invited lectures given at the *International Symposium on Perceptual Multistability and Semantic Ambiguity* held at the University of Bremen, Germany, March 22–25, 1993.

For research on self-organization in brain dynamics and in cognition multistability certainly is a crucial case. Confronted with a constant stimulus pattern perception fluctuates between two or more possible interpretations. The stimulus is a boundary condition to the autonomous process of order formation in cognition. Multistability is a paradigmatic tool for investigating cognitive order formation. Multistability opens a window to self-organized brain dynamics. The scientists dealing with multistable processes come from physics, biology, psychology, linguistics and mathematics. The concepts used for investigation of multistable phenomena on different levels of analysis are mainly synergetics, neural network theory, and gestalt theory. The papers included in this volume reflect the joint effort of the invited scientists. Convergent experimental results and theoretical ideas concerning cognitive multistability are presented and discussed. The circular causality of elements forming global order and of global order influencing elementary dynamics is shown for different levels of analysis. The volume gives a new look on an old cognitive phenomenon.

Unfortunately Gaetano Kanizsa, the great artist and creative phenomenologist, died the week before the beginning of the symposium, which he had planned to attend. His contribution could nonetheless be included in this book. The volume is dedicated to his memory.

The preparation of the volume would not have been possible without the indefatigable help of Petra Kralemann, Ulla Lohmann, Sabine Pfaff, Thomas Fabian, Silke Katterbach and Christina Schlase. Furthermore we wish to thank the Deutsche Forschungsgemeinschaft, Bonn, for the financial support of the symposium.

Our special thanks go to the editor of the Springer Series of Synergetics, Hermann Haken, for his continuous encouragement on the topic of multistability in cognition.

Bremen
January 1994

P. Kruse
M. Stadler

Table of Contents

Introduction
G. Roth 1

I Basic Concepts

The Function of Meaning in Cognitive Order Formation
M. Stadler and P. Kruse 5

Some Basic Concepts of Synergetics with Respect
to Multistability in Perception, Phase Transitions
and Formation of Meaning
H. Haken 23

II Psychophysical and Phenomenological Approaches

Multistability as a Research Tool
in Experimental Phenomenology
G. Kanizsa and R. Luccio 47

The Significance of Perceptual Multistability
for Research on Cognitive Self-Organization
P. Kruse, D. Strüber, and M. Stadler 69

Task, Intention, Context, Globality, Ambiguity:
More of the Same
C. van Leeuwen 85

Multistability – More than just a Freak Phenomenon
A.C. Zimmer 99

**Recognition of Dynamic Patterns
by a Synergetic Computer**
R. Haas, A. Fuchs, H. Haken, E. Horvath, A.S. Pandya,
and J.A.S. Kelso 139

III Attractors in Behavior

**Multistability and Metastability
in Perceptual and Brain Dynamics**
J.A.S. Kelso, P. Case, T. Holroyd, E. Horvath, J. Rączaszek,
B. Tuller, and M. Ding 159

Self-Organizational Processes in Animal Cognition
G. Vetter 185

Dynamic Models of Psychological Systems
H. Schwegler 203

IV Semantic Ambiguity

**Ambiguity in Linguistic Meaning
in Relation to Perceptual Multistability**
W. Wildgen 221

The Emergence of Meaning by Linguistic Problem Solving
A. Vukovich 241

V Models of Multistability

A Synergetic Model of Multistability in Perception
T. Ditzinger and H. Haken 255

**Concepts for a Dynamic Theory of Perceptual Organization:
An Example from Apparent Movement**
G. Schöner and H. Hock 275

**Artificial Neural Networks
and Haken's Synergetic Computer:
A Hybrid Approach for Solving Bistable Reversible Figures**
H.H. Szu, F. Lu, and J.S. Landa 311

Multistability in Geometrical Imagination:
A Network Study
H.-O. Carmesin 337

VI Brain Processes

A Psychophysiological Interpretation of Theta and Delta
Responses in Cognitive Event-Related Potential Paradigms
M. Schürmann, C. Başar-Eroglu, T. Demiralp, and E. Başar 357

Slow Positive Potentials in the EEG
During Multistable Visual Perception
C. Başar-Eroglu, D. Strüber, M. Stadler, P. Kruse, and F. Greitschus   389

Brain Electric Microstates, and Cognitive
and Perceptual Modes
D. Lehmann 407

The Creation of Perceptual Meanings in Cortex
Through Chaotic Itinerancy and Sequential State Transitions
Induced by Sensory Stimuli
W.J. Freeman 421

VII Symmetry Breaking in Nature

Multistability in Molecules and Reactions
P.J. Plath and C. Stadler 441

Perception of Ambiguous Figures: A Qualitative Model
Based on Synergetics and Quantum Mechanics
G. Caglioti 463

Name Index 479

Subject Index 489

Introduction

G. Roth

Institute of Brain Research and Center for Cognitive Sciences,
University of Bremen, D-28334 Bremen, Germany

The research group "Interdisciplinary Cognition Research" which organized the International Symposium "Perceptual Multistability and Semantic Ambiguity" at the University of Bremen tries to bring together a number of disciplines including experimental and theoretical neurobiology and neural network theory, cognitive psychology, learning research and didactics, computer science and robotics. It is very appropriate that this symposium is focused on the central topic of our work, namely the development of a naturalistic theory of meaning, i.e. the origin of semantics in the brain. Until very recently, such an endeavor was seen by almost every scientist and philosopher as vain from the very beginning. The brain was viewed by neurophysiologists and neurochemists as a purely physicochemical system and the processes going on inside the brain as nothing but electrochemical events. What can be measured are action potentials and transmitter release, but no meaning. The behaviorist dogma was that this was sufficient to explain behavior and cognitive acts. On the other hand, psychologists, philosophers and computer scientists believed and to a large extent still believe that meaning or "information" constitutes a domain in itself, with its own laws and phenomena that can be described and understood *independently* of brain processes.

Recently, this view has begun to change for several reasons. One is the development of new tools for measuring both local and global brain activity (such as positron emission tomography and advanced EEG and evoked potential technique) which suggest an often strict parallelism between cognitive operations and brain processes as well as a correlation between brain dysfunctions and cognitive dysfunctions, a parallelism that can no longer be ignored by psychologists. A second reason is the development of new concepts in natural science, particularly in theoretical physics and biosystems research dealing with the phenomena of self-organization, synergetics, chaos theory and the origin of order and ordered functions in complex systems. Multistability, the topic of the present symposium, is one of the most interesting phenomena arising from these systems. A particularly fruitful new field in this framework is the neural network theory. This theory gives evidence that tasks that are commonly considered cognitive by computer scientists can be simulated by different kinds of parallel distributed networks.

However, in the recent past it has been stated that such networks are still very far from natural neuronal networks. Therefore, a primary task is to develop neural networks that include principles related to cognition that have been identified in nervous systems of man and animals. A second and equally important criticism is that those networks are cognitive only insofar as they have been developed and their achievements been interpreted by natural cognitive systems, i.e. by human beings. However, at least in our brain, the domain of physicochemical neuronal events and the domain of semantics coincide; in other words: we are the proof that meaning is constituted by neuronal activity. For us, the solution of this seeming paradox cannot lie in a neurobiological reductionism, which is popular among many of my colleagues from neurobiology. Rather, it can be shown that a naturalistic or physicalistic approach is consistent with the view that meaning is a phenomenal domain of its own that originates from the complex interaction of neurons and assemblies of neurons within the context of *self-evaluation*. Of course, I will not go on and discuss the present state of our hypothesis. I hope that this conference will give us stimulating new ideas about the relationship between the physical and the semantic domain inside the brain and our cognitive system. I am very happy that parallel to this conference there is an exhibition about a personality who in my eyes is one of the greatest scientists and thinkers of this century, Wolfgang Köhler. His concept of "psychoneural identism" represented the seed of what has yet to be developed in greater detail in the future by cognitive neuroscience. As a neurobiologist, I must say that the neuroscience of Köhler's days has played a dubious role with respect to his physicalistic and at the same time non-reductionist view of the brain-mind relationship.

The papers collected in these proceedings give an impression of the success of the conference and bear witness to the important advances that have recently been achieved and to others which are on the horizon.

Part I

Basic Concepts

The Function of Meaning in Cognitive Order Formation

M. Stadler and P. Kruse

Institute of Psychology and Cognition Research and Center for Cognitive Sciences, University of Bremen, D-28334 Bremen, Germany

Abstract: The following theses on the function of meaning in multistable perception are discussed: (1) Every stimulus pattern allows more than one interpretation. Thus, every percept has more than one state of stability. (2) Multistable perception is switching between attractors in the neural network. (3) Meaning is constituted in a neural network by relations between attractors. Meaning does not refer to external things, but it is generated within the system. (4) Meaning is an order parameter of complex brain processes. Its function is the reduction of information and the stabilization of established attractors. (5) Meaning may influence the elementary structure from which it has emerged. It may cause decisions at bifurcation points.

1 Introduction

Multistability has often been judged as a freak phenomenon. Without any stimulus change two or more different phenomena are experienced. The case where there is a stable environment and a working cognitive system is a paradigm case for self-organization in the brain. There is, however, no reason to assume that there is no self-organization in a cognitive system surrounded by a changing environment, but the process of multistability is especially interesting for three reasons:

(1) It allows one to refute all simple stimulus–response theories of cognitive processes;

(2) it is hard to explain by the theory of direct perception and

(3) it is a perfect test case for the investigation of self-organizing processes in the brain.

In this general introduction to the main topic of this volume a more theoretical view will be presented, while the experimental data of our research group will be described in the contributions of Kruse, Strüber & Stadler and Basar-Eroglu et al. in this volume. For the illustration of the theoretical arguments we shall be using the method of experimental phenomenology, which was invented by the gestaltists and refined to a special art by Gaetano

Kanizsa (cf. 1979), who died in his 80th year one week before this symposium took place and to whom this volume is dedicated.

Multistability is a phenomenon that exists on all levels of matter and organization:

- There are multistable physical phenomena like the Bénard instability in fluid dynamics (Haken, 1977) and optical bistability (DelleDonne, Richter & Ross, 1981);
- there is a variety of bistable molecules and chemical reactions as shown by Plath and Stadler, (1994);
- biological processes, especially in highly developed brains often show different stable states (Freeman, 1991) or a continuous change of stable states (Lehmann, 1994) as a response to an unchanging stimulus situation;
- very well known are the various examples of multistable visual and auditory perceptions detected and constructed since the first description of Necker's cube in 1832;
- it is an everyday experience that there are multistabilities of the meaning of words and sentences (Vukovich, 1993; Wildgen, 1994);
- even animal and human thinking can be understood as a sudden restructuring of one and the same problem space leading to a solution (Köhler, 1917; Wertheimer, 1945);
- social processes have been described as bistable concerning for instance the predominance of one or the other animal species: the snowhare/lynx example of McLulich (1937); and the social development was described by Marx and Engels as a sequence of stable formations of society interrupted by highly instable phases of social revolutions;
- and finally scientific progress itself passes phases of instability and change in scientific revolutions – the so called paradigm shifts. Kuhn (1962) used reversible figures in perception as an analogue for these nonlinear changes in the structure of the sciences.

In all these examples of autonomous changes in the non-living and living matter and in the mental and social world we find certain kinds of phase transitions that have much in common and whose characteristics are described in the interdisciplinary research field of synergetics that was founded by Herman Haken (1977, 1983). Two fundamental types of transition between two stable states with an intermediate phase of instability can be distinguished:

Spontaneous reversion: $order_1 \longrightarrow$ instability $\longrightarrow order_2$
Evolution: order $\longrightarrow$ instability $\longrightarrow$ higher order.

Demonstrations of multistability, especially the reversible figures, were not only a favorite toy of the psychologists (see the summary of Kruse, 1988), but have fascinated physicists (Ditzinger & Haken, 1989, 1990, 1994; Caglioti, 1992; Schöner, 1994), neural network theorists (Kawamoto & Anderson, 1985; Kienker et al., 1986; Szu, 1994) and especially artists like Salvadore Dali and Mauritz C. Escher (Fisher, 1967; Nicki, Forestell & Short, 1979). Among the philosophers it was Wittgenstein in his Tractatus (1921) and his

Philosophical Investigations (1952), who paid attention to the importance of reversible figures for the analysis of mind. He used the duck/rabbit-pattern (see Figure 1) to underline the following arguments: there are two meanings of "to see". Figure 1 is in a physical sense a picture but in a subjective sense, he argued, I see this picture as a rabbit or I see this picture as a duck. We can see it as a rabbit or as a duck, i.e. we see it as we interpret it (1949, IIxi).

In the epistemology of radical constructivism the multistability in perception is a crucial case which gives serious reason to doubt, whether knowledge about reality is possible (Stadler & Kruse, 1990).

2 What is multistability in perception?

There is much evidence that every stimulus pattern offers the possibility of more than one interpretation. The perceptual systems, however, have developed certain criteria which favor one interpretation. But there is no certainty that this more probable choice fits reality. Perception gives, if at all, a viable picture of the world and as long as an animal does not fall down a cliff or is eaten by a well camouflaged tiger it may take its interpretation of the world for real. Hens, for instance, often show flight reactions when they are confronted with the trap shown in Figure 2 moving to the right resembling a bird of prey. The same trap moving to the left resembling a goose evokes no reaction (Lorenz, 1939). Hens are better acquainted to geese than with birds of prey, which they observe more seldom, so the symmetry of the bistable pattern of Figure 2 is broken with respect to probability (Schleidt, 1961).

There are many different types of patterns that induce a continuous change of perceptual interpretations, spontaneous reversions or at least bear the possibility of different interpretations:
1) Fluctuations of complex patterns (Figure 3). Patterns like this show directly the activity of the perceptual system in search of order and stability.
2) Figure-ground tristability. Figure 4 shows (i) a simple line, which may be (ii) the boundary of a figure to the right (female body) or (iii) to the left (profile of a face). It is impossible to see the two figures at a time, i.e. the line exerts its boundary function always to one side only (Metzger, 1975b).
3) Multistability of symmetry axes: Every equilateral triangle, for instance, has three symmetry-axes which determine where the triangle is pointing and where its phenomenal base is (Attneave, 1971). Every figure with more than one symmetry axes suffers from this kind of instability because the perceptual system can obviously only realize one at a time.
4) Multistability of 2-dimensional projections of 3-dimensional bodies: The famous Necker-cube belongs together with all other geometric projections to this category (Schröder's staircase, Wundt's ring, Mach's book and many others, see especially van Leeuwen, 1993; Zimmer, 1994).

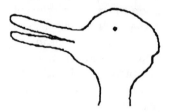

Fig. 1 The duck/rabbit-pattern

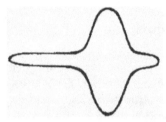

Fig. 2 The trap moving to the right is perceived as a bird of prey; moving to the left it looks like a goose

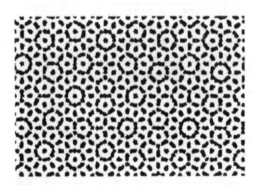

Fig. 3 Multistable pattern. It shows a permanent change of innumerable different structures

5) Multistability of actually 3-dimensional objects: The Ames rotating trapezoidal window is an eminent representative of this category. Most transparent rotating objects switch the direction of movement after some time of observation. Metzger's (1933) rotating rod platform is another example.
6) Multistability of motion direction in apparent movement. The stroboscopic alternative movement introduced by von Schiller (1933) and the circular apparent movement are demonstrated by Kruse et al. (1991, 1994).

Fig. 4 Figure-ground-tristability. A simple line (i) may be seen as the boundary of a figure to the right (ii) or as the border of a figure to the left (iii)

7) Multistability of meaning attribution: The change between duck and rabbit in Figure 1 is the most simple of these demonstrations. Many others are to be found in this volume.

All the examples given above in the seven categories are multistabilities in the visual system. But there are switches and spontaneous reversions in auditory perception, too. Especially in music there are often different structural realizations of one and the same notation (Stadler, Kobs & Reuter, 1993). Spontaneous changes in meaning appear in language perception (Wildgen, 1994; Vukovich, 1994).

What do all these kinds of multistable perception have in common and in what respect do they differ? In the categories (1) to (6) there are always spontaneous reversals. Usually the first reversal takes some time, up to three minutes, but then the reversion rate continuously increases until it is stabilized at a rate that is specific to different personalities. Actually the individual reversion rate seems to be a cognitive-style variable to be called **cognitive flexibility** which includes originality in thinking (Klintman, 1984) and ability to imagine (Kruse et al., 1992). The individual differences of the reversion rates range from about 60/min. to about 1/min. Spontaneous reversions usually do not appear in the multistable patterns of the category (7). An aspect first seen by a person is kept for a long time, maybe forever, if the alternative aspect is not learned. Spontaneous reversions appear seldom, usually the switches are controlled voluntarily. In contrast to this the spontaneous reversions of the patterns of type (1) to (6) are mostly not accessible to voluntary control if perceived under conditions of eye fixation.

It can be said for multistable patterns in general that there are always more than only two aspects. Even if, at the first moment, only two aspects are realized, more aspects appear after some time of observation and even further aspects may be learned. However, usually not all theoretically possible aspects are seen. Especially in patterns that are very complex and for which the possibilities number some power of ten, usually only about four aspects may be perceived (Metzger, 1933; Hoeth, 1966). This is due to the fact that in complex patterns gestalt organization plays a dominant role. By gestalt organization the most simple and regular interpretations are selected and other non-salient interpretations are supressed. Furthermore, we have argued that every pattern, in a way, is multistable, but in everyday life probability constitutes certain criteria for the selection of an aspect of the pattern that is assumed to best fit reality.

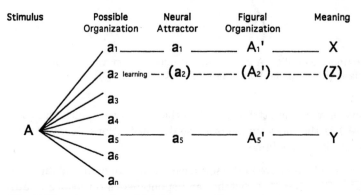

Fig. 5 The stages of multistable perception; see text

Figure 5 gives an overview of the different stages of multistable perception: One stimulus (A) has a certain number of possible organizations $(a_{1,n})$. Only some of these meet a neural attractor (e.g. a_1, a_5) which result in a figural organization (A_1, A_5) that is created by attention. There is switching of the alternatives as a function of the varying strength of the attractors. The experienced alternatives of certain figural organizations may be linked to certain meanings (X, Y). If the meaning switches from one to the other, the figural organization changes, too. Certain organizations that do not come to attention spontaneously may be learned: e.g. a_2, A_2, Z. The concept of multistable perception as switching between attractors will be discussed in the next chapter.

3 Switching between attractors

In mathematics, certain repeated operations always lead to the same eigenvalues. These values are called **attractors**. There exist fix-point attractors, periodic, quasi-periodic and chaotic attractors. In psychophysics we usually find fix-point attractors in perception (Stadler et al., 1991) and periodic attractors in behavior (Vogt, Stadler & Kruse, 1988; Stadler, Vogt & Kruse, 1991). In psychophysics attractors are identified by the method of **serial**

reproduction. The repeated reproduction of patterns always leads to very simple results that do not change thereafter. These attractors are stable states that show **hysteresis**, i.e. they tend to be resistant to further change. Stable states in perception resemble good gestalts (Kanizsa & Luccio, 1990). In psychophysics the attracting forces can even be measured by systematic differences between stimulus and perception. An angle, for instance, exhibits a strong attractor at 90°. Differences to angles of 89° or 91° are very obvious. At a distance, in the range of 30° to 60° an angle is perceived as very instable and changes of about 10° are not even realized. But there is always a systematic bias towards 90°, i.e. angles are always estimated bigger if the stimulus is below 90°, and smaller if the stimulus is above 90°.

In conclusion, attractors are ordered states of high stability surrounded by instability. In neurophysiology, attractors are identified with **coherent oscillations** (Skarda & Freeman, 1987; Gray et al., 1990; Eckhorn & Reitboeck, 1990). These attractors are surrounded temporally and spatially by chaotic activity in the brain. The Singer-group has identified scene segmentations in perception on the neurophysiological level with the coherency of locally distributed oscillations. If there are locally distributed oscillations in phase with one another, then they represent one and the same object.

Multistable perception can be understood as a switching between two or more different attractors. The attractors of the resulting system dynamics can be modelled as a potential landscape, as shown in Figure 6. If the parameter V is changed, the percept becomes more or less stable.

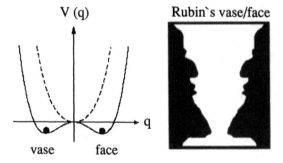

Fig. 6 Potential of a bistable pattern

If attractors in cognitive systems are stable states, and if instability is a prerequisite for the generation or finding of new attractors, then there must be a counteractive process against the tendency to stability. Two kinds of processes may be taken into account for that: **Chaos generation**, which has been demonstrated by Skarda and Freeman (1987) to be a process that precedes the formation of new attractors in the neural network. The second is **self-satiation** of figural processes in the brain as first proposed by Köhler (1940). Although Köhler's hypothesis about the brain processes that evoke self-satiation does not agree with today's knowledge about neural network functioning, his fundamental idea of field-like processes in the brain has gained new interest among brain researchers. The parallel distributed-processing approach of neural networks characterises brain processes in a way that was proposed by Köhler: every local change in the network has an effect on the whole. This is basically the same as expressed in the idea that every stimulus is a system stimulus. In such a general sense, self-satiation, i.e. the self-inhibition of attractors, may be the cause of the permanent change in multistable perceptual situations. In the model of a potential landscape the active attractor basin (one side of the potential landscape of Figure 6) becomes progressively flattened, as long as one figural version is represented. At a certain time the particle will fall into the neighboring attractor, which means a switch to the alternative figure. There the process of self-inhibition starts anew. The concept of flattening attractor basins is well supported by empirical results, as most studies show an acceleration of the switching rate in the first minutes of observation (Köhler 1940; Kruse, Stadler & Strüber, 1991). Ditzinger and Haken's (1989) synergetic model of multistable perception on the basis of self-satiation agrees with these empirical results. On the other hand Hock, Kelso and Schöner (1993) and Schöner (1994) have modelled the same bistable situations without the concept of self-satiation. Schöner assumes only stochastic processes underlying the changes and the hysteresis of the two attractors, which comes closer to the above mentioned concept of active chaos generation. Although there is agreement that multistability is switching between attractors, the question of the cause of the switching can, at the moment, not be decided.

4 What does meaning mean?

In the classic work of Ogden and Richards on "The meaning of meaning" (1923) sixteen different definitions are presented. The most adequate concept for our purpose is the following: meaning is "the place of anything in a system", which is furthermore explained by the authors: "the meaning of anything has been grasped if it has been understood as related to other things or as having its place in some system as a whole" (Ogden & Richards, 1923, p. 196).

This definition is interesting, because it does not refer to any objects or facts outside the system. If meaning is the place of anything in a system, its content is attributed by its surrounding, i.e. the place of other things in the same system. Meaning is constituted by meaningful contexts. Change in the system changes the meanings of anything in this system. This theory agrees very well with the fundamental assumptions of the **radical constructivistic view**. This philosophical approach claims that no semantic information enters the brain or the cognitive system from the outside. From that it follows that meaning must be created in the system itself by self-reference. Therefore it is not possible to refer from cognitive objects to real objects. Real objects surely are the cause of certain stimulus patterns for the sensory systems, but these patterns are, as we have argued, always ambiguous. So the external stimulus patterns are only boundary conditions that stimulate the self-organizing activity of the brain. In other words, cognitive systems are closed systems with respect to semantics and open systems with respect to energy flow.

5 Meanings and attractors

Cognitive systems are, as we have seen, not reacting to external objects and facts, but they develop their own picture of a world, rich in meaningful objects, emotional living beings and goal-directed behavior. How can we understand the way the brain works and how it creates such a rich phenomenal world? Surely, brains are evolving dynamical systems of extremely high complexity. If they are structured at all, their dynamics consists of networks of attractors and repellers that form huge potential landscapes (Amit, 1989). Many authors claim that attractors in the dynamics of the neural network are the places where meanings are located and generated. But how do attractors and meanings come together? To explain this three aspects are discussed:

1) Meanings are attributed to attractors: Amit (1989, p.42) states that "input into an **attractor neural network** (ANN) may, or may not, lead within a biologically reasonable short time, to one of the network's attractors. If it does, it can lead to a biological function which endows it with meaning. Accordingly, we classify stimuli entering the ANN as cognitively meaningful if they lead the network quickly to an attractor. Otherwise, the input is classified as meaningless and ignored. This is the coarse division between the meaningful and the meaningless. It refers to the inputs into the ANN. The specific meaning is in the particular attractor, and it depends, of course, on the level of cognitive processing in which the ANN is situated, i.e., it may be a major retrieval, a recognition of the result of an arithmetical computation. If the unfamiliar stimulus is imposed persistently enough, then it may become a candidate for learning, which is the process of formation of new attractors". This kind of learning is usually called associative learning, which is modelled by Kohonen's (1984) associative neural networks.

Meanings cannot only be attributed to attractors by associative learning but also be detached from attractors. This is exemplified by the well known phenomenon of **lapse of meaning**, which is often explained by semantic satiation (Wertheimer & Gillis, 1958) – the same process that seems to cause multistable switching of meaning. Lapse of meaning is usually observed, when a word or sentence is repeated very often.

Another example of the detachment of meaning is **blindsight**, a neuropsychological disturbance. Patients suffering from this disease are able to see a pattern, even to give a graphic description of it, but are not able to name and recognize an object. The association of the percept with an earlier experience is lacking, i.e. its meaning cannot be established.

2) Meanings are a priori related to attractors: In this view meanings and attractors are identical or, in a weaker formulation, have an isomorphic relation. Zaus (1992) named these preexisting relations **prerepresentations**. An example is the relation between certain figural and phonetic properties as shown in Köhler's famous Maluma and Takete-pattern (Figure 7).

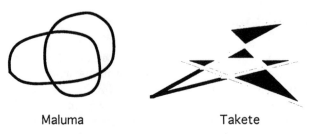

Maluma   Takete

Fig. 7 Common gestalt qualities of words and figures

The links between the figures and their names are gestalt qualities that constitute intermodal relations. A similar case to this, **phonetic symbolism**, is the fact that there is a good chance for subjects to attribute the correct meaning to contrasting notions in foreign languages, even if they belong to another language family (Ertel & Dorst, 1965).

The prerepresentations of attractor neural networks may result in new representations in the course of learning processes. This learning is not arbitrary in direction, as most associative learning is, but it follows certain goals, which are the prerepresented meaningful attractors. Learning in this sense resembles **learning by insight** (Köhler, 1917; Wertheimer, 1945). The trajectories of cognitive processes itinerate until they reach attractors, which represent a necessity for certain meanings (Köhler, 1938). Or in the words of Metzger (1975a): Those things belong together which have a natural relation.

3) Meanings are generated in the attractor neural network, i.e. they emerge from the relations between the attractors. In other words, meanings are a function of the attractor network. In Schwegler's (1992) non-substancialistic network theory there exist relaters and relations. The relaters as the nodes between the relations have no spatial or temporal persistence. The meaning is generated in the relations between these nodes, they are a function of the relations between them. Schwegler's system theory of network relations meets on the one hand the semiotic relation theory of Ch.S. Peirce (cf. Stadler & Wildgen, 1993), and on the other hand von Ehrenfels' theory of gestalt qualities (1890). The latter claims that for instance a melody does not consist of the tones, which are the sensory data for musical experience, but of the intervals between the tones and the relations between the

intervals. So a melody can be transposed in such a way that it remains the same with all the tones being changed. The property of **transposibility** is besides **non-summativity** the criterion for the existence of gestalt qualities. Actually all perceptions apparently obey these two criteria, i.e. they are all relative. Their invariants consist of the relations between the stimuli and if there is need for absolute judgements, the cognitive system constructs its own **frames of reference** from earlier experience (Witte, 1975).

Walter Freeman comes to the same conclusions (see Freeman in this volume) saying that perception depends dominantly on expectation and only marginally on sensory input. "The perceptual process ... allows the brain to 'know' only its own experience of the object and not the 'reality'". Skarda and Freeman (1987) have outlined a theory, how "sense of the world" might be constructed on the basis of chaotic brain activity.

6 Meanings as order parameters

In this chapter we shall assume that meanings are generated as attractors in the neural network and that meanings may have **weak causal effects** on the network process itself. Hermann Haken's synergetics (1977, 1983, 1991) is a formal theory that explains such **top-down influences** in the cognitive system. In synergetics one distinguishes between microscopic and macroscopic states of a system. For instance there are the molecules as constituents of an ideal gas. They may be described by the extent and direction of their motions. But at the same time there is the macroscopic view, that is the temperature of the gas, which is a function of the motions of the molecules, but represents at the same time an entirely new quality, which is found in none of the elements. This is a rather simple example of the emergence of new qualities by scale transition. This concept was applied to the emergence of cognitive processes from elementary brain interactions. Haken described the process of interaction between the microscopic and the macroscopic level in more detail using the same concepts which he first applied to the explanation of laser light: stimulated by a continuous increase of energy input from the sense organs (the so called **control parameter**), the neural elements of the brain system begin to interact in a nonlinear way with each other, comparing and competing with various possible collective behaviors of the structure. Under the condition of a further increase in the control parameter at a certain point, the critical moment of decision approaches. Now the system is in a state of high instability and a minimal fluctuation would suffice to cause a phase transition to a collective, i.e. highly synchronized, behavior. This collective behavior is called an order parameter (Figure 8).

The order parameter, once established, has a backward effect on the activity of all the elements from which is has emerged. This is the so called slaving-process. The collective behavior is now in a state of high stability

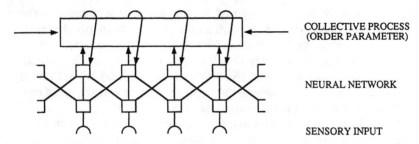

Fig. 8 Emergence of an order parameter in synergetics; see text

and consequently demonstrates hysteresis before change. An attractor has been born. Haken himself (1990) identified order parameters with the phenomenal states of the brain: ... the order parameters represent the meanings of the microscopic activity in the neural network of the brain. The circular causal process of the formation of an ordered pattern from the chaotic elementary brain activity has, as Haken has pointed out, a deep analogy to the process of pattern recognition. So here again we find that the formation of an attractor is analogous to the generation of meaning.

If meaning is an order parameter that influences the microscopic brain activity from which it has emerged, we should be able to confirm this from experience. On the one hand, there is surely the everyday experience that we control our actions towards achieving goals that we have set up on the phenomenal level. But there should also be experimental proof that there exist top-down influences from the meaning and interpretation to the structure of the percepts. These are demonstrated in the contribution of Kruse, Strüber and Stadler (1994). Multistable perceptions are a crucial case for such a type of experiments, for they allow the variation of cognitive processes without any stimulus change. Thus, the effects of self-organized cognitive processes may be demonstrated.

The effects of meaning as an order parameter can also be shown in learning. Many experiments demonstrate that the memory for complex verbal chains is organized by meaning (for a summary see Kruse & Stadler, 1993; Stadler, Kruse & Strüber, 1993). In an experiment that is nearly a hundred years old two teachers of the telegraphic language (Morse code) published very illustrative data confirming our main thesis. They presented to their scholars letters in Morse code that formed no words, furthermore letters that formed words but no sentences and finally letters that formed words that formed sentences as used in normal connected discourse. The data of the learning curves over 8 months are shown for one subject in Figure 9.

The letter-condition without word meanings exhibits a normal linear learning curve. The "word"-curve shows a slight phase transition and the "sentence"-curve a very distinct phase transition, where the meaning of the Morse signs acts as an order parameter increasing the learning effect. The

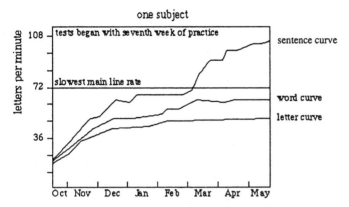

Fig. 9 Learning telegraphic language: Letter, words, sentences (Bryan & Harter 1899)

function of meaning in complex learning processes seems to lie in the reduction of complexity, i.e. the compression of statistical information by the formation of global structures.

The top-down approach to cognitive systems has often been criticized. Especially the classical argument of Höffding (1887) is still brought into the discussion today (Kanizsa & Luccio, 1987, 1990). The argument is as follows: An association of an element b to an element a cannot take place unless b is already recognized as similar to a. That means any identification or interpretation of perceptual data is necessarily bound to the basic perceptual object constitution. When it is possible to show that basic perceptual properties like figure-ground discrimination or direction of movement in multistable situations can be influenced by meaning (Davis, Schiffman & Greist-Bousquet, 1990; Kruse, Strüber & Stadler, 1994) a paradox appears: The identification and cognitive processing of a percept seems to be prior to the precategorial order-formation. This means, as we have pointed out, that the interpretation of a percept is already part of the process of finding its attractor. The paradox – actually very similar to that of Zenon – results from the implicit assumption that perceptual processes take place in discrete steps of cause and effect. However, in the perspective of self-organizing neural networks one can no longer discriminate between single linear causal steps, because causality is assumed to be circular and encompasses the global dynamics as well as the elementary neural interactions. In this sense the process of pattern formation and pattern recognition in the brain cannot be discriminated.

7 Conclusions

Our main thesis was that meaningful cognitive processes may influence elementary brain processes. It is often asked how this might be possible without violating the physical conservation laws. Our hypothesis is that the mental effort influences the brain activity at the moment when it is in a highly instable state. In other words, when the brain process has reached the top of a repeller, the mind may decide into which of a variety of attractor basins it should fall. Thus, from the mental point of view we experience the power to direct our attention to certain details of a percept and to control our actions towards attaining a self-set goal. We experience that our attention is pulled towards the interesting aspect in the perceptual field and that our actions are pulled towards the goals. This is exactly what the attractors of the neural network seem to do. On the other hand, subjective experience is limited to the macroscopic section of the circular causal brain process. We do not experience the elementary nervous interactions which cause the emergence of the macroscopic order parameter. From the brain's point of view, the direction of causality goes from the elementary nervous activity to the macroscopic collective processes.

If it is a good hypothesis that meaning influences brain processes at the most instable points of the brain attractor landscape, then there is a good reason to begin the investigation of the relations between brain and cognitive processes with the problem of multistabilty, as we are doing in this volume. Multistability may be the point of departure for an experimental analysis of the problems of semantics, intentionality, and the mind–brain relation.

References

Amit, D.J. (1989): Modelling Brain Function: The World of Attractor Neural Networks. Cambridge MA: Cambridge University Press.

Attneave, F. (1971): Multistability in perception. Scientific American 225, 62 – 71.

Basar-Eroglu, C., Strüber, D., Stadler, M., Kruse, P. & Greitschus, F. (1994): Slow-positive potentials in the EEG during multistable visual perception. In this volume.

Bryan, W.L. & Harter, N. (1899): Studies on the telegraphic language. The acquisition of a hierarchy of habits. Psychological Review 6, 345 – 375.

Caglioti, G. (1992): The Dynamics of Ambiguity. Berlin: Springer.

Davis, J., Schiffman, H.R. & Greist-Bousquet, S. (1990): Semantic context and figure–ground organization. Psychological Research 52, 306 – 309.

DelleDonne, M., Richter, P.H. & Ross, J. (1981): Dynamic fluctuations in optical bistability. Zeitschrift für Physik B - Condensed Matter 42, 271 – 283.

Ditzinger, T. & Haken, H. (1989): Oscillations in the perception of ambiguous patterns. A model based on synergetics. Biological Cybernetics 61, 279 – 287.

Ditzinger, T. & Haken, H. (1990): The impact of fluctuations on the recognition of ambiguous patterns. Biological Cybernetics 63, 453 – 456.

Ditzinger, T. & Haken, H. (1994): A synergetic model of multistability in perception. In this volume.
Eckhorn, R. & Reitboeck, H.J. (1990): Stimulus-specific synchronization in cat visual cortex and its possible role in visual pattern recognition. In: H. Haken & M. Stadler (Eds.), Synergetics of Cognition, pp. 99 – 111. Berlin: Springer.
Ehrenfels, Chr.von (1890): Über Gestaltqualitäten. Vierteljahresschrift für wissenschaftliche Philosophie 14, 249 – 292.
Ertel, S. & Dorst, R. (1965): Expressive Lautsymbolik. Zeitschrift für experimentelle und angewandte Psychologie, 12, 557 – 569.
Fisher, G.H. (1967): Measuring ambiguity. American Journal of Psychology 80, 541 – 547.
Freeman, W.J. (1991): The physiology of perception. Scientific American 264, 78 – 87.
Freeman, W.J. (1994): The creation of perceptual meanings in cortex through chaotic itinerancy and sequental state transitions induced by sensory stimuli. In this volume.
Gray, C.M., König, P., Engel, A.K. & Singer, W. (1990): Synchronization of oscillatory responses in visual cortex: A plausible mechanism for scene segmentation. In: H. Haken & M. Stadler (Eds.), Synergetics of Cognition, 82 – 98. Berlin: Springer.
Haken, H. (1977): Synergetics - An Introduction. Berlin: Springer.
Haken, H. (1983): Advanced Synergetics. Berlin: Springer.
Haken, H. (1990): Synergetics as a tool for the conceptualization and mathematization of cognition and behavior - how far can we go? In: H. Haken & M. Stadler (Eds.), Synergetics of Cognition, pp. 2 – 31. Berlin: Springer.
Haken, H. (1991): Synergetic Computers and Cognition. Berlin: Springer.
Haken, H. (1994): Some basic concepts of synergetics with respect to multistability in perception, phase transitions and formation of meaning. In this volume.
Hock, H.S., Kelso, J.A.S. & Schöner, G. (1993): Bistability and hysteresis in the organization of apparent motion patterns, Journal of Experimental Psychology: Human Perception and Performance 19, 63 – 80.
Höffding, H. (1887): Psychologie in Umrissen auf der Grundlage der Erfahrung. Leipzig: Fues Verlag.
Hoeth, F. (1966): Gesetzlichkeit bei stroboskopischen Alternativbewegungen. Frankfurt/Main Kramer.
Kanizsa, G. (1979): Organization in Vision. New York: Praeger.
Kanizsa, G. & Luccio, R. (1987): Formation and categorization of visual objects: Höffding's never refuted but always forgotten argument. Gestalt Theory 9, 111 – 127.
Kanizsa, G. & Luccio, R. (1990): The phenomenology of autonomous order formation in perception. In: H. Haken & M. Stadler (Eds.), Synergetics of Cognition, pp. 186 – 200. Berlin: Springer.
Kawamoto, A.H. & Anderson, J.A. (1985): A neural network model of multistable perception. Acta Psychologica 59, 35 – 65.
Kienker, P.K., Sejnowski, T.J., Hinton, G.E. & Schumacher, L.E. (1986): Separating figure from ground with a parallel network. Perception 15, 197 – 216.
Klintman, H. (1984): Orgininal thinking and ambiguous figure reversal rates. Bulletin of the Psychonomic Society 22, 129 – 131.
Köhler, W. (1917): Intelligenzprüfungen an Anthropoiden I. Abhandlungen der preussischen Akademie der Wissenschaften, physikalisch- mathematische Klasse (whole no. 1).
Köhler, W. (1938): The Place of Value in a World of Facts. New York: Liveright.
Köhler, W. (1940): Dynamics in Psychology. New York: Liveright.

Kohonen, T. (1984): Self-Organization and Associative Memory. Berlin: Springer.
Kruse, P. (1988): Stabilität-Instabilität-Multistabilität. Selbstorganisation und Selbstreferentialität in Kognitiven Systemen, Delfin 6, 35 – 57
Kruse, P. & Stadler, M. (1993): The significance of nonlinear phenomena for the investigation of cognitive systems. In: H. Haken & A.S. Mikhailov (Eds.), Interdisziplinary Approaches to Nonlinear Complex Systems. pp. 138 – 160, Berlin: Springer.
Kruse, P., Stadler, M., Pavlekovic, B. & Gheorghiu, V. (1992): Instability and cognitive order formation: Self-organization principles, psychological experiments, and psychotherapeutic interventions. In: W. Tschacher, G. Schiepek & E.J. Brunner (Eds.), Self-Organization and Clinical Psychology, pp. 102 – 117. Berlin: Springer.
Kruse, P., Stadler, M. & Strüber, D. (1991): Psychological modification and synergetic modelling of perceptual oscillations. In: H. Haken & H.P. Koepchen (Eds.), Rhythms in Physiological Systems, pp. 299 – 311. Berlin: Springer.
Kruse, P., Strüber, D. Stadler, M. (1994): The significance of perceptual multistability for research on cognitive self-organization. In this volume.
Kuhn, T.S. (1962): The Structure of Scientific Revolutions. Chicago, IL: University of Chicago Press.
Lehmann, D. (1994): Brain electric microstates, and cognitive and perceptual modes. In this volume.
Lorenz, K. (1939): Vergleichende Verhaltensforschung. Verhandlungen der deutschen zoologischen Gesellschaft 1939, 69.
McLulich, D.A. (1937): Fluctuations in the Numbers of VarYing Hare. Toronto: University of Toronto Press.
Marx, K. & Engels, F. (1969): Die deutsche Ideologie. Werke, Vol. 3. Berlin: Dietz.
Metzger, W. (1933): Beobachtungen über phänomenale Identität. Psychologische Forschung 19, 1 – 60.
Metzger, W. (1975a): Psychologie, Darmstadt: Steinkopff (5th ed.).
Metzger, W. (1975b): Gesetze des Sehens, Frankfurt/Main: Kramer (3rd ed.).
Necker, L.A. (1832): Observations on some remarkable phenomenon which occurs on viewing a figure of a crystal or geometrical solid. The London and Edinburgh Philosophical Magazine and Journal of Science 3, 329 – 337.
Nicki, R.M., Forestell, P. & Short, P. (1979): Uncertainty and preference for "ambiguous" figures, "impossible" figures and the drawings of M.C. Escher. Scandinavian Journal of Psychology 20, 277 – 281.
Ogden, C.K. & Richards, I.A. (1923): The Meaning of Meaning. London: Routledge & Kegan Paul.
Plath, P. & Stadler, C. (1994): Multistability in molecules and reactions. In this volume.
Schiller, P.von (1933): Stroboskopische Alternativversuche. Psychologische Forschung 17, 179 – 214.
Schleidt, W.M. (1961): Über die Auslösung der Flucht vor Raubvögeln bei Truthühnern. Naturwissenschaften 48, 141 – 142.
Schöner, G. (1994): Concepts for a dynamic theory of perceptual organization: An example from apparent movement. In this volume.
Schwegler, H. (1992): Systemtheorie als Weg zur Vereinheitlichung der Wissenschaften? In: W. Krohn & G. Küppers (Eds.); Emergenz: Die Entstehung von Ordnung, Organisation und Bedeutung, pp. 27 – 56. Frankfurt/Main: Suhrkamp
Skarda, C.A. & Freeman, W.J. (1987): How brains make chaos to make sense of the world. Behavioral and Brain Science 10, 161 – 195.

Stadler, M., Kobs, M. & Reuter, H. (1993): Musik – Hören, Verstehen und Spielen. Kognitive Selbstorganisation und Einfühlung. In: K.E. Behne, G. Kleinen & H. de la Motte-Haber (Eds.), Musikpsychologie. Jahrbuch der Deutschen Gesellschaft für Musikpsychologie 9, 7 – 24.

Stadler, M. & Kruse, P. (1990): Über Wirklichkeitskriterien. In: V. Riegas & C. Vetter (Eds.), Zur Biologie der Kognition, pp. 133 – 158. Frankfurt/Main: Suhrkamp.

Stadler, M., Kruse, P. & Strüber, D. (1993): Struktur und Bedeutung in kognitiven Systemen, in press.

Stadler, M., Richter, P.H., Pfaff, S. & Kruse, P. (1991): Attractors and perceptual field dynamics of homogeneous stimulus areas. Psychological Research 53, 102 – 112.

Stadler, M., Vogt, S. & Kruse, P. (1991): Synchronization of rhythm in motor actions. In: H. Haken & H.P. Koepchen (Eds.), Rhythms in Physiological Systems. Berlin: Springer., pp. 215 – 231.

Stadler, M. & Wildgen, W. (1993): Semiotik und Gestalttheorie, in press.

Szu, H. (1994): Artificial neural networks approach for bistable reversible figures. In this volume.

van Leeuwen, C. (1993): Task, intention, context, globality, ambiguity: More of the same. In this volume.

Vogt, S., Stadler, M. & Kruse, P. (1988): Self-organization aspects in the temporal formation of movement gestalts. Human Movement Science 7, 365 – 406.

Vukovich, A. (1994): The emergence of meaning by linguistic problem solving. In this volume.

Wertheimer, M. (1945): Productive Thinking. New York: Harper & Brothers.

Wertheimer, M. & Gillis, W.M. (1958): Satiation and the rate of lapse of verbal meaning. Journal of General Psychology 59, 79 – 85.

Wildgen, W. (1994): Ambiguity in linguistic meaning in relation to perceptual multistability. In this volume.

Witte, W. (1975): Zum Gestalt- und Systemcharakter psychischer Bezugssysteme. In: S. Ertel, L. Kemmler & M. Stadler (Eds.), Gestalttheorie in der modernen Psychologie, pp. 76 – 93. Darmstadt: Steinkopff.

Wittgenstein, L. (1921): Tractatus logico-philosophicus. Ostwalds Annalen der Naturphilosophie.

Wittgenstein, L. (1952): Philosophical Investigations. London: Basil Blackwell.

Zaus, M. (1992): Kritische Thesen zur Bedeutungsentstehung. Berichte aus dem Institut für Kognitionsforschung, Universität Oldenburg, No. 7.

Zimmer, A.C. (1994): Multistability – More than a freak phenomenon. In this volume.

Some Basic Concepts of Synergetics with Respect to Multistability in Perception, Phase Transitions and Formation of Meaning

H. Haken

Institute for Theoretical Physics and Synergetics, University of Stuttgart,
Pfaffenwaldring 57/IV, D-70569 Stuttgart, Germany

Abstract: We first remind the reader of the goals and the basic concepts of synergetics in a nonmathematical fashion. In order to understand ambiguity of perception and the occurrence of meaning, we treat the brain as a synergetic system, or, in other words, as a self-organizing system close to instability points. As will be outlined, ambiguities may be resolved by oscillations or by a hierarchy of order parameters. The general concepts are used to devise a computer model for perception that reveals the fundamental role of attention, in particular in the resolution of ambiguous patterns. Finally, we discuss semantic information and the self-creation of meaning.

1 A reminder of the goals of synergetics

In practically all scientific disciplines we are dealing with systems that are composed of many individual parts, components or subsystems. These components represent the microscopic level. By means of their interaction the components may quite often produce a total action on the macroscopic level. This total action may consist of the formation of spatial, temporal, or functional structures or behavior. In particular, in natural systems, such as plants and animals, these structures are not imposed on the systems from the outside, but they are produced by the systems themselves via self-organization. The spontaneous formation of structures can be observed in quite simple systems of physics and chemistry, but such structures can also occur in complex systems, such as the economy or ecology. The question I asked some twenty years ago was: are there general principles governing self-organization irrespective of the nature of the parts? That question may have sounded rather strange at that time, because the individual parts can be as diverse as molecules in a liquid or in chemical reactions, cells in organs, or humans in a society. But over the past decades, it was shown that such general principles can be found provided we focus our attention on those situations where qualitative changes on macroscopic scales occur. Under such circumstances, we observe the emergence of new qualities. To treat these problems, we developed a rather comprehensive mathematical theory that has been presented in several books of mine (Haken, 1983, 1987, 1988). In the present article I shall try to present the underlying concepts in a nonmathematical language. I wish to remind the reader, however, of the three different approaches we have now developed to deal with these problems. In the microscopic or mesoscopic approach we start from the individual components of the system. Their behavior is described by evolution equations for their microscopic variables (Haken, 1983, 1987). We then study how these evolution equations allow the description of the spontaneous formation of macroscopic structures and patterns. In very complex systems, such as those of biology, quite often the microscopic variables are not known nor are their evolution equations. Then the macroscopic approach applies, in which so-called unbiased estimates based on available data are made; this is essentially a stochastic theory, where distribution functions and stochastic evolution equations for macroscopic observables are derived (Haken, 1988)[3]. Finally, we mention the important phenomenological approach. As the microscopic approach reveals, close to points where new qualities emerge, the dynamics of a complex system is governed by few variables, the so-called order parameters. Therefore, under such circumstances, it becomes possible to describe and then to model a system in terms of these order parameters (cf. Haken, Kelso & Bunz, 1985).

2 Synergetics of the brain

The heading of this section is identical with the title of a book that was edited in 1983 by Başar, Flohr, Haken and Mandell. This book is the proceedings of a conference held in 1983 at Schloß Elmau, Bavaria. The conference, organized by one of these editors (H.H.), brought together scientists from different disciplines ranging from physics to neurobiology. I believe that in several articles of that book the foundation was laid to approaches that are now quite well-known and to concepts that have got acquired an important place in modern studies on the behavior of the brain.

When we deal with perception and the origin of meaning, we may start with a study of the properties of the brain, where we take as a starting point a material point of view. The brain is a complex system, because it is composed of, say, a hundred billion neurons with many synapses connecting the neurons in a complicated fashion. It is also a dynamical system, because it hosts uncountably many electrical and chemical processes. It becomes a synergetic system, because it exhibits the emergence of new qualities in the processes of perception, motor-action, thinking, speech-control, and so on. Two approaches to deal with these phenomena may be compared according to Table 1:

macroscopic	cells
collective	individual
network	grandmother cells
	steering cells
delocalized	localized
self-organization	programmed computer
close to instability points	stable

The right side represents the older view in which for instance perception is attributed to the action of individual cells, in particular to the so-called grandmother cells. In motor-action specific steering cells are claimed. This picture may be called a localized one. According to a somewhat different paradigm, the brain is considered as a computer. Furthermore it is thought that the state of the brain is stable. In our more modern view, that is intimately connected with the concepts of synergetics, the new qualities mentioned above, such as perception, are produced at a macroscopic level by collective effects in which many cells participate in the frame of a network. Brain activity can no longer be localized at individual cells, but is more or less delocalized and may even cover large regions of the brain. The most important issue brought forward by synergetics is the idea that the brain is a self-organizing system that keeps itself close to instability points. Under these circumstances, the complex system "brain" can switch quickly from one state to another and is thus able to react quickly to new situations.

3 Basic concepts of synergetics

In this field one considers systems composed of many components; the behavior of each component with index j is described by a variable q_j where q_j may be also a vector comprising several components. The state of the total system is described by a state vector $\mathbf{q} = (q_1, q_2, ..., q_N)$. The evolution of the system in the course of time is described by the temporal change of the state vector $\mathbf{q}$. This change is determined by the present state $\mathbf{q}$, by so-called control parameters, and by chance events.

Let us discuss some examples of control parameters. When a fluid is heated from below, a temperature difference between its lower and upper surface is generated. This temperature difference may act as a so-called control parameter. In the brain control parameters may be represented by the concentrations of neurotransmitters, drugs, hormones, or other agents. A great variety of systems may be described by evolution equations of different forms. At first sight it seems hopeless to make general statements about the solutions of such evolution equations. However, the following strategy is suggested by experiments (Fig. 1). For instance, when we consider a fluid that is not heated from below, there is no temperature difference; the control parameter α_0 vanishes. In such

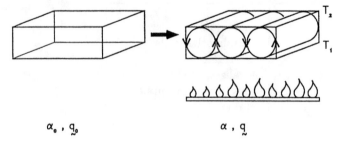

$\alpha_0, \mathbf{q}_0$ $\qquad\qquad\qquad\qquad \alpha, \mathbf{q}$

Fig. 1 Left: A fluid in a box at rest. For a given control parameter α_0, i.e. no temperature difference between a lower and upper surface, a homogeneous steady state is present. The state is described by the state vector $\mathbf{q}_0$. Right: When the liquid is heated from below, a temperature difference $T_1 - T_2$ is established with its corresponding control parameter α. The state in the form of rolls is described by a state vector $\mathbf{q}$

a situation, the state of the system $\mathbf{q}_0$ is known. When we change the control parameter by progressively heating up the liquid, suddenly a point is reached where a qualitatively new state occurs, e.g. a roll pattern, where the fluid undergoes a macroscopic motion with a well-ordered up- and down-welling. Thus our strategy consists in considering those situations, where the state vector undergoes a qualitative change, or, in other words, where it undergoes an instability. As is revealed by the mathematical methods of synergetics, the macroscopic motion is governed by a new quantity, the order parameter ξ. Once the order parameter ξ is determined, the behavior of all the components of a system is determined uniquely maybe except for microscopically small fluctuations. This is the content of the slaving principle of synergetics (Fig. 2).

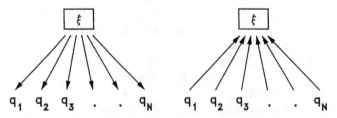

Fig. 2 Left: Illustration of the slaving principle of synergetics. Once an order parameter is given, it determines the action of the individual components. Right: By their action, the individual components sustain the order parameter

While a huge amount of information is required if we have to describe the behavior of the system by means of the variables q_j of the components, an enormous information compression is achieved by means of the order parameter. A particular phenomenon should be observed here. While on the one hand an order parameter enslaves the subsystems, i.e. prescribes the behavior of the subsystems, in all synergetic systems the subsystems determine in turn the behavior and existence of the order parameter (Fig. 2). This effect is called "circular causality". As the mathematical analysis reveals, order parameters are slowly varying quantities in time. When they are disturbed, they return slowly to their original state. On the other hand, the individual components, once they are disturbed, will relax on a short time scale. This time scale separation allows us in many cases to identify order parameters. For instance in the central nervous system, the individual cells have typical response times of milliseconds, whereas perception, motor-action, and so on take place on time scales of a hundred milliseconds. Thus it seems reasonable to describe processes of perception, motor-action, and so on by means of adequate order parameters. As may be again exemplified by fluid dynamics or laser physics, it is also possible for several order parameters to appear. They may compete such that, eventually, one order parameter wins the competition. Or they may coexist, for instance in a fluid two roll patterns may be superimposed on each other. Finally, order parameters may cooperate, e.g., in fluid dynamics, three order parameters determining roll patterns moving in different directions can cooperate. Then, in this case, the three order parameters stabilize each other and by a superposition of the individual motions a honeycomb pattern may evolve. A warning should be added at this point: While our explicit example is taken from fluid dynamics, it can be shown that the laws derived by synergetics are based on mathematical structures and relationships rather than on physics. Thus synergetics does not imply any physicalism.

4 Dynamics of order parameters I

As mentioned above, close to instability points, the dynamics of a system is determined by its order parameters. Let us now discuss that dynamics of order parameters for the special case of a single order parameter. In such a case, the behavior of the order parameter can be easily visualized by identifying the variable ξ of the order parameter with the coordinate of a ball sliding down within a landscape of Fig. 3. For a certain control parameter value α_0, the landscape is represented by the left-hand side of Fig. 3.

When the control parameter reaches a value where instability sets in, the bottom of the valley becomes very flat (Fig. 3, middle). As one may show, order parameters are always subjected to random fluctuations, similar to a ball kicked by a football team entirely at random. Because the valley is very flat, the random forces have a considerable effect on the motion of the ball and its motion fluctuates quite heavily. This phenomenon is called "critical fluctuations". Because the valley is flat, the restoring forces are very small and the ball relaxes to its position only very slowly. This effect is called "critical slowing down". Critical fluctuations and critical slowing down are typical

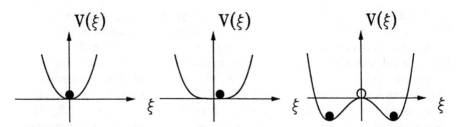

Fig. 3 l.h.s.: For a control parameter α_0 the potential landscape shows only one minimum. Middle part: If the control parameter reaches its critical value, the valley becomes very flat. R.h.s.: Beyond the critical value of the control parameter α, the potential landscape exhibits two minima

effects that occur when a system approaches its instability point. In this context, it is worth noting that such effects have been discovered not only in physical systems, such as lasers, but also in movement coordination, in particular by the now famous experiments of Kelso on qualitative changes in finger movement coordination (cf. Schöner & Haken, 1986, where further references may be found). When the control parameter is still further increased, the landscape is deformed to that of Fig. 3 right-hand side. Here instead of the former minimum two new minima occur. The system now has the choice between two macroscopic states that are attached to the minima. These two new states are stable so that we can speak of bistability. Since the system can choose one of these two states, one speaks of a symmetry breaking instability. Which macroscopic state will be occupied by the system is determined by a small chance event. Therefore, as an important consequence, it follows that such small causes can generate large macroscopic effects. The symmetry can also be broken artificially from the outside by giving the ball a push to the right or to the left so that it rolls down to the corresponding minimum. The phenomena just described are connected with the deformation of the landscape and are denoted as "nonequilibrium phase transition". This word has been coined, because there are systems in thermal equilibrium in which such transitions occur; an example is a ferromagnet, where the disordered motion of the individual elementary magnets becomes entirely ordered when the ferromagnet is cooled down beyond a critical temperature. In the case of synergetics, however, we are not dealing with systems in thermal equilibrium, such as a ferromagnet, but with systems far from equilibrium, which are called open systems.

Their state is maintained by a continuous influx of energy, matter and/or information. While the present author (Haken, 1983) observed analogies between perception (bistability, hysteresis) and that of general synergetic systems, Kruse and Stadler (1990a,b) stressed profound analogies between Gestalt theory (Köhler, 1920) and the behavior of synergetic systems. Their discovery marked the beginning of what I believe to be a very important development in psychology. I believe that we are still only at the beginning

Basic Concepts of Synergetics   29

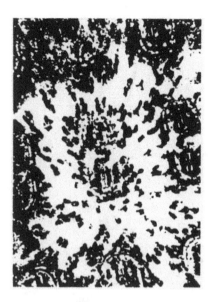

Fig. 4 Depending on our preknowledge we recognize a microscopic picture of a ferromagnet or a scene on the stock market

of a very fruitful development as is already witnessed by the contributions to this volume. The deformation of the so-called potential landscape need not always occur in an entirely symmetric fashion as exhibited in Fig. 3, rather potential landscapes may also be deformed in a way that is depicted in the upper row of Fig. 5.

Here, when a control parameter is changed, a single minimum is replaced by a single minimum plus a local minimum and then, eventually, the first minimum disappears again and only the new minimum survives. When we change the control parameter value in the opposite direction, the state of the system first remains in the new minimum, even when the same control parameter value α_1 is reached. Only later is the old state acquired again. Here we observe the fact that even for the same control parameter value quite different macroscopic states of the system may be acquired depending on the history of the system. This effect is hysteresis. An example in perception is shown in Fig. 6.

When we read the words from top to bottom, we read CHAOS, CHAOS, CHAOS, and then, eventually, ORDER. When we read the same words from bottom to top, we read ORDER, ORDER, ORDER, and then CHAOS. Thus we observe that the switching from one meaning to another one occurs at a different position depending on the history.

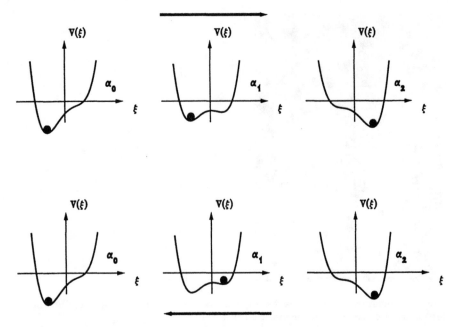

Fig. 5 Visualization of the effect of hysteresis. The state vector switches at different situations depending on the history

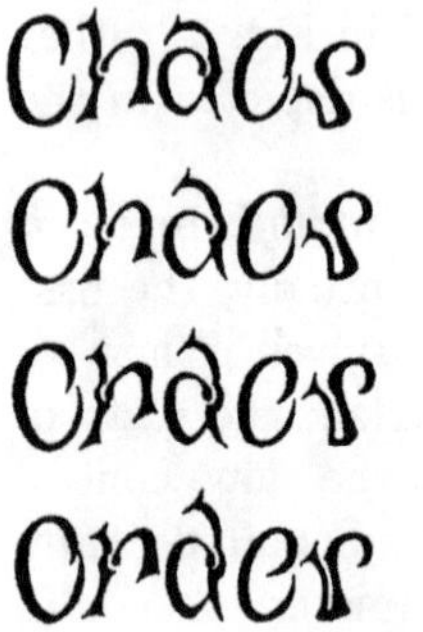

Chaos
Chaos
Chaos
Order

Fig. 6 An example for hysteresis and visual perception. If we read the picture from above to bottom, we read CHAOS, CHAOS, CHAOS, then only ORDER, in the opposite direction we perceive ORDER, ORDER, ORDER and then only CHAOS

5 Dynamics of order parameters II

Let us briefly consider a few typical effects in cases where several order parameters occur. Then we may again draw a potential landscape, but now as a landscape in a multidimensional space. Here again local minima may occur, but instead of two minima several such minima may exist. This is the case of multistability. Another example connected with two or more order parameters is that of oscillations in which two order parameters change their magnitude periodically in the course of time. Finally, when there are three or more order parameters, in addition to cases of multistability and oscillations, so-called deterministic chaos will appear. An example of the occurrence of oscillations in perception is provided by ambiguous figures, such as those of Figs. 7 and 8.

In the case of the Necker cube, the percept switches in this way: At a particular time one interprets the one face of the Necker cube as front face and then as the rear face. A similar phenomenon can be observed with the figure of Rubin: vase or two faces? The percepts "vase" and "face" change periodically in the course of time. A detailed representation of a model of these oscillations will be given in the contribution by Ditzinger and Haken (1994) to this volume.

6 Resolution of ambiguities

As we have just seen, our brain resolves conflict situations provoked by ambiguous figures by means of oscillations. It is impossible for the brain to recognize "vase" and "faces" simultaneously, rather the percept switches between these two interpretations. Another resolution of ambiguities is performed by the brain with respect to the occurrence of ambiguous words. Scrutinizing a language reveals quickly that there are numerous ambiguous words (cf. the contribution by Wildgen (1994)). Their meaning may be made unique by means of the context or by means of a hierarchy. I suggested at early instances to consider this disambiguation by means of a dynamic process. Consider the string of words of Fig. 9, where the middle word is ambiguous.

The words of that string generate the order parameter which, in turn, reacts on the individual words (cf. Fig. 2). According to the slaving principle, the meaning of these words is now uniquely determined so that the ambiguity is removed. In this interpretation disambiguation is done by means of a dynamic process, where the brain can be considered as a synergetic system governed by order parameters.

As is shown by Wildgen and others not only words but whole sentences may be ambiguous. In such a case the ambiguity may again be removed by means of a context in which these sentences are embedded into a text which, in its entirety, again acts as an order parameter. In a number of cases ambiguity may also be removed by the experience that one interpretation is highly unlikely.

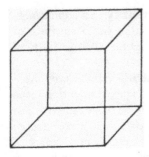

Fig. 7 The Necker cube

Fig. 8 Vase or faces?

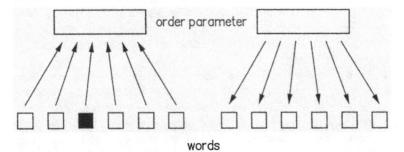

Fig. 9 Disambiguation of the meaning of a word by means of the order parameter concept in a dynamical process

7 Pattern recognition

So far we have given a qualitative description of synergetics. In order to show how these concepts can be put into an operational form, we consider the problem of pattern recognition by humans and machines. Our approach to pattern recognition is based on three ingredients: First and in accordance with the generally accepted belief, we identify pattern recognition with the action of an associative memory. An example for an associative memory is provided by a telephone book. For instance when we look up the name Adam Miller, the telephone book tells us, in addition to the name, his telephone number. Thus associative memory serves the completion of a set of data. The second ingredient of our approach is the assumption that the associative memory is realized by means of a dynamics of order parameters within a potential landscape in analogy to landscapes of the form of Fig. 3, right-hand side, where, however, we now have to deal with a multistable landscape. The third and most essential ingredient of our approach is the idea that pattern recognition is nothing but pattern formation (Haken, 1991). Let us consider to this end a simple example from fluid dynamics (compare Fig. 10). Here we simulate the behavior of a liquid that is heated from below so that roll patterns may be formed.

First an initial state in the form of a single up-welling roll is prescribed. Then, according to the computer calculations depicted by Fig. 10, left-hand side, a complete roll pattern evolves. If we prescribe a different single initial roll, a correspondingly completed roll pattern emerges (Fig. 10, middle). Finally, when we prescribe two up-welling rolls, one a little bit stronger than the other one, a competition between these two roll patterns sets in and the originally stronger roll pattern wins the competition. In terms of synergetics the following happens: The initially partially ordered state of the fluid gives rise to a set of order parameters. These order parameters compete, whereby the originally largest one wins. It acts on the system by the slaving principle and thus, eventually, forces the whole system to go into its corresponding ordered state. In more abstract terms we may say that a partially ordered system generates its corresponding order parameters that react on the system and force the system into the totally ordered state. Quite the same happens in pattern recognition. Once a number of features are given, the features generate the order parameters that force the total system to complete all the features so that a total figure is reconstructed. For instance when we prescribe, say, eyes and nose of a person, according to this process the whole face will be reconstructed.

To describe our procedure more explicitly, let us consider a set of faces that have been photographed (Fig. 11).

In order to process these patterns on a computer, we put a grid over the individual photographs and contribute to each pixel a grey value. The set of all grey values of the stored patterns represents a prototype vector $\mathbf{v}$. Because several prototype vectors are stored, we distinguish them by an index $u, \mathbf{v}$. Once now a test pattern vector $\mathbf{q}$, e.g. eyes and nose, is presented, we have to decide to which stored face this test pattern vector belongs. To this end we

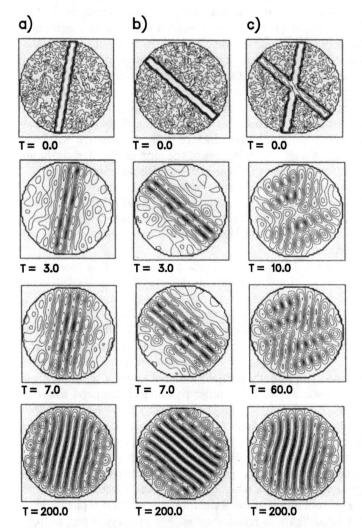

Fig. 10 Computer simulation of the behavior of a liquid heated from below. Left column: Initially a single up-welling roll is prescribed, then the full pattern evolves. Middle part: A different direction of the up-welling roll is prescribed. Right: Competition between two initially given rolls

construct a dynamics that is described by the following differential equation for the temporal evolution of the test pattern vector

$$d\mathbf{q}/dt = \sum_u \lambda_u (\mathbf{v}_u \cdot \mathbf{v}^u \mathbf{q}) - \sum_{uu'} B_{u,u'} (\mathbf{v}^u \mathbf{q})^2 (\mathbf{v}^{u'} \mathbf{q}) \mathbf{v}_u. \tag{1}$$

Basic Concepts of Synergetics 35

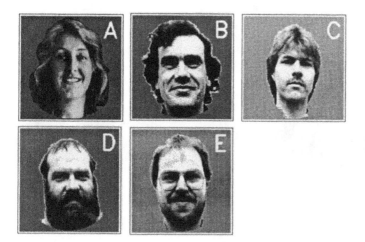

Fig. 11 Examples of prototype patterns stored in a synergetic computer

According to this equation, the temporal change of the test pattern vector is described by so-called attention parameters λ_u and a learning matrix $\mathbf{v}_u \cdot \mathbf{v}^u$. The second term on the right-hand side serves for the discrimination between patterns and for saturation, i.e. that the test pattern vector values do not exceed a certain value. The temporal evolution may be influenced by some chance events; however, these can be neglected in our general procedure. As it turns out, the temporal evolution of $\mathbf{q}$ can be again described by means of a ball rolling down in a potential landscape. As a consequence of this dynamics, the originally given test pattern vector is pulled in the course of time into one of the prototype patterns, which are represented by means of the minima of the potential landscape that is depicted in Fig. 12.

Examples for stored prototype patterns and the completion process are given in Figs. 11 and 13. As it turns out, our procedure is very sensitive and allows the computer to distinguish between facial expressions, where examples are provided by Figs. 14 and 15.

The abstract algorithm described by equations (1) can be realized, in particular, by a parallel computer, where each pixel j is connected with a neuron with an activity q_j. Equations (1) can then be interpreted in the following way: The neuron index j receives inputs from other neurons q_k, or from three other neurons with activities q_k, q_ℓ, q_m. The neuron under consideration then sums up over these contributions by means of synaptic strengths λ_{jk} or $\lambda_{jk\ell m}$. This can easily be seen when we rewrite (1) in terms of its components:

$$dq_j/dt = \sum_k \lambda_{jk} q_k + \sum_{k\ell m} \lambda_{jk\ell m} q_k q_\ell q_m. \tag{2}$$

Fig. 12 Example of a potential landscape, where the minima represent stored prototype patterns

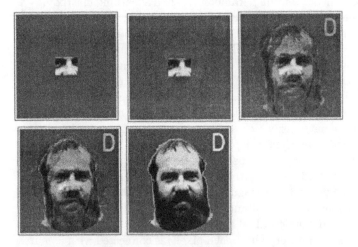

Fig. 13 Pattern recognition process from the initially given part of a face to a full face including the family name

8 The role of attention

It is not difficult to develop a preprocessing of the computer so that it recognizes faces that are shifted in their position in space. When we then show the computer a scene of Fig. 16, it recognized first the lady in the foreground. Then the attention parameter for the lady was put equal to zero, either from

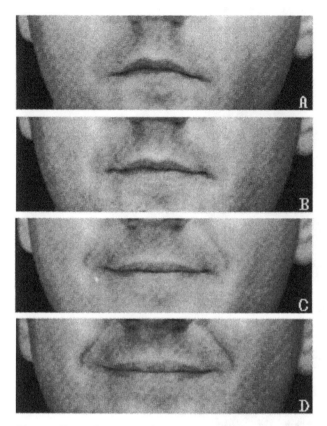

Fig. 14 Examples of facial expressions that can be distinguished by the synergetic computer

the outside or by the computer. When then the picture was shown again to the computer, it recognized the man behind. Similarly, by setting the corresponding attention parameters equal to zero once an individual face had been recognized, the computer was able to recognize scenes of Fig. 17.

These findings lead to the idea that, in humans too, our perception is significantly influenced by attention parameters. This view is strongly reinforced by models developed by Ditzinger and the author (1990) in which a saturation of attention parameters is included. If this is taken into account, the oscillations found in the perception of ambiguous figures can easily be explained. An example for an ambiguous figure is presented in Fig. 18: Einstein's face or ...?

The time during which one specific interpretation of an ambiguous figure is perceived, is called reversion time. Because in our approach it is connected with attention parameters, one may try to influence the reversion times by

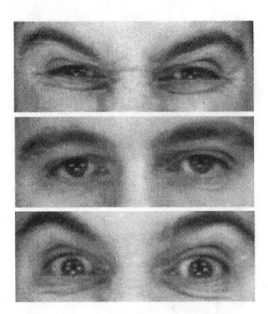

Fig. 15 Same as Fig. 14, but with respect to expressions of eyes

Fig. 16 Example of a complex scene

means of attention parameters. One possibility that I suggested a couple of years ago was to influence the attention parameters by drugs. In addition one may try to use reversion times to get insight into the emotional state of the test person.

Fig. 17 Another example of a complex scene with 5 patterns

While in our approach (Ditzinger & Haken, 1990) the saturation of attention parameters plays a crucial role for the explanation of oscillations of perception, two other approaches deserve attention. Caglioti (1994) stresses the analogous behavior of some quantum systems, while Kelso et al. (1994) can explain their experimental results by means of discrete maps in the sense of intermittency.

9 Expert systems of AI, heuristics of human actions, and synergetics

The concepts of order parameters, slaving principle and circular causality allow us to make contact with other fields dealing with human (and artificial) intelligence and behavior. Let us start with an approach to an expert system for medicine, that came to my attention recently by a talk by M. Stefanelli (1993). The philosophical basis for this approach is provided by the concepts of the American philosopher Pierce, who distinguished between deduction, induction, and abduction. The relationships between data (e.g., on a patient) and the hypothesis (e.g., diagnosis or therapy) are shown in Fig. 19.

The problem is, of course, to draw conclusions for the diagnosis (or therapy) from known data. This is done by a cyclic procedure: From incomplete data directly by induction or indirectly – by some kind of filtering via abstraction and abduction – hypotheses are formed that lead to the prediction of additional data (to be found by an investigation of the patient). Now the process can be repeated. Here it becomes useful to establish a correspondence with basic concepts of synergetics. The "data" correspond to the

Fig. 18 Einstein's face or ... ?

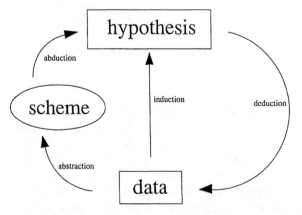

Fig. 19 Relationship between given data and hypotheses formed on account of these data via various channels

partially ordered subsystems, whereas the "hypotheses" correspond to the order parameters. In both cases, cyclic causality is present. The incompletely "ordered" subsystems, e.g., parts of a face, generate a set of order parameters of different sizes. The competition dynamics of the order parameters selects one definite order parameter so that in this case the "correct" hypothesis was found. In the case of the synergetic computer for pattern recognition, the order parameter dynamics is given by equation (1). It will be an interesting task for the future to study the relationship between this explicit algorithm and the scheme of Fig. 19 more closely and to generalize our algorithm. This may have a direct impact on how to devise expert systems.

In this context it is of great importance to know how the process of hypothesis building goes on in humans and how they react in specific situations. Here we may refer to comprehensive studies by W.A. Wagenaar (1993). For illustration consider the following situation: Someone comes home and switches the light switch, but no light appears. Then he or she will first try to switch it on and off several times. If this does not work, he or she may try to exchange the bulb. Eventually – in some cases – the person will come to the conclusion that the power station had broken down. According to Wagenaar, persons develop their own "heuristics" to react to given situations with incomplete information. They have a list (or program) of possible reactions. On top of their list is a situation, which had previously occurred most often. If their corresponding reaction fails, they try the next most probable reaction, and so on. It is easy to make contact with concepts of synergetics. The subsystems are here represented both by sensory inputs and motor actions. A set of sensory inputs generates order parameters, where one wins the competition and calls upon the corresponding motor action. But if this action is inadequate, the process is repeated whereby the "false" attention parameter is put equal to zero, the evolution of the former order parameter prevented and a new order parameter will cause a new motor action. In the recognition of scenes (Figs. 16, 17) the attention parameters of all the previously recognized faces were put equal to zero and all the others were left unequal to zero and of equal size, whereas in the present case, the size of the attention parameters is chosen according to the relative frequency with which the corresponding cases had happened previously. As one may show, the system then let that order parameter win which belongs to the biggest attention parameter. If the procedure must be repeated as described above, that attention parameter is put equal to zero and the next largest one comes into play, and so on.

10 Recognition of movement patterns

In section 7 we represented our approach to the recognition of static figures. Our approach may be generalized to the recognition of movement patterns, e.g., to distinguish between different kinds of motions of humans. An example that was treated explicitly (Haas et al., 1994) is that of Johanssons' experiments (1973), where lights were attached to the joints of persons that were moving in the dark. When the movie of these persons was shown to observers, they could not say what these light points meant when the person did not move. On the other hand, when the person moved, the observers could immediately tell what kind of movement was being performed and they could even distinguish between the movement of men and women. In our approach the periodic movement of the joint angles was decomposed into Fourier series with amplitudes and phases. Then a potential landscape for amplitudes and phases was constructed. Once the potential landscape has been constructed, the procedure prescribed in section 7 or by Ditzinger and Haken (1990) on the recognition of ambiguous figures can be applied.

11 Meaning and semantic information

In this section I to wish to give a brief outline of our interpretation of the occurrence of meaning. To be more precise, we shall speak here of semantic information when a message is transmitted. In our interpretation, semantic information is characterized by its specific effect on a dynamical system that is represented by a state vector, whose dynamics is described by nonlinear evolution equations with control parameters and possibly fluctuating forces. The effect of the message is to fix the initial values of the state vector and/or of control parameters α. Thus the information initiates a temporal change of the state vector, which then approaches an attractor state of the system. We may speak of semantic information if there is a specific correspondence between each such information and the dynamics of a receiver leading to a specific attractor state. A simple example is provided by an attractor landscape composed of fixed points, or in a more popular description, by a golf course. Quite evidently, such a system is self-correcting; even an incomplete or noisy message will be completed and made noiseless. An example was provided in the section on pattern recognition. Then sensitisation will correspond to an opening, habituation to a closing of attractors. Such an opening or closing of perception attractors can be achieved by changing attention parameters. One should note that attractors need not only be fixed points represented by valleys in a landscape, but attractors may be also stable oscillations (limit cycles), or chaotic attractors. As we know from synergetics, attractors are governed by one or few order parameters. Therefore, we may say that meaning is attached in each case to a specific order parameter. Such order parameters govern in each case a whole configuration – in our case now that of a neural network or synergetic computer.

Let us now consider the process of generation of meaning, or what one may call the self-creation of meaning. To this end it is useful to compare the learning process of a neural or synergetic computer with human or animal learning. For these computers one may distinguish between supervised and unsupervised learning. In supervised learning the following happens: A set of examples is presented to the computer, which then processes the input and, eventually, produces an output. The desired output is known to the teacher, but in the learning phase the output of the system may be different from the correct one. Then the system is told by learning rules how to change its internal synaptic strengths. After a certain training period, the system is (hopefully) able to reproduce the right output for the given test set. In unsupervised learning, the system learns just from the frequency of the presented pictures or data. One may say that it digs holes into its potential landscape, where the holes are the deeper the more often the corresponding prototype pattern is offered. In this kind of learning, the system learns on frequency of occurrence. Such a learning certainly occurs in a number of cases in human life, when for instance a school teacher presents the same letter to the pupils again and again. On the other hand, there are quite important learning processes which are not based on frequency but based on a single event, e.g., when a child touches a hot stove. It seems to be important to incorporate such learning processes in computers. At any rate we realize that the meaning of, say, hot stove is connected here with a single, but very important event. In this case we realize that the establishment of meaning of certain events or information is based on the context. In general, we may expect a whole network of such contexts.

References

Başar, E., Flohr, H., Haken, H. & Mandell, A.J. (Eds.) (1983): Synergetics of the Brain. Berlin: Springer.
Caglioti, G. (1994): Perception of ambiguous figures: a qualitative model based on synergetics and quantum mechanics. In this volume.
Ditzinger, T. & Haken, H. (1990): The impact of fluctuations on recognition of ambiguous patterns. Biological Cybernetics 63, 453 – 456.
Ditzinger, T. & Haken, H. (1994): A synergetic model of multistability in perception. In this volume.
Haas, R., Fuchs, A., Haken, H., Horvath, E., Pandya, A.S. & Kelso, J.A.S. (1994): Recognition of dynamic patterns by a synergetic computer. In this volume.
Haken, H. (1983): Synergetics. An Introduction. Berlin: Springer (3rd ed.).
Haken, H. (1987): Advanced Synergetics. An Introduction. Berlin: Springer (2nd ed.).
Haken, H. (1988): Information and Self-Organization. Berlin: Springer.
Haken, H. (1991): Synergetic Computers and Cognition. Berlin: Springer.
Haken, H., Kelso, J.A.S. & Bunz, H. (1985): A theoretical model of phase transitions in human hand movements. Biological Cybernetics 51, 347 – 356.
Johansson, G. (1973): Visual perception of biological motion and a model for its analysis. Perception and Psychophysics 14, 201 – 211.
Kelso, J.A.S., Case, P., Holroyd, T., Horvath, E., Rączaszek, J., Tuller, B. & Ding, M. (1994): Multistability and metastability in perceptual and brain dynamics. In this volume.
Köhler, W. (1920): Die physischen Gestalten in Ruhe und im stationären Zustand. Braunschweig: Vieweg.
Kruse, P. & Stadler, M. (1990): Stability and instability in cognitive systems: multistability, suggestion, and psychosomatic interaction. In: H. Haken & M. Stadler (Eds.). Synergetics of Cognition, pp. 201 – 215. Berlin: Springer.
Schöner, G., Haken, H. & Kelso, J.A.S. (1986): A stochastic theory of phase transitions in human hand movement. Biological Cybernetics 53, 247 – 257.

Stadler, M. & Kruse, P. (1990): The self-organization perspective in cognition research: historical remarks and new experimental approaches. In: H. Haken & M. Stadler (Eds.), Synergetics of Cognition, pp. 32 – 52. Berlin: Springer.

Stefanelli, M. (1993): The Artificial Intelligence Approach to Modelling Medical Reasoning. Talk given at the Symposium: Natural Sciences and Human Thought, Villa Vigoni, Italy, 29.3. – 2.4.1993.

Wagenaar, W.A. (1993): Heuristics: Simple Ways for Dealing with Complex Problems. Talk given at the Symposium: Natural Sciences and Human Thought, Villa Vigoni, Italy, 29.3. – 2.4.1993.

Wildgen, W. (1994): Ambiguity in linguistic meaning, in relation to perceptual multistability. In this volume.

Part II

Psychophysical and Phenomenological Approaches

Multistability as a Research Tool in Experimental Phenomenology

G. Kanizsa and R. Luccio

Università degli Studi di Trieste, Facoltà di Lettere e Filosfia,
Corso di Laurea in Psicologia, Via dell'Universita 7, I-34123 Trieste, Italy

Abstract: Multistable situations can help us in understanding the processes in the visual system and in identifying rules, constraints, limits, ways in which extra perceptual elements contribute to the formation of visual objects. An example of the use of multistable situations is given by the research on stroboscopic alternative motions. Another example is the series of investigations aimed at identifying the factors determining figure–ground separation. In both cases we have studied the very first perceptual organization, and we have always found an absolute prevalence of the autochthonous factors. We conclude that perception (always considered as a *primary process* leading to the constitution of visual objects) is essentially encapsulated from top-down influences.

1 Multistable displays: Exceptional but utterly natural phenomena

The phenomenal world is a stable world, mostly made up of objects maintaining their features and above all maintaining their identity in a sufficiently constant manner in time. We have adjusted to living in such a world, and this explains why reversible figures like Necker's cube or Rubin faces, which appear at first like one thing, and then suddenly change and turn into something else, are considered only to be amusing tricks, having little to do with the normal vision that enables us to have an unequivocal knowledge of the external world.

It is easy to be ironic about the interest that students of perception have for such rare and exceptional phenomena, and this indeed occurred. For instance, Gibson (1979), in proposing his approach to the study of perception, mockingly named the other existing approaches "psychology of the snapshot". With this phrase he intended an artificial, laboratory psychology, as opposed to a psychology that should be natural, in harmony with life, and, to use a fashionable word, "ecological". We should point out that the term "snapshot" only makes sense if applied to experiments using *tachistoscopic* presentation and that reversible figures have nothing to do with snapshots, nor do optical illusions which are subject to similar denigration. They do not need to be presented with a rigid predetermination of fixation, lighting and exposure time; they can be observed at ease in every environment and for as long as one wishes. In any case they continue to be reversible or illusory. They are definitely peculiar visual situations, but not owing to their being produced or observed in unnatural or artificial conditions as the term "snapshots" induces us to believe. Their peculiarity lies in the fact that they are very rare in ordinary everyday perception. One could claim that they are not to be found in nature, but this would be inaccurate: The very moment they enter an observer's visual field they gain full status of visual objects. The frequency of this event occurring is irrelevant. (In the same way, the temperatures close to absolute zero or the particles obtained by nuclear scission do not belong to the category of ordinary physical events; to produce them, we must create suitable conditions. Under normal circumstances they either do not occur at all or they occur only exceptionally. But this does not mean they are not to be found in nature or they are less important for understanding the laws of physics.)

Even accepting these qualifications, one could insist that situations of multistability (but also optical illusions, anomalous surfaces, apparent motions, impossible figures, phenomena of amodal completions, or visual embedding) are mere peculiarities or errors of nature (such as a hand with six fingers or a fruit without seeds) that are of little or no value for science. According to this point of view, science should deal with quite different matters and should not waste time with abnormal phenomena of no "ecological validity". In our view, such a reductive opinion is totally incorrect. However, it seems useless to insist on this point here. We shall confine ourselves to recalling that in our opinion situations of multistability, as other anomalous phenomena, have essentially two useful functions:

1.1 A rhetoric function

According to common sense things appear as they do because they are what they are. Man is unable to see where the problem lies. Multistable situations undermine this widespread naïve conviction: Man has proof of the existence and the extent of the problem. In this way we realize that even in everyday ordinary perception things that appear so straightforward could also take on a different appearance; this, however, happens so rarely that the real problem is to discover the reasons of such stability.

1.2 A cognitive function

Secondly, plurivocal situations can help us in understanding the workings of the visual system, discover rules, constraints, limits, ways in which extra perceptual elements (top-down subjective and cognitive factors) contribute to the formation of visual objects. (To say that our visual system picks up the useful information from an optical array or that it 'resonates' to the invariants that are contained in it is, on one hand, an important claim. It implies that it is not necessary "to go beyond the information given", because the information *is all in the array*. On the other hand, however, it tells us very little about *how* the system picks out information from a mass of potential data much richer than the one actually utilised. If one does not clarify the "how", in other words if one does not identify the factors that determine precisely this specific selection excluding thousands of other possible solutions, this claim runs the risk of remaining unfounded, that is to say that perception is as it is because it is as it is.) In this perspective, multistable situations and other perceptual anomalies are or may become valuable tools for investigating ordinary perception with ecological validity, an apparently obvious issue, which is however still mysterious. Among others, Ramachandran and Anstis (1983), Kruse (1988) and Kruse and Stadler (1990) have stressed this methodological aspect of multistability.

Let us now see how the use of the paradigm of multistability has in some cases furthered the research on the functioning of visual system. We have chosen two problems which we have contributed in this context.

2 Multistability as a research tool

2.1 The stroboscopic alternatives

An example of the use of multistable situations to identify the organizing principles ruling the visual system is given by the research on stroboscopic alternative motions carried out by the Hungarian psychologist P. von Schiller sixty years ago (1933).

When a cross (like the one of Fig. 1) is presented and replaced by a similar cross rotated by 45°, in which direction will the stroboscopic motion be perceived? Linke (1907), Wundt's pupil, had tried to answer this question and the results of his experiments had led him to conclude that here, as it is generally the case with stroboscopic motion, it was not possible to find laws: The cross could rotate indifferently clockwise or anticlockwise. It depends on the situation or on the observer's intentions. We are well aware of the fact that Wertheimer (1912) had a different opinion: His classic research on stroboscopic motion demonstrated that it is a highly regular phenomenon that obeys precise laws and that, even in Linke's case, there is a lack of predictability only in conditions of equidistance. By modifying the length of

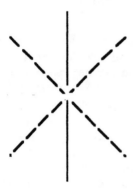

Fig. 1 A cross (vertical-horizontal) is presented, then replaced by a similar cross rotated by 45°. (Linke, 1907)

the path, the event gets more predictable. This enabled Wertheimer to state the "law of minimal distance".

Von Schiller re-examined Linke and Wertheimer's problem; he had realized that such multistable situations could be very useful to establish the factors which determine direction and shape of motion. His analysis begins with a situation similar to Linke's. Instead of two crosses he used the simple tachistoscopic device illustrated in Fig. 2. In the first phase he projected the two dots in A simultaneously on a screen, and then, after turning them off, the two dots in B. The four stimuli were located at the vertexes of a square. The distance between each pair of contiguous vertexes was equal. In this case the phenomenal displacements occur indifferently either clockwise or anti-clockwise. However, the results become immediately more univocal when one introduces a differentiating element, for instance a difference in colour. When in the first phase the upper dot is black and the lower is white and in the second phase the right dot is black and the left one is white, the prevailing motion seen will be clockwise. Similar results are obtained when, instead of chromatic equality, visual objects similar in shape, size, or brightness are used, as can be seen from the examples in Fig. 3. In all these cases the alternative disappears according to the principle of similarity: Similar goes with similar and the tendency of visual objects to maintain their own identity and to persist in time without changing, takes control of the situation.

Von Schiller then examined the strength of the factor of similarity and he contrasted it with the tendency to favour the shorter path. He found that similarity (of shape, size, colour, or brightness) is decisive in determining an alternative direction of motion. All these kinds of similarities succeed in counteracting the tendency to prefer the shorter path; at least as long as the difference between the lengths of the paths is less than 20°. When they act together they can overcome the shorter distance up to a difference of 60° between the path's lengths. Using stroboscopic alternative motions, v. Schiller

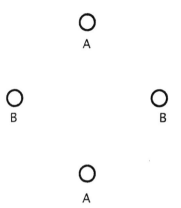

Fig. 2 Von Schiller's display. First the two dots A are projected simultaneously onto a screen and then the two dots B, after turning the first ones off

succeeded in establishing a hierarchy between these factors of similarity. From series of his experiments, in which all the possible pairs of factors were contrasted, the following relationships emerged. The similarity of *shape* clearly has a greater weight than the similarity of *size* in deciding the alternative kind of motion. With closed figures, as in the example of Fig. 4, the motion between figures with equal shape still prevails when the ratio between sizes is 1 to 2. When open and closed figures are contrasted, as in Fig. 5, once again the shape prevails over the size when the ratio between sizes is 1 to 3.5. The shape also wins over colour if the hues are not complementary. However, when the colours are complementary, shape no longer prevails and the multistable situation is restored (see Fig. 6).

The other combinations have demonstrated that *colour* always prevails over *size*, and that *brightness* is more influential than either *colour* or *size* in determining the direction of the motion. Besides similarity and proximity, closeness and good continuation are other factors that affect the direction of the motion and that can emerge through the method of stroboscopic alternatives (Figs. 7 and 8). In the display illustrated in Fig. 7 the clockwise motion prevails, each arc moves forward and gains the position of the partner on its left. The counterclockwise motion is very rare. In conclusion, it can be remarked that factors favouring grouping in static situations also condition the successive relationship in kinematoscopic situations.

These are the main results of the work of v. Schiller. Most of his papers published in *Psychologische Forschung* in those years, report a number of other data and observations. There is also a chapter entitled "*Einstellung und subjective Faktoren*" which is devoted to the influence of the observer's attention, voluntary set, and expectancies on the phenomenon. It is difficult to control such factors. They affect the outcome of the alternatives only when the gestalt factors are in unstable equilibrium.

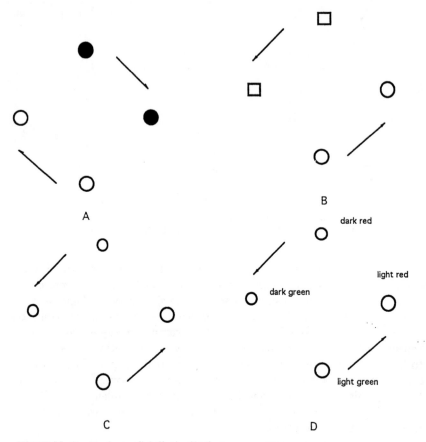

Fig. 3 Variants of von Schiller's display

(The classical papers on *Psychologische Forschung* are mostly known through the English translations, briefly summarized and published in the reader edited by Ellis, or through the synthesis presented in the textbooks by Koffka or Köhler. It is worth reading them in the original, not only because they are written in a non-standard style, today rather uncommon, but they also contain many interesting remarks. They are written in such a way as to challenge a widespread prejudice: namely, that the gestaltists' works are characterized by little rigor or experimental scrupulousness, with little, if any, statistical elaboration. They are written mostly in German, but we advise reading even to avoid perpetuating superficial and often wrong judgements, and the malpractice of presenting well-studied phenomena as "new findings". The same is true for *Zeitschrift für Psychologie*. The theoretical frame and the interpretations may be old but the studied phenomena, the facts and the discoveries survive the eclipse of the theories.)

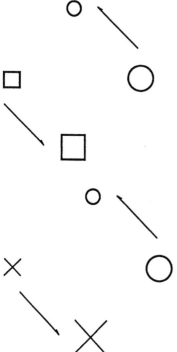

Fig. 4 With closed figures, the apparent motion between figures with equal shape prevails

Fig. 5 When open and closed figures are contrasted, the apparent motion between figures with equal shape prevails over the motion between figures with equal size, when the ratio between sizes is 1 to 3.5

Further research carried out by P. Kolers (1964), F. Hoeth (1966, 1968), H. Erke and H. Gräser (1972), and V. Ramachandran (1985) has essentially confirmed the results obtained by v. Schiller. The true novelties are few and include the following observation by Hoeth, confirmed by Ramachandran and Anstis. In a bistable display like v. Schiller's (Fig. 9) dots on two diagonally opposite corners are first shown, then switched off and followed by two spots appearing simultaneously on the remaining two corners. The entire procedure is repeated in a continuous cycle. The two possible percepts that are equally probable and mutually exclusive are a vertical oscillation and a horizontal oscillation (indicated in the diagram). A third theoretically possible percept is *continuous clockwise* (or *anticlockwise*) motion, but this is *hardly ever seen*.

2.2 Our experiments

To this effect, we have performed two experiments. Our aim was not to find something particularly new; we intended to focus on two aspects of the results that we felt to be somewhat neglected by previous authors. The point that we have repeatedly made in the last few years is that the properly visual process in perception is characterized by the segmentation of the visual field in different units (gestalten). According to us, this *primary process* is fully

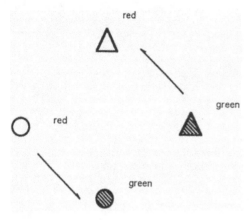

Fig. 6 When the hues are not complementary, the apparent motion between figures with equal shape prevails over the motion between figures with equal colour. When the colours are complementary, shape no longer prevails

Fig. 7 Clockwise motion prevails

"informationally encapsulated" (to use Fodor's expression, 1983), and the extra perceptual factors can act only at a later stage. The two aspects on which we have focused our attention were: (a) The very first motion the subjects see; (b) the ease with which they can pass from one kind of motion to the alternative one.

In the first experiment, we used a display very similar to v. Schiller's – the difference was that we presented the stimuli on the screen of a Macintosh and the animation was generated by the Macromind program. We had 7 kinds of stimuli that the subjects (Ss) saw in random order, different for each S: (1) stimuli equal in size, shape, and colour (Fig. 10a); (2) stimuli equal in size and shape but not in colour (Fig. 10b); (3) stimuli equal in size and colour but not in shape (Fig. 10c); (4) stimuli equal in shape and colour but not in size (Fig. 10d); (5) stimuli equal only in shape (Fig. 10g); (6) stimuli equal only in size (Fig. 10f); (7) stimuli equal only in colour (Fig. 10d). In each condition we presented the stimuli on the vertical first (a and c) and then on the horizontal (b and d). The possible motions were i) a-b-a-b+ and c-d-c-

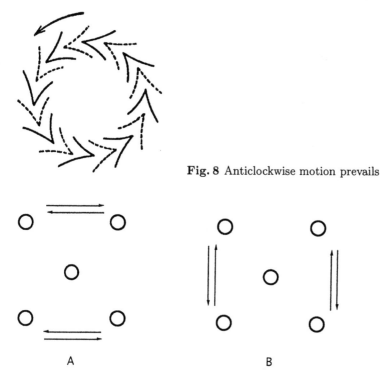

Fig. 8 Anticlockwise motion prevails

Fig. 9 The displays used by Ramachandran and Anstis

d+ , v. ii) a-d-a-d+ and c-b-c-b+ . The Ss had to notice which motion they saw first (i or ii) and following this they had to rate (on a seven-point scale) the ease with which they could invert the motion (pass from i to ii, or vice versa). In other words, the Ss were asked to change their set and try to change voluntarily the direction of perceived motion; for most Ss this task was easy to understand. Two Ss reported perception of a circular motion (a-bc-d+ or a-d-c-b+); for 3 Ss it was impossible to invert the direction of apparent motion in any situation and they always rated 1. 3 Ss on the contrary always rated 7. The scores of all these Ss were excluded.

The results of the first experiment might appear a little disappointing. There is no significant influence of shape, colour or size in determining the appearance of the first perception of motion i) or ii). Only colour appears to be a significant factor: It was significantly easier to move from one motion to the other when the second motion unified stimuli of the same colour, and significantly more difficult when the stimuli of the same colour were already unified in the first motion. Note that most Ss reported that as the observation proceeded they had spontaneous alternation of direction. The unification for colour, shape or size prevailed when there was only one factor present (stimuli b, c, and d). When two factors were present (stimuli e, f,

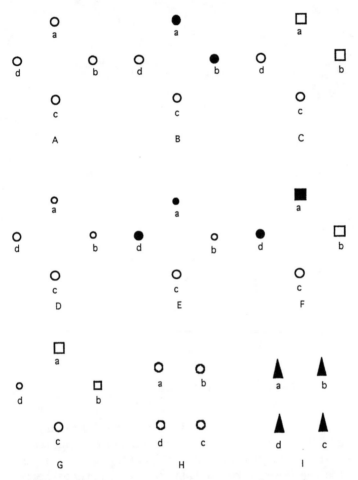

Fig. 10 The displays used in our experiments (see text)

and g), colour prevailed over shape and distance. But between the last two factors neither prevailed.

In a second experiment we added the factor of proximity. The stimuli were like the stimuli 1a-d of the first experiment, but the four points were at the vertexes of a rectangle. So the distances a-b (or respectively a-d) and c-d (or respectively c- b) were shorter than a-d (or respectively a-b) and c-b (or respectively c-d). Everything else was like in the first experiment (Fig. 10 h).

In this case, the factor proximity turns out to be dramatically effective. For all Ss the very first motion perceived is in direction of the nearer point. It proves very hard for them to invert voluntarily the direction of the motion. In fact, for about half of the Ss this was almost impossible. The spontaneous inversions are less frequent and the alternative lost only very short time before the Ss returned to the first motion again. This is not the case, however, if we contrast proximity with directionality of the figure (Figure 10 i). Here the two alternative directions are equiprobable.

Taken together, the results of the two experiments are hardly surprising, and essentially confirm all previous findings. What is important to stress is the dramatic difference between proximity and other factors like colour, shape, and size (with the exception of directionality, which, however, is a tertiary quality). Proximity reveals itself as an autochthonous (autonomous) factor, preceding any other factor in organizing movement and highly resistant to *Einstellung* or other cognitive factors. The other factors prevail only when they are recognized.

3 The figure-ground articulation

3.1 Perceptual and extra perceptual factors

Let us now turn to another series of investigations based on the utilisation of multistability. We shall begin with a short history of the steps that have allowed to identify the factors determining figure-ground articulation. Using a typical reversible display similar to the one in Fig. 11, E. Rubin in his classical essay of 1921 tried to establish the conditions determining that some parts of the field assume the role of "figure" with respect to others that are seen as "ground". His accurate investigation showed the most important conditions to be: The *relative size* of the parts, their *topological relationship,* and the kind of their *contours.* Ceteris paribus, the area that is *smaller, included,* with *convex* contours will tend to emerge as a figure (Fig. 12). In Fig. 12 all these factors concur to favour the central zone as a figure. The situation is not ambiguous nor is there any spontaneous reversibility. When, on the other hand, the factors are conflicting and no single one prevails on the others, a typical situation of ambiguity occurs, where instability and the continuous reversibility of the figure-ground relationship dominate. This is the case in Fig. 13.

In 1928 P. Bahnsen, a pupil of Rubin, pointed out the role played by *symmetry* in figure-ground articulation. Some of his displays showing the influence of this new factor are well known, having been presented in many textbooks on visual perception (Fig. 14).

A factor analogous to symmetry, identified and studied by S. Morinaga (1942), is *Ebenbreite* or *constancy of width* of a region of the field. As one can see in Figs. 15 and 16, this factor tends to favour the assumption of the role of "figure" by the regions whose contours are phenomenally parallel. In the same vein one can consider as a sort of symmetry also the condition at work in Fig. 17. In this condition the vertical stripes bounded by contours of the same style are more likely to stand out as a "figure".

Fig. 11 Rubin's classical figure

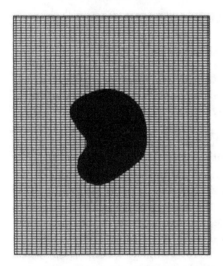

Fig. 12 The area smaller, included, with convex contours, will tend to emerge as a figure

In 1975 Kanizsa noticed that in Bahnsen's displays this factor had been not sufficiently controlled, and he set up some situations in which convexity was put in contrast to symmetry (Figs. 18–21). Observing this material, it seemed to him that symmetry could be overcome or at least counterbalanced by preference for a convex form. A quantitative research carried out by W. Gerbino has confirmed this impression (G. Kanizsa and W. Gerbino, 1976). In the case of Figs. 18 and 19, the asymmetrical but convex pillars stand out as "figure" in 70 % of cases. The result is more clearcut with displays 20, where the convexity of the asymmetrical areas is more accentuated; it follows that the contours of symmetrical regions are more concave. In 90 % of cases convexity prevails over symmetry. The symmetrical regions have no independent existence, but they combine to form a black or white ground (Fig.20). According to Biederman, Hilton and Hummel (1991) this union of symmetrical regions in a unitary ground should be favoured in Kanizsa's displays by the fact that collinearity of the top and bottom edges is stronger

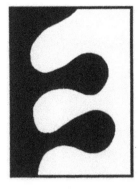

Fig. 13 The factors are conflicting and no single one prevails over the others

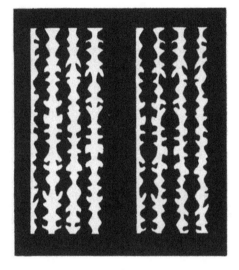

Fig. 14 The regions whose contours are symmetrical tend to assume the role of "figure" (Bahnsen, 1928)

for the symmetrical sections than for the asymmetrical ones. In fact when one interrupts collinearity of contours of symmetrical regions these become more visible as figures (Fig. 21).

At this point it was inevitable to consider the use of bistable figure-ground situations as a methodology well suited to give an empirically founded answer to a question that arises each time one attempts to establish which factors are responsible for a given class of perceptual phenomena. Granted that figure-ground articulation is a fundamental operation in the process of constitution of visual objects, the question is: Is such segmentation of the visual field entirely due to a bottom-up process internal to the visual system or does it also result from top-down processes external to the visual system, such

Fig. 15 Ebenbreite or constancy of width: The regions whose contours are phenomenally parallel tend to assume the role of "figure" (Morinaga, 1942)

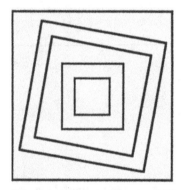

Fig. 16 Another example of Ebenbreite

as attention, intentions, and especially cognitive variables like familiarity, likelihood, and meaning?

Rubin had already claimed that prior experience might influence figure-ground organisation. He observed that, once a region of a bistable display is singled out as a figure, it is maintained as a figure for longer durations than the other potential alternative. However, according to Rock and Kremen (1957), the procedure used by Rubin contained so many serious flaws that one could not really be sure that the figural aftereffect actually occurred. So they repeated Rubin's experiment removing the questionable aspects of the procedure and making sure that the effect, if obtained, could be attributed to the spontaneous influence of past experience on figure-ground organisation. Their observers first participated in a cover task in which they viewed one of

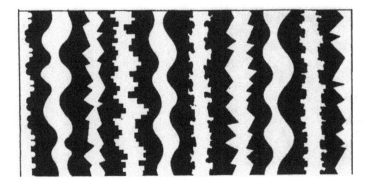

Fig. 17 The vertical stripes bounded by contours of the same style tend to assume the role of "figure" (Metzger, 1975)

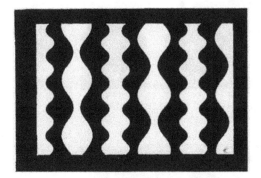

Fig. 18 The convex pillars stand out as "figure"

the two regions of the figure-ground displays. Afterwards they reported the organization perceived in brief presentations of the full figure-ground displays. The figural aftereffect was not obtained.

However, other research by Epstein and Rock (1960) seems to support Rubin's results. Their experiment aimed to examine the reliability of the widespread conviction that the expectancy set is a determinant of perception. Their findings demonstrated that it is the most recent perceptual experience that controls the finally attained percept.

Why are the results of Rubin and Epstein and Rock so interesting? They contrast with an argument that in the history of psychology is known as Höffding's step, from the name of the Danish philosopher who first stated it clearly (1887). According to this argument, the identification of a form presupposes the previous existence of the *identificandum*, of what has to be recognized. Thus the determination of figure and ground relationships ought to precede the operations involved in recognition processes. In fact it is only in virtue of this previous parsing that the field articulates in a series of distinct

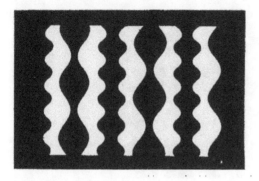

Fig. 19 See Figure 18

Fig. 20 Convexity prevails over symmetry

Fig. 21 According to Biederman et al. (1991), when one interrupts collinearity of contours of symmetrical regions, these become more visible as figures

visual objects that stand out on an undifferentiated ground. It would seem obvious to conclude that one cannot recognize a shape until one has perceived it (see Kanizsa and Luccio, 1987).

Not all authors agree with this opinion. Among those dissenting are, for instance, P. Kruse and M. Peterson. P. Kruse (1986), in research aimed at challenging the cogency of Höffding's step, has proposed a modification of the display in Fig. 20 to demonstrate that meaning can influence the determination of perceptual alternatives. That is to say that a process of identification can take place before the field is segmented into visual objects with a character of a figure or of a ground. In Fig. 22 A and B, the greater meaningfulness of one of the possible perceptual alternatives determines its phenomenal prevalence. In this case the impression of a certain number of heads that face one another imposes itself. A slight modification eliminating the physiognomic appearance is sufficient for vases to be seen more easily (Fig. 22 C and D). According to Kruse (1988), these findings demonstrate the inadequacy of a unidirectional model of visual processing in terms of stages, and suggest the theoretical opportuneness to replace it with the idea of a stimulation-cognition continuum characterized by circularity and by mechanisms of parallel analysis (Shepard, 1984).

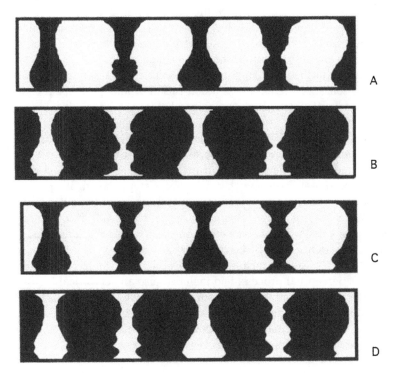

Fig. 22 According to Kruse (1986), in A and B the impression of a certain number of heads that face one other imposes itself. In C and D a slight modification is sufficient for the series of vases of the other alternative to be seen more easily

More recently, M. Peterson has reopened the discussion, resorting to experiments with bistable displays (Peterson, Harvey, and Weidenbacher, 1991). She begins with the remark that even though many investigations have undoubtedly demonstrated that figure-ground relationships can be determined without shape recognition input. These demonstrations have not ruled out the possibility that figure-ground computations can weigh inputs from shape recognition analysis as well. Her observers viewed both upright and inverted versions of reversible figure-ground stimuli. The central area was favoured by the factors of relative size, convexity, symmetry, inclusion. In the upright version the surround had a meaningful shape, in the inverted condition it was a far less denotative shape (Fig. 23).

Peterson found that when the surround was upright rather than inverted it was maintained as a figure for longer durations, and was more likely to be obtained as a figure by reversal out of the centre-as-figure interpretation. A series of four experiments shows that these effects reflect contributions of a shape recognition route entailing access to orientation-specific memory representations. Alternative interpretations in terms of eye movements or motivation were ruled out. Peterson, confirming and extending a previous finding of Hochberg and Peterson (1987) that observers' intentions can influence perceived organization, is persuaded that the evidence of her experiments indicate that figure-ground computations weigh shape recognition inputs, as well as inputs from routines assessing other variables such as symmetry, relative area, and so on. She also proposes a non-ratiomorphic mechanism that permits shape recognition processes to operate before the discrimination of figure-ground relationships.

The results of Kruse and Peterson's experiments confirm the Epstein and Rock's findings and they seem to show that the visual system is not a module absolutely impermeable to the influence of higher-order cognitive processes. Meanwhile, they also seem to show that Höffding's step, though it appears irrefutable from a logical point of view, cannot reflect all reality, when facts contradict it. As for the rest, in reports on perceptual research it is not rare to find remarks on the possibility of some effect on the phenomena under observation by prior experience or subjective factors as voluntary sets. We must not forget that gestalt psychologists themselves, claiming for a bottom-up conception of the perceptual process, had included also an "empirical factor" among their organizing principles (Wertheimer, 1923). Moreover we all know that, after finding an embedded figure in a drawing with great difficulty, or after integrating the lacking parts of a Street figure, we find it very easy to see them in successive exposures. On this issue W. Köhler (1913) says: "Wir erkennen also, dass frühere Gestaltbildung spätere beeinflussen kann, und entnehmen daraus, dass im Lebensgang einmal realisierte *Gestaltung* ebenso Dispositionen für ähnliche Hergänge in Zukunft hinterläßt ..." (p. 52).

It should be noted that the action of these extra perceptual (intentional or cognitive) factors is always described as weak, fleeting, unpredictable, asystematic, and it exhibits itself in conditions that one can hardly define with precision. However, it is an effect that can be ascertained only through complex procedures or hypothesized on the basis of sophisticated statistical computations. Of course, if one is unwilling to accept the thesis of the impenetrability of early visual processing, one single case of penetrability is enough to falsify the thesis.

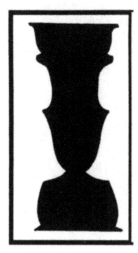

Fig. 23 According to Peterson et al., the surround is maintained as a "figure" for longer durations when it is upright rather than inverted

3.2 Our experiments

In this case too we performed two experiments with Kruse's figures. As in the experiments on apparent motion, our interest did not bear on the prevalence of one configuration over the other during the spontaneous reversal, but in the very first configuration perceived. In the first experiments, carried out by A. Santisi and S. Paluzzi in the Department of Psychology of the University La Sapienza of Rome, 96 Ss observed only one of the 4 figures of Kruse (24 Ss for each figure), after having seen 20 slides with animals, objects and persons, with no possibility of a figure-ground inversion. The exposure time was 180 msec. The Ss had to say what they had seen, and press a key to indicate the colour (black or white) of what they had seen. (As a matter of fact, the Ss saw 8 slides with Kruse's figures, each one repeated twice as the importance of the effects of sequence was investigated as well. The results were very weak and not significant, so here we are interested only in the first slide.)

The results are extremely straightforward. Nearly all Ss reported having seen black over white (over 95 % for each slide). With the figures D they were unable to say what they saw: Only one S said that he had recognized the white vases. In the second experiment, carried out by C. Primi in the Department of Psychology of the University of Florence, we used a "double blind" procedure, to eliminate the risk of a Rosenthal effect; moreover, we avoided presenting other slides before the critical one and requesting other tasks, like describing what they had seen and pressing some key. Ss were simply told that they had to look at the figure and to say immediately if they were seeing a black figure on white or a white one on black. Only later, were they asked to describe what they had seen. The exposure time was longer, 1 second, so the Ss could look at the figure with a certain ease.

112 Ss participated in the experiment, 28 for figures A and B, and 27 for figures C and D, and the results are quite similar to those of the previous experiment. In figure A, 23 Ss saw black on white and five white on black; in figure B, 24 Ss saw black on white and four white on black; in figure C, 23 Ss saw black on white and four white on black; in figure D, 26 Ss saw black on white and one white on black. In other words, there was no difference at all between the four figures, with an overwhelming prevalence of black on white.

We do not contest that *later*, once the meaningful part has been recognised, it is easier to see this part again. As we said earlier, Rubin, but Köhler too, admitted this fact that for instance Wertheimer's famous example of the figure-ground organization of the letter E clearly demonstrates.

Our results do not contradict Kruse and Peterson's findings. We studied only the very first perceptual organization, not the prevalence of one alternative over the other in the first period of observation. On the basis of our results, however, the meaning of Kruse and Peterson's demonstrations does not appear to be completely convincing. In any case, we think it is unwise to open a debate on this issue - which would risk appearing captious - to decide on the possible existence of effects whose manifestations are admittedly weak even in the eyes of those who claim their existence. Faced with these demonstrations obtained with such difficulty we can cite innumerable robust experimental facts illustrating the rigidity of our perceptual deliverances even in cases in which we know perfectly that they are not veridical. These perceptual events remain quite unalterable in spite of every effort, of logic, of common-sense and of everything known about experience. Comparing the weight of these facts, which are unquestionable and which everybody observes continuously, with the exiguity and the uncertainty of the latter ones, it seems to us wise to conclude that perception (always considered as a *primary* process leading to the constitution of visual objects) is essentially independent, a process that is encapsulated from top-down influences. In our opinion, the visual system is endowed with sets of built-in rules which impose constraints on early visual processing. This primary process is mainly unidirectional and delivers outputs to the higher cognitive centres, but it is in any interesting sense impenetrable to inputs from them. What is important is to learn more about these wired-in rules and constraints. This is a task that Wertheimer began, obtaining the first identification and systematisation of organising factors. This task must be continued.

References

Bahnsen, P. (1928): Symmetrie und Asymmetrie bei visuellen Wahrnehmungen. Zeitschrift für Psychologie, 108, 129 – 154.

Biederman, I., Hilton, H. J. & Hummel, J.E. (1991): Pattern goodness and pattern recognition. In: G.R. Lockhead & J.R. Pomerantz (Eds.), The Perception of Structure. Washington, DC: APA, 73 – 96.

Ellis, W.D. (Ed.) (1938): A Source Book of Gestalt Psychology. London: Routledge & Kegan Paul.

Epstein, W. & Rock, I. (1960): Perceptual set as an artifact recency. American Journal of Psychology 73, 214 – 228.

Erke, H. & Gräser, H. (1972): Reversibility of perceived motion: Selective adaptation of the human visual system to speed, size and orientation. Vision Research 12, 69 – 87.

Fodor, J. (1983): The Modularity of Mind. An Essay on Faculty Psychology. Cambridge, MA: MIT Press.

Gibson, J.J. (1979): The Ecological Approach to Visual Perception. Boston, MA: Houghton Mifflin.

Hochberg, J. & Peterson, M. (1987): Piecemeal organization and cognitive components in object perception: Perceptually coupled responses to moving objects. Journal of Experimental Psychology, General 116, 370 – 380.

Hoeth, F. (1966): Gesetzlichkeit bei stroboskopischen Alternativbewegungen. Frankfurt/Main: Kramer.

Hoeth, F. (1968): Bevorzugte Richtungen bei stroboskopischen Alternativbewegungen. Psychologische Beiträge 10, 494 – 527.

Höffding, H. (1887): Psychologie in Umrissen auf Grundlage der Erfahrung. Leipzig: Fus's Verlag.

Kanizsa, G. (1975): The role of regularity in perceptual organization. In: G. Flores d'Arcais (Ed.), Studies in Perception, pp. 48 – 66. Firenze: Martello-Giunti.

Kanizsa, G. (1979): Organization of Vision. New York: Praeger.

Kanizsa, G. & Gerbino, W. (1976): Convexity and symmetry in figure-ground organization. In: M. Henle (Ed.), Vision and Artifact. New York: Springer.

Kanizsa, G. & Luccio, R. (1986): Die Doppeldeutigkeiten der Prägnanz. Gestalt Theory 8, 99 – 135.

Kanizsa, G. & Luccio, R. (1987): Formation and categorization of visual objects: Höffding's never confuted but always forgotten argument. Gestalt Theory 9, 111 – 127.

Kanizsa, G. & Luccio, R. (1990): The phenomenology of autonomous order formation in perception. In: H. Haken & M. Stadler (Eds), Synergetics of Cognition, pp. 186 – 200. Berlin: Springer.

Köhler, W. (1913): Über unbemerkte Empfindungen und Urteilstauschen. Zeitschrift für Psychologie 66, 51 – 80.

Kolers, P. (1964): The illusion of movement. Scientific American 211, 98 – 106.

Kruse, P. (1986). Wie unabhängig ist das Wahrnehmungsobjekt vom Prozeß der Identifikation? Gestalt Theory 8, 141 – 143.

Kruse, P. (1988): Stabilität - Instabilität - Multistabilität. Selbstorganisation und Selbstreferenzialität in kognitiven Systemen. Delfin 11, 35 – 57.

Kruse, P. & Stadler, M. (1990): Stability and instability in cognitive systems: Multistability, suggestion and psychosomatic interaction. In: H. Haken & M. Stadler (Eds.), Synergetics of Cognition, pp. 201 – 217. Berlin: Springer.

Linke, P.F. (1907): Die stroboskopische Täuschung und das Problem des Sehens von Bewegung. Psychologische Studien 3, 393 – 545.

Metzger, W. (1975): Gesetze des Sehens. Frankfurt/Main: Kramer.

Morinaga, S. (1942): Beobachtungen über Grundlagen und Wirkungen anschaulich gleichmässiger Breite. Archiv für die gesamte Psychologie 110, 310 – 348.

Peterson, M.A., Harvey, E.M. & Weidenbacher, H.J. (1991): Shape recognition contributions to figure-ground reversal: which route counts? Journal of Experimental Psychology: Human Perception and Performance 17, 1075 – 1089.

Ramachandran, V.S. (1985): The neurobiology of perception. Perception 14, 97 – 103.

Ramachandran, V.S. & Anstis, S.M. (1983): Perceptual organization in moving patterns. Nature 304., 529 – 531

Rock, I. & Kremen, I. (1957): A re-examination of Rubin's figural after-effect. Journal of Experimental Psychology 53, 23 – 30.

Rubin, E. (1921). Visuell wahrgenommene Figuren. Teil 1. Kopenhagen: Gyldenalske.

Shepard, R. (1984): Ecological constraints on internal representation: Resonant kinematics of perceiving, imagining, thinking, and dreaming. Psychological Review 91, 417 – 447.

Stadler, M. & Kruse, P. (1990): The self-organization perspective in cognition research: Historical remarks and new experimental approaches. In: H. Haken & M. Stadler (Eds.), Synergetics of Cognition, pp. 32 – 52. Berlin: Springer.

von Schiller, P. (1933): Stroboskopische Alternativversuche. Psychologische Forschung 17, 179 – 214

Wertheimer, M. (1912): Experimentelle Studien über das Sehen von Bewegung. Zeitschrift für Psychologie 62, 371 – 394.

Wertheimer, M. (1923): Untersuchungen zur Lehre der Gestalt 2. Psychologische Forschung 4, 301 – 350.

The Significance of Perceptual Multistability for Research on Cognitive Self-Organization

P. Kruse, D. Strüber and M. Stadler

Institute of Psychology and Cognition Research and Center for Cognitive Sciences, University of Bremen, D-28334 Bremen, Germany

Abstract: Perceptual order formation is a process of self-organization in a complex neural network and not a pick up of external information. In this view any stimulus condition is multistable. Stimuli are only boundary conditions of the autonomous process of perceptual order formation. Stability in perception is the result of a fast converging process of autonomous order formation which normally acts on a time scale far beyond conscious realization. Multistability in perception is an exceptional case in which the process of order formation (confronted with one constant boundary condition) spontaneously oscillates between two or more attractors established in the system dynamics. The spontaneous reversions in perceptual multistability show characteristics of nonlinear phase transitions and can be simulated with a high degree of correspondence on basis of self-organizing networks and synergetic modelling. In everyday experience multistability in perception is a relatively irrelevant curiosity but for investigating the process of order formation in cognition multistability is a paradigmatical research tool. Perceptual multistability can be used as a window to the underlying neural system dynamics. In a variety of different experiments the possibility is shown to change the potential landscape of the system dynamics in multistable perception by learning, context and meaning. The reversion process is discussed as an indicator for innersystemic fluctuations. Some hypothetical links to pathological phenomena in cognition are outlined.

1 Introduction

Like multistable perception itself, the scientific interest in this topic is also characterized by a high degree of instability and ambiguity. Multistable perception is, on the one hand, one of the most intensively investigated psychological phenomena, whilst, on the other, often judged to be an irrelevant and marginal curiosity. It has attracted, irritated, and amused perceptual scientists ever since the first more systematic descriptions of spontaneous perceptual reversions were given by the physicist Necker (1832). The interest in multistable perception was most of the time due to the subjective quality of the experience of the process of spontaneous perceptual reversion and the devaluation of the significance of the phenomenon was often mainly theoretically based. Understanding perception as pick up of external information and as stimulus driven informational processing necessarily classifies multistable perception as a rare break down of the normal functioning of the perceptual system. In any strictly hierarchical stimulus-oriented approach of perception spontaneous changes on the basis of unchanged stimulus conditions simply have to be a marginal curiosity. In contrast to this, perceptual multistability is of central importance when the inner dynamics of the visual system itself is seen as a major determinant of perception. Consequently in gestalt theory perceptual multistability was evaluated totally differently right from the beginning. In the gestalt-theoretical perspective the reversion of figure and ground, the change in boundary function of a contour, the multistability of perspective or of direction of apparent motion, and the ambiguity of meaning is understood as a paradigmatical tool to investigate the process of perceptual order formation (see Köhler, 1940; Kanisza, 1979; Kruse, 1988). Recently this old gestalt-theoretical idea occupies again an important position on the background of the concept of cognitive self-organization (see Kruse & Stadler, 1990, 1993; Kruse, Stadler & Strüber, 1991). Although originally limited to the theoretical assumptions of linear thermodynamics, in many respects gestalt theory is a legitimate precursor of the modern theory of self-organizing dynamic systems (Kruse, Roth & Stadler, 1987; Stadler & Kruse, 1990).

2 Perceptual multistability and cognitive self-organization

In the concept of cognitive self-organization the brain is understood as a complex system which operates close to instability points (see Haken, 1994; Haken & Stadler, 1990). Stable macroscopic states of order are explained as attractors in the dynamics of the elementary neural network. Stimuli are only boundary conditions. Stability in perception is the result of a fast converging process of autonomous order formation which acts on a time scale far beyond conscious realization. The symmetry of figure and ground, the symmetry of the boundarying function of any discontinuity or contour in the stimulus array, and the symmetry of different simultaneously responding receptive fields has to be actively broken for the basic constitution of perceptual objects, figural identity, and motion direction. In this sense any stimulus condition is multistable (see Kruse, 1988; Stadler & Kruse, 1994; Zimmer, 1994). What is usually termed perceptual multistability is only an exceptional case in which – even when confronted with one constant boundary condition – the process of order formation sometimes spontaneously oscillates between two or more attractors established in the dynamics of the visual system. A lot of examples for perceptual multistability are presented in this volume.

To understand perception as a process of autonomous order formation leads to a fundamental methodological problem. In experience perception is normally characterized by stability. In contrast to the process of order formation in learning or in motor behavior, there is usually no way to directly realize the assumed system dynamics (see Kruse & Stadler, 1993). Only in the case of perceptual multistability does the fast converging process of order formation in the visual system become evident. In view of the theory of self-organization, perceptual multistability therefore is an outstanding methodological window for the measurement of the dynamics of the visual system. Following this line of argumentation, it is a necessary step to demonstrate that the reversion process in perceptual multistability can be characterized as a nonlinear phase transition. There are two complementary ways of investigating perceptual multistability in the perspective of self-organization theory: (i) Synergetics and network modelling and (ii) phenomenological synergetics. In the first approach (see Haken, 1977) the system dynamics is modelled on basis of differential equations and the resulting simulation is compared with the behavior of the system to be investigated. In the second approach (see Haken, Kelso & Bunz, 1985; Schöner, Haken & Kelso, 1986) relevant parameters of system behavior are measured and analysed. By demonstrating the existence of behavioral indicators like hysteresis, critical fluctuations, and critical slowing down, the existence of a nonlinear phase transition is proven. The potential landscape of the system dynamics is estimated by indirect measures of stability.

On the basis of synergetic modelling Haken and Ditzinger were able to simulate a variety of properties typical for multistable perception with very good correspondence to experimental data (see Ditzinger and Haken, 1994) and already a variety of different neural network architectures have been designed which are able to create spontaneous reversions when confronted with one stable stimulus condition (see e.g. Szu, Lu & Landa, 1994). Clear indicators have been demonstrated which support the idea of understanding the reversion process in multistable perception as a nonlinear phase transition (especially in connection with the hysteresis effect; see Kelso et al., 1994; Schöner & Hock, 1994). Further evidence in the sense of phenomenological synergetics can be expected on the basis of EEG-studies on perceptual multistability. Because of its rapidity on the level of experience, direct behavioral reactions to the reversion process seem to be inappropriate to reliably measure critical fluctuations or critical slowing down. But the identification of EEG-correlates of the spontaneous perceptual switches allows measurement on a faster time scale (see Basar-Eroglu et al., 1993). Therefore, to analyse the reversion process on a neurophysiological level is especially promising for investigation of the dynamics of perceptual multistability.

Psychological analysis of perceptual multistability is more or less restricted to indirect measures of stability. In the following empirical strategies are outlined which use psychological influences to introduce changes in the dynamics of the visual system or which use properties of perceptual multistability as global indicators for hypothetical situational and interpersonal differences in the process of perceptual order formation. In a first part some experimental data are presented which show that it is possible to change the potential landscape of the dynamics of different situations of perceptual multistabilty by learning, context and meaning. These data are discussed in context of the basic idea that in view of self-organization theory any perception is multistable and in context of the general notion that meaning is an order parameter of the process of cognitive self-organization (see Stadler & Kruse, 1994). In a second part the reversion process is used as a possible indicator for situational and interpersonal differences in the amount of spontaneous inner-systemic fluctuations. Different cognitive styles are assumed as a possible result of the interpersonal variances and some hypothetical consequences for understanding psychosomatic and pathological phenomena are discussed.

3 Perceptual multistability and psychological influences

Facing the diversity of the phenomenon, the first necessary step for designing experiments on multistable perception is to choose appropriate stimulus configurations. Reviewing the existing experimental data, multistable apparent motion (AM) patterns seem to be most adequate for several reasons: (i) Due to the early stage of visual integration AM patterns are usually very reliable in their perceptual behavior. (ii) AM patterns are easily modifiable in various ways. It is e.g. possible to introduce continuous change by manipulating the distance between distinct stimuli triggering the motion impression. (iii) Any AM-pattern with more than two alternating stimuli is multistable in the sense of theoretically possible motion directions. (iv) For AM-patterns it is especially easy to construct situations with three or more possible stable states.

The AM-displays used in the experiments to be described are the stroboscopic alternative motion (SAM) and the circular apparent motion (CAM). Both displays and the main alternative stable states occurring spontaneously during their presentation are explained schematically in figure 1.

Viewing SAM or CAM a continuous change between the different possible stable states is perceived. The time for which one stable state persists (residence time) decreases systematically during prolonged observation (see figure 2a). Consequently the rate of apparent change (RAC), the frequency

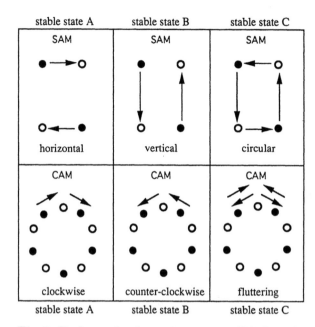

Fig. 1 Stroboscopic alternative motion (SAM) and circular alternative motion (CAM). The white and the black dots are presented in alternation. Due to the general temporal characteristics of AM, succession and simultaneity can be additionally observed as stable states in both displays. In SAM the circular alternative is relatively seldom perceived

with which the visual system shifts between the perceptual alternatives, increases over time (see figure 2b).

During prolonged observation the possible stable states seem to destabilize more and more. Physiologically this can be understood as a process of satiation and in the theoretical framework of synergetics it is the result of a flattening of potential landscape of the underlying system dynamics (see Kruse, Stadler & Strüber, 1991). This synergetic interpretation is further substantiated by the observation that the more improbable perceptual alternatives tend to appear spontaneously only after a longer time of observation. The attractor basins of the dominating alternatives have to become less pronounced before the stochastic fluctuations can move the system into the more improbable states. In this view the cumulative residence times of the different perceptual alternatives during prolonged observation can be seen as a global estimate of the shape of potential landscape. By measuring the cumulative residence times, therefore, the effect of different psychological influences on the potential landscape can be analyzed. In the perspective of self-organization theory it should be possible to stabilize and further destabilize a multistable perception by introducing contextual information or meaning. When e.g. a multistable AM-display is surrounded by some mono-directional

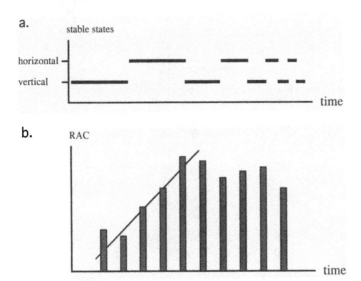

Fig. 2 Change in residence time (a) and rate of apparent change (b) during prolonged observation of SAM (data from Kruse, Stadler & Wehner, 1986)

displays the multistable display shows a strong bias in the direction contextually supported (see Kruse, Stadler & Strüber, 1991). The same effect can be demonstrated when a bias of figural identity is introduced. For this a frame can be added around one of the alternating light-points forming the CAM. After every cycle of alternation the frame is moved one light-point in the counter-clockwise direction. As a result a invariant frame is perceived which moves clearly mono-directionally over the light-points. Although all the different possible motion alternatives of the CAM still appear, the cumulative residence time of the biased stable state is significantly enhanced under these conditions (see figure 3).

Interpreted as a change in potential landscape of the underlying system dynamics, the data show that it is possible to stabilize a multistable display in favour of one attractor state. The next experiment was designed to demonstrate a further destabilization of multistable perception. In this experiment the SAM was used. Normally the SAM appears to be bistable. Most of the time untrained people only perceive the horizontal and the vertical alternative (see figure 1). During a special training session perceivers learned to be aware of the third circular motion alternative. In the training session the SAM was biased towards cicular motion using figural identity, as already described above for the CAM (see figure 4a). After some training sessions for most of the perceivers the circular motion alternative appeared spontaneously and the cumulative residence time of this alternative was significantly enhanced (see figure 4b). This effect did not occur in a control condition where only an unbiased SAM was presented in between. The perception of the AM-display

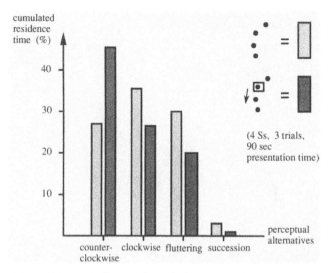

Fig. 3 Effect of a bias of figural identity on the residence times of the different perceptual alternatives of CAM (unbiased residence times: left bars; biased residence times: right bars)

was fundamentally changed by the special training procedure. A new attractor was added and the bistable display has been further destabilized.

The experiments presented up to now support the assumption that stability in perception has to be understood as mainly determined by the inner dynamics of the visual system. The stimulus only constrains the self-organized order formation. But nevertheless the demonstrated experimental effects were closely related to the stimulus configuration and therefore have to be interpreted as part of the primary process of perceptual order formation. In this sense the effects were more or less bottom up. But also the existence of clearly top down oriented effects on primary perceptual processes can be demonstrated on the basis of the CAM. Introducing meaning by verbal suggestion or by constructing meaningful AM-displays are also able to change the potential landscape as the introduction of the clearly stimulus triggered biases. When e.g. the CAM is integrated in a display which suggest the functioning of a water-wheel (see figure 5a) the first perceived motion direction follows mainly the interpreted causality (see figure 5b).

Also the introduction of meaning by verbal suggestion is able to influence the basic constitution of motion direction as already shown by Sherif on basis of the autokinetic phenomenon (see Sherif, 1935). In the case of SAM or CAM one motion direction happens to be seen more often and with longer residence times when supported by verbal suggestions. Even if the verbal suggestions are presented subliminally a similar effect can be demonstrated. During the perception of SAM perceivers listened to different audio tapes. On one audio tape the suggestion 'up and down like bouncing balls' was

a.

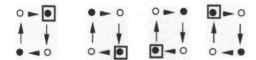

b.

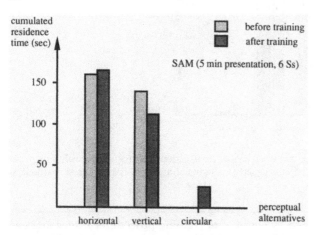

Fig. 4 Spontaneous appearence of the circular motion alternative in SAM before and after training (before training: left bars; after training: right bars)

masked by music and natural noises. The masked suggestive content of the audio tape was not consciously realized by the perceivers. On an other audio tape only the music and the natural noises were recorded and presented to a control group during perception of SAM. The subliminal suggestion was able to significantly enhance the cumulative residence times of the supported vertical motion alternative of the SAM compared to the control (see figure 6).

Obviously meaning directly influences the basic constitution of motion direction. Here the question arises whether it is possible to generalize the idea that meaning functions as an order parameter in the dynamics of the visual system also to the basic constitution of visual objects. A positive answer would be in contradiction to Höffding's argument that the identification of an object cannot be prior to its perceptual constitution (see the chapter by Kanizsa & Luccio in this volume). In the perspective of synergetic modelling, pattern recognition and pattern formation are essentially identical processes. Therefore in this perspective it is no logical contradiction to assume that meaning can influence the basic object constitution (see Stadler & Kruse, 1994). To empirically demonstrate the possibility of a top down influence on the perception of objects, two experiments were designed. In a first exper-

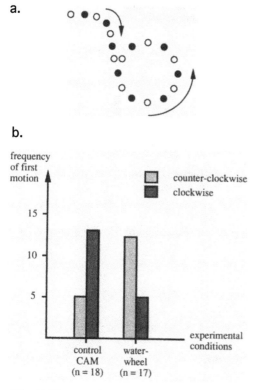

Fig. 5 First motion impression of CAM with and without additional meaning

iment a modification of the vase-face figure of Rubin was used (see Kruse, 1988). By changing the shape of the contours either the face or the vase interpretation was supported. When confronted with the figure-ground pattern, perceivers reports of the first impression showed a stronger preference for the face-parts of the picture when the contours were shaped like a normal human profile (see figure 7).

In the second experiment the top down influence on the perception of objects was analyzed on basis of Jastrow's duck-rabbit figure. It was suggested that the perceived position of the eye in this ambiguous pattern is influenced by the attributed meaning. To measure this difference in perceived position the duck-rabbit figure was presented ten times to each perceiver. The perceivers were instructed to concentrate always on one of the possible interpretations of the figure (five times 'duck' and five times 'rabbit' in alternation). The task of the perceivers was to mark the perceived position of the eye between two lines above the duck-rabbit figure only when the instructed alternative was stable (see figure 8a). Results show that there is a significant difference in perceived position of the eye depending on which alternative of

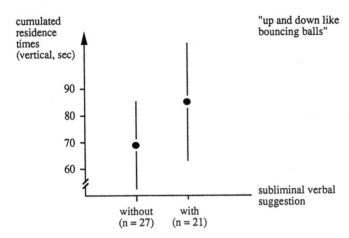

Fig. 6 Effect of subliminal verbal suggestion on perception of SAM (3 min presentation)

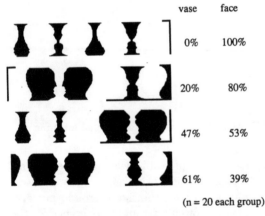

Fig. 7 Top-down influence of meaning on object constitution in figure-ground separation

the ambiguous figure is realized (see figure 8b). This result could be replicated using a different psychophysical measurement technique (method of constant stimuli).

Obviously it is a promising approach to analyze the process of perceptual order formation by use of the phenomenon of multistable perception. Multistability opens a window to the system dynamics underlying the perceptual process. Complicated perceptual interactions and subtle influences from higher psychological functions can be demonstrated and analyzed with the help of patterns which induce spontaneous reversions (see also Kanizsa & Luccio, in this volume).

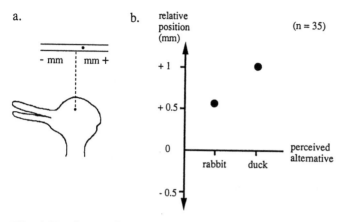

Fig. 8 Top-down influence of meaning on perceived position of the eye in Jastrow's duck-rabbit figure

4 Perceptual multistability and system state

During the long history of investigations on perceptual multistability it has often been reported that there are significant interpersonal differences in frequency of the spontaneous changes between perceptual alternatives. Some people tend to display low rates of apparent change (RAC) and for others the perception oscillates very quickly between different stable states. The level of RAC produced by a person shows a high degree of re-test reliability (see e.g. Bergum & Bergum, 1981). In view of self-organization theory differences in RAC can be interpreted as differences in the amount of innersystemic fluctuations. The higher the amount of fluctuations in a system the easier the system state can pass from one attractor basin to another. Therefore the RAC may function as a measure for how close an individual cognitive system works to instability points. The nearness to instability points may vary situationally or as a personal characteristic. Following predictions of self-organization theory, systems which operate more closely to instability points should be more sensitive and able to adapt to changing boundary conditions than systems which are more stable. Concerning habitual individual differences this is strongly reminiscent of Pavlov's theory of types of higher nervous activity (Pavlov, 1955). In his typology, individual nervous systems are mainly classified according to nervous system strength and mobility of the nervous processes. Strength and sensitivity are in an inverse relationship to each other. The strongest system is the most stable and at the same time the least sensitive. The weakest system is the least stable and the most sensitive (see also Nebelytsyn, 1964).

On the psychological level, these ideas suggest a relationship between differences in innersystemic instability and adaptive mental functions. Following the assumption that the readiness to change in multistable perception (e.g. quantified by RAC or amount of hysteresis) is a measure of innersystemic instability, a variety of correlations between individual differences in multistable perception and other cognitive phenomena may be assumed. Reviewing the experimental literature a complex and interesting network of intercorrelations can be found (see figure 9).

In the situation of symmetry breaking, systems are sensitive to very subtle influences (small causes with large effects). Theoretically a relationship between the readiness to change in multistable perception and the readiness to react on suggestive influences (suggestibility) can be predicted. This has been empirically substantiated using the SAM and a sensory suggestibility score (see Gheorghiu & Kruse, 1991; Kruse et al., 1992). Inner-systemic instability is a basic determinant of self-organized order formation. Therefore the readiness to change in multistable perception should be correlated to cognitive flexibility, creativity, originality in thinking, and imagination. These relationships, too, have already been empirically substantiated, too (see Gräser, 1977; Klintman, 1984). Consistent with these findings comparable correlations were also found for these adaptive cognitive functions and suggestibility (see Crawford, 1989). Suggestibility as well as RAC are reduced with age (Feingold, 1982; Heath & Orbach, 1963; Stukat, 1958). It seems to be possible to judge individual cognitive systems on a scale between two extreme cognitive strategies just in the way Pavlov did from a more neurophysiological perspective. The higher the degree of instability, the higher is the readiness for new processes of order formation triggered spontaneously by stochastic inner-systemic fluctuations or by changes in boundary conditions. A stable system is very reliable concerning its inner states but not very adaptive, while a system which is more often in an unstable state is less reliable but normally highly adaptive. On growing old the individual cognitive systems seem to develop in the direction of higher stability.

In addition to the correlations mentioned, there are some interesting empirical findings concerning pathological system states which can be interpreted in terms of the relation between cognitive order formation and innersystemic instability. In psychotic episodes schizophrenia often is characterized by symptoms like bizarre delusions, prominent hallucinations, disordered thoughts, incoherence, or loss of normal association between ideas. Most of the symptoms indicate instability in the process of cognitive order formation. Consistently Calvert et al. (1988) found that patients with a diagnosis of schizophrenia show a higher readiness to change in perceptual multistability. In the experiments the ambiguous Schröder staircase was used. Schizophrenic patients were significantly better able to see the staircase from below (normally a seldom realized alternative) and they showed a tendency to have a higher RAC than controls. Emrich performed a number of comparable experiments using inversion illusion (Emrich & Dirlich, 1993). Normally it is nearly impossible for perceivers to correctly interpret a picture of a negative cast of

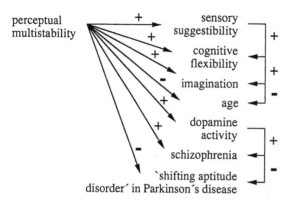

Fig. 9 Network of intercorrelations of perceptual multistability and other cognitive and physiological phenomena

a human face when seen from the inside. The picture is always misinterpreted as a normal face in frontal perspective. There is a strong bias for the ecologically more valid interpretation. Schizophrenic patients have significantly less difficulty in realizing the uncommon possible stable state of order.

In a mathematical analysis of neural network behavior, Carmesin recently demonstrated a relationship between the degree of spontaneous fluctuations and the formation of stable states of order in a neural network which comes quite close to the described characteristics of schizophrenia (for details see Kruse, Carmesin & Stadler, 1993). Enhancing the level of fluctuations in the network produced a phase transition in the ability of the network to approach stable states of order and to solve tasks. Up to a critical level of fluctuations the network was able to create stable solutions with an increasing level of complexity. But above the critical level the potential landscape progressively lost its structure and previously reached attractor states broke down. When the level of fluctuations in the network was lowered again, it was no longer able to solve already learned tasks. Depending on the level of fluctuations the network produced positive symptoms and residual disturbances (negative symptoms) comparable to a psychotic episode in schizophrenia.

One neurophysiological parameter discussed in the context of schizophrenia is the activity of the dopaminergic system. It has been proposed that the positive symptoms of schizophrenia result from overactivity in parts of this system (see e.g. Carlsson, 1974). Biochemical blocking of the dopaminergic receptors has a strong antipsychotic effect. Again a close relationsship to innersystemic instability can be found. Application of chlorpromazine, which has dopamine receptor blocking properties, to healthy persons lowers the RAC and the ability to perceive the less preferred alternatives in multistable perception (see Phillipson & Harris, 1984). Generally the application of dopamine receptor blocking drugs impairs cognitive flexibility (see Berger et al., 1989). Consistently with this, people suffering from Parkinson's disease show a disorder of cognitive shifting aptitude (see Cools et al., 1984). The upper end of the scale of cognitive instability seems to turn from high complexity and flexibility of order formation to the risk of pathological inability to stabilize at all, as described for the psychotic episodes of schizophrenia. And the lower end of the scale of cognitive instability seems to develop from reliability of simple order formation to rigidity and cognitive shifting aptitude disorder, as described for Parkinson's disease. Recently Elbert et al. (1992) demonstrated that patients with a diagnosis of schizophrenia show a higher dimensional complexity in EEG (Fz and Cz) compared to healthy controls. Therefore dimensional complexity of EEG may function as a physiological measure of innersystemic instability. Here the hypothesis arises that readiness to change between different stable states in multistable perception should be positively correlated with dimensional complexity, i.e. chaoticity, of the EEG.

The suggested relationship between innersystemic instability and cognitive order formation opens interesting research perspectives also in the fields of psychosomatics and investigation of hypnosis. Because in instability the effect of top down influences is enhanced, psychosomatic reactivity should increase with the degree of innersystemic instability (see Kruse & Stadler, 1990). Hypnosis can be characterized as a state of enhanced psychosomatic reactivity and after hypnotic induction the ability to imagine, cognitive flexibility, and suggestibility are higher. Therefore, the state of hypnosis may be understood as a state of enhanced innersystemic instability (see Kruse & Gheorghiu 1992). Generally, the measurement of situational and habitual differences in multistable perception is a promising tool for emipirical analysis of complex interactions, different system states and pathological phenomena in cognition.

Acknowledgements

The authors would like to thank Drs. H.-O. Carmesin and V. Gheorghiu for helpful discussions.

References

Basar-Eroglu, C., Strüber, D., Stadler, M., Kruse, P. & Basar, E. (1993): Multistable visual perception induces a slow positive EEG wave. International Journal of Neuroscience 73, 139 - 151.

Berger, H.J.C., van Hoof, J.J.M., van Spaendonck, K.P.M., Horstink, M.W.I., van den Bercken, J.H.L., Jaspers, R. & Cools, A.R. (1989): Haloperidol and cognitive shifting. Neuropsychologica 27, 629 - 639.

Bergum, J.E. & Bergum B.O. (1980): Reliability of reversal rates as a measure of perceptual stability. Perceptual and Motor Skills 50, 1038.

Calvert, J.E., Harris, J.P., Phillipson, O.T., Babiker, I.E., Ford, M.F. & Antebi, D.L. (1988): The perception of visual ambiguous figures in schizoprenia and Parkinson's disease. International Clinical Psychopharmacology 3, 131 - 150.

Carlsson, A. (1974): Antipsychotic drugs and catecholamine synapses. Journal of Psychiatrical Research 11, 57 -64.

Crawford, H.J. (1989): Cognitive and physiological flexibility: Multiple pathways to hypnotic responsiveness. In: V.A. Gheorghiu, P. Netter, H.J.Eysenk & R. Rosenthal (Eds.), Suggestion and Suggestibility: Theory and Research, pp. 155 - 167. Berlin: Springer.

Cools, A.R., van den Bercken, Horstink, M.W.I., van Spaendonck, K.P.M. & Berger, H.J.C. (1984): Cognitive and motor shifting aptitude disorder in Parkinson's disease. Journal of Neurology, Neurosurgery, and Psychiatry 47, 443 - 453.

Ditzinger, T. & Haken, H. (1994): A synergetic model of multistability in perception. In this volume.

Emrich, H.M. & Dirlich, G. (1993): Systemtheoretische Modellierung psychotischer Bewußtseinsstörungen. Paper presented at the Symposion Gehirn, Geist, Bewußtsein. GMD, Bonn.

Elbert, T., Lutzenberger, W., Rockstroh, B., Berg, P. & Cohen, R. (1992): Physical aspects of the EEG in schizophrenics. Biological Psychiatry 32, 181 - 193.

Feingold, E. (1982): Untersuchungen zur sensorischen Suggestibilität sowie zum Zusammenhang zwischen sensorischer Suggestibilität und der Placebo-Ansprechbarkeit im Schmerzbereich. Doctoral dissertation, Mainz.

Gheorghiu, V.A. & Kruse, P. (1991): The psychology of suggestion: An integrative perspective. In: J.F. Schumaker (Ed.), Human Suggestibility. Advances in Theory, Research, and Application, pp. 59 - 75. New York: Routledge.

Gräser, H. (1977): Spontane Reversionsprozesse in der Figuralwahrnehmung. Doctoral dissertation, Trier.

Klintman, H. (1984): Original thinking and ambiguous figure reversal rates. Bulletin of the Psychonomic Society 22, 129 - 131.

Haken, H. (1977): Synergetics. Berlin: Springer.

Haken, H. (1994): Some basic concepts of synergetics with respect to multistability in perception, phase transitions and formation of meaning. In this volume.

Haken, H. & Stadler, M. (Eds.) (1990): Synergetics of Cognition. Berlin: Springer.

Haken, H., Kelso, J.A.S. & Bunz, H. (1985): A theoretical model of phase transitions in human hand movements. Biological Cybernetics 51, 347 356.

Heath, H.A. & Orbach, J. (1963): Reversibility of the Necker cube: IV. Responses of elderly people. Perceptual and Motor Skills 17, 625 - 626.

Kanizsa, G. (1979): Organization in Vision. New York: Praeger.

Kelso, J.A.S., Case, P., Holroyd, T., Horvath, E., Raczaszek, J., Tuller, B. & Ding, M. (1994): Multistability and metastability in perceptual and brain dynamics. In this volume.

Köhler, W. (1940): Dynamics in Psychology. New York: Liveright.

Kruse, P. (1988): Stabilität, Instabilität, Multistabilität: Selbstorganisation und Selbstreferentialität in kognitiven Systemen. Delfin 6, 35 - 57.

Kruse, P. & Gheorghiu, V.A. (1992): Self-organization theory and radical constructivism: A new concept for understanding hypnosis, suggestion, and suggestibility. In W. Bongartz (Ed.), Hypnosis: 175 Years after Mesmer. Recent Developments in Theory and Application, pp. 161 - 171. Konstanz: Universitätsverlag Konstanz.

Kruse, P. & Stadler, M. (1990): Stability and instability in cognitive systems: multistability, suggestion, and psychosomatic interaction. In: H. Haken & M. Stadler (Eds.), Synergetics of Cognition pp. 201 - 215, Berlin: Springer.

Kruse, P. & Stadler, M. (1993): The significance of nonlinear phenomena for the investigation of cognitive systems. In: H. Haken & A.S. Mikhailov (Eds.), Nonlinearity in Complex Systems, pp. 138 - 160. Berlin: Springer.

Kruse, P., Carmesin, H.-O. & Stadler, M. (1993): Stabilisierung adaptiver Dynamik durch Bewertung: Schizophrenie als Korrespondenzproblem plastischer neuronaler Netzwerke? Paper presented at the 3rd Herbstakademie: Selbstorganisation in Psychologie und Psychiatrie, Sozialpsychiatrische Universitätsklinik Bern.

Kruse, P., Roth, G. & Stadler, M. (1987): Ordnungsbildung und psychophysische Feldtheorie. Gestalt Theory 9, 150 - 167.

Kruse, P., Stadler, M. & Strüber, D. (1991): Psychological modification and synergetic modelling of perceptual oscillations. In: H. Haken & H.P. Koepchen (Eds.), Rhythms in Physiological Systems, pp. 299 - 311. Berlin: Springer.

Kruse, P., Stadler, M. & Wehner, T. (1986): Direction and frequency specific processing in long-range apparent movement. Vision Research 26, 327 - 335.

Nebelytsyn, V.D. (1964): An investigation on the connection between sensitivity and strength of the nervous system. In: J.A. Gray (Ed.), Pavlov's Typology, pp. 402 - 245. Oxford: Pergamon Press.

Necker, L.A. (1832): Observations on some remarkable phenomenon which occurs on viewing a figure of a crystal or geometrical solid. The London and Edinburgh Philosophical Magazine and Journal of Science 3, 329 - 337.

Pavlov, I.P. (1955): Selected Works. Moscow: Foreign Language Publishing House.

Phillipson, O.T. & Harris, J.P. (1984): Effects of chlorpromazine and promazine on the perception of some multi-stable visual figures. The Quarterly Journal of Experimental Psychology 36A, 291 - 308.

Schöner, G. & Hock, H. (1994): Concepts for a dynamic theory of perceptual organization: An example from apparant movement. In this volume.

Schöner, G., Haken, H. & Kelso, J.A.S. (1986): A stochastic theory of phase transitions in human movement. Biological Cybernetics 53, 247 - 257.

Sherif, M. (1935): A study of some social factors in perception. Archives of Psychology 187, 1 - 60.

Stadler, M. & Kruse, P. (1990): The self-organization perspective in cognition research: historical remarks and new experimental approaches. In: H. Haken, & M. Stadler (Eds.), Synergetics of Cognition, pp. 32 - 52. Berlin: Springer.

Stadler, M. & Kruse, P. (1994): The function of meaning in cognitive order formation. In this volume.

Stukat, K.G. (1958): Suggestibility: A Factorial and Experimental Analysis. Stockholm: Almqvist and Wiksell.

Szu, H.H., Lu, F. & Landa, J.S. (1994): Artificial neural networks and Haken synergetic computer hybrid approach for solving bistable reversible figures. In this volume.

Zimmer, A.C. (1994): Multistability, more than a freak phenomenon. In this volume.

Task, Intention, Context, Globality, Ambiguity: More of the Same

C. van Leeuwen

University of Amsterdam, Faculty of Psychology, Department of Psychonomics, Roetersstraat 15, NL-1018 WB Amsterdam, The Netherlands

Abstract: Network Ecology Theory (NET) is introduced as a general processing principle, applying to both semantic and perceptual ambiguity. It is proposed that the ambient activity of networks may be characterized in terms of local precedence yielding over time to global preference. An asymptote is set to the global preference reached by the single parameter of this approach, the reset rate. I review examples from semantic and perceptual ambiguity research to illustrate the applicability of this processing principle.

1 Introduction: Ambiguity and context

I will discuss two examples of ambiguity research, from the two domains of the present symposium. The first is from semantic ambiguity: The disambiguation of word meaning in context as investigated by Swinney (1979). The second is from perceptual ambiguity of the Necker cube as investigated, among others by Peterson and Hochberg (1983). Each study introduces a phenomenon that I think is pervasive in information processing. I claim these phenomena possess a basic theoretical significance. Once noticed, analogues will readily be visible in several domains. I will present a theory of context information usage that accounts for the phenomena illustrated.

General processes need a general, qualitative theory. A theory should disregard specific assumptions about network architecture, activation functions, learning rule and parameters. I will therefore not present my theory in the form of, for instance, a specific neural network model, but as a set of assumptions specifying some general qualitative constraints on neural networks.

2 Swinney (1979): Disambiguation not used initially

The first experiments to be discussed are from the domain of semantic ambiguity and how it is resolved by context. Swinney (1979) performed experiments, in a cross-modal primed lexical- decision paradigm. Sentences like the following were auditorily presented (Anderson, 1990):

Rumor had it that, for years, the government building had been plagued with problems. The man was not surprised when he found several spiders, roaches, and other bugs in the corner of the room.

The prime in this sentence was the ambiguous word bug. Targets were visually presented with varying delay from prime onset. The word ant is related to the meaning of bug that is primed by the preceding sentence context, while spy is related to the unprimed meaning. If the target is presented within 400 ms of the prime, facilitation of both ant and spy occurs. Thus, the presentation of **bug** immediately activates both of its meanings and their associations. After 700 ms, there was facilitation only for the context-related word ant. Thus, two meanings of an ambiguous word rapidly become activated, but context operates soon afterwards to select the appropriate meaning.

An alternative way of stating this is the following: Initially the listener uses only cues locally available to identify the meaning of a certain word (i. e. the word itself). Later, more global cues (i. e. the preceding sentence context) are brought to bear on the interpretation. This phenomenon I therefore call **local precedence** giving way over time to **global preference**. There are many analogues of this process in perception and cognition. They can be observed at different time scales ranging from 200 ms (Hogeboom, van Leeuwen & Bakker, 1993) to hours of study (van Leeuwen, Buffart & van der Vegt, 1988). For a review, see van Leeuwen (1993).

3 The first NET principle:
Local precedence/global preference

To explain these phenomena, I will present the first assumption of a theory called "Network ecology theory" (NET). The "ecology" part of NET is a tribute to Gibson's "scientific revolution" that allowed him to focus on the **higher order** invariant patterns in the ambient array, see Figure 1. Figure 1A shows light radiating from a source. The light conveys first-order information (intensity, colour). Figure 1B shows light reaching the eye from the ambient array. The light is reflected from surfaces, and thus contains higher order information (intensity and colour patterns).

A similar revolution is attempted here for network activity. Take an arbitrary fragment of the network, i.e. a single unit, or a group, and call this a **node**. Traditionally network theories have used radiant (spreading) activation from a node to its neighbours (Figure 2A) to model important behaviour,

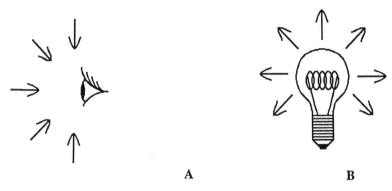

Fig. 1 Ambient light (A) and radiant light (B)

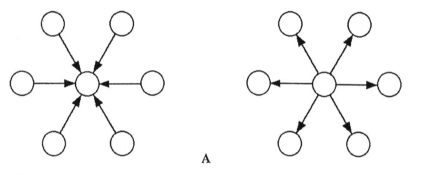

Fig. 2 Radiant (A) vs. ambient (B) network activation

such as memory search and priming effects. The paradigm case for such behaviour is first-order information, the expected value (mean) of the output activation of a node. I will, however, draw attention to network ecology, i.e. the ambient array of incoming signals in the neighbourhood of the node (Figure 2B). In the ambient array of a node, the paradigm is **higher-order information**, i.e. the expected values of activation **patterns**. The higher the order of the information conveyed, the stronger constraint is represented in the ambient array of a node.

There is a remarkable invariant about how the order of information in network ecology develops as a function of time. For instance, assume some units in the environment of a node contain information not earlier transmitted and start transmitting this information at t_0. I say these units are **reset** at t_0. The energy of signals transmitted from the receptors may may achieve this; units also may be reset at random intervals. At t_1, the node central in our analysis receives information from its neighbours only. These, in turn, receive at t_1 input from their neighbours, and may start to transmit a modified version of their information, combined with what they received at t_1. This

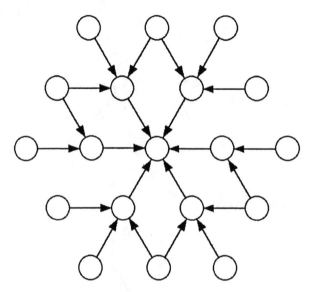

Fig. 3 Direct vs indirect ambient influence

leads to an increase of the order of the information in the signal transferred. At t_2, the modified signal arrives at the node central in our analysis.

If I compare the signals incoming at this node at t_1 and t_2, the incoming information at t_1 is a combination of signals from the direct neighbourhood. There are few constraints on the activation pattern established at t_1, but those constraints which operate, are specific (in the sense that the source of information is specified by the local connectivity pattern). I call this state of receiving, relatively, lower order information a state of local organization. Thus in this initial state of the information process, information is local. I therefore call this a state of **local precedence**.

By contrast, at t_2 the information is more **global**. The signals arriving are influenced by more distant nodes, that through the direct neighbours convey a higher order pattern of information. They therefore more strongly constrain the state of the central node at t_2. This is called **global preference**. This tendency is independent on the number of steps. At t_3, this will be more strongly so. And this could go on, in principle, indefinitely. In sum, the characteristic process according to this description is local precedence leading to global preference.

Global information processing is diffuse, in the sense that the sources of information are no longer specified individually through the local connectivity pattern. If the activity of two indirect neighbours of a node is combined, it is likely that whatever is common in the two is strengthened, whereas what is individual to each one is suppressed. In this sense, the incoming information from more distant sources is increasingly abstract.

The local precedence and global preference in the theory could be used to explain the results of Swinney. The information of the prime initially (> 400 ms) is not related to the information of the global sentence context, so the prime's meaning is not disambiguated locally. Later (< 700 ms) the preceding sentence context is combined with the prime information to disambiguate the prime globally.

Similar considerations apply to several other experiments in the literature on visual perception and cognition, where local information precedes global information. For instance, in perception, the microgenesis studies by Sekuler and Palmer (1992); the experiments on visual search by Nattkemper and Prinz (1984), or the eye-movement research of de Graef, Christiaens and d'Ydewalle (1990), as well as studies on perceptual learning (van Leeuwen, Buffart & van der Vegt, 1988) and development (van Leeuwen, L., Smitsman & van Leeuwen, C., 1993).

Several recent studies on the perception of structure have conflated the issue of local precedence with the global precedence discussed by Navon (1977). These are, however, not opposing results, but two diferent issues. Our study deals with late, focal perception and cognition, Navon with early and peripheral vision. Our local and global are in terms of lower order and higher order information, Navon's local and global relate to low vs high spatial frequencies. It is likely that Navon's results should be explained in terms of the relative speed of early processing channels for different spatial frequencies, whereas NET applies to endogeneous factors.

4 The Necker cube

For the introduction of the second part of my theory, I now turn to the issue of figure reversal. One of the best known ambiguous figures is the Necker cube, a wire figure with at least two interpretations. See Figure 4 (middle), the flanking figures are alternative interpretations of this cube. The cube can be seen in two alternative orientations, viewed obliquely either from above or from below. Spontaneous reversal is known to occur.

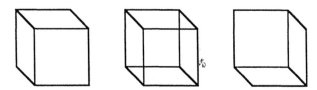

Fig. 4 Middle: Necker cube; left and right: alternative interpretations

5 Peterson and Hochberg (1983):
Cues within a figure are not globally used

Peterson and Hochberg (1983) studied reversal in Necker cubes. They used a technique in which local orientation cues were added (Figure 5) and people must indicate which orientation they perceive.

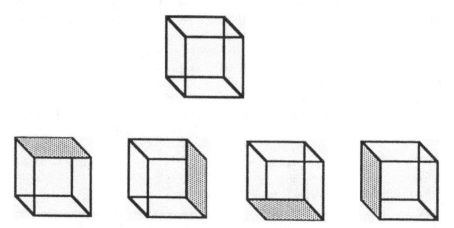

Fig. 5 Local disambiguations of the Necker cube (after Peterson & Hochberg, 1983)

In addition the location of attention was varied. It could be near the locally disambiguating cue, or further away (Figure 6). When a disambiguated area was attended to, people perceive the orientation according to the disambigation cue. When, however, an unbiased location was attended, the local disambiguation elsewhere in the figure had less, or no influence. Yet these figures were small enough (2 degrees of visual angle) for the whole figure to be captured in one glance.

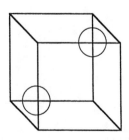

Fig. 6 Alternative focus of spatial attention (after Peterson & Hochberg, 1983)

6 Perception is piecemeal?

Peterson and Hochberg concluded from these, and related studies, that perception is piecemeal. A percept is not a strong, global whole but a mere loose construction out of elements.

Both Gestalt psychology (Koffka, 1935) and ecological realism (Gibson 1979) are committed to the claim that the interpretation of local parts of the visual field is determined by properties of its overall structure. Peterson and Hochberg's results seem therefore problematic for these theories. In our view, however, there is no need to assume that a global organization principle is violated by these studies. It could be assumed that subjects perform local (i.e. piecemeal) information processing **for strategic reasons**.

7 Global context influence

We figured that the strategy to process the Necker cube locally could have been induced by the absence of context within Peterson and Hochberg's procedure. We therefore replicated the studies of Peterson and Hochberg with context added (Figure 7). Each cube was surrounded by eight other ones, which could either be ambiguous themselves, or in one of the two possible orientations. It was shown in our experiments that context orientation had a strong overall influence on the perceived orientation in the central, ambiguous Necker cube. This illustrates, supporting Gibson and the Gestaltists against Hochberg's claims, that piecemeal perception is not necessary; global processing is possible. A global perceptual strategy could replace the local, piecemeal strategy as soon as helpful information is presented in the surrounding context of the Necker cube.

8 The strategic character of local vs. global processing

Peterson and Hochberg, together with our replication illustrate the second phenomenon relevant for our discussion. I shall call it the strategic character of local vs. global information processing. We have found recent support also from other areas of perception and cognition (Stins & van Leeuwen, 1993; Hogeboom & van Leeuwen, 1993).

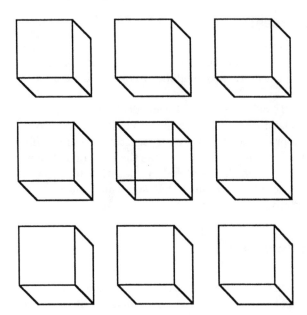

Fig. 7 Necker cube with context added (after van Leeuwen, Weitenberg, & Smit, 1993)

9 The second NET principle: Reset rate equilibria

To explain the strategic character of local or global processing of ambiguous figures and elsewhere, I will introduce an important further assumption of NET. This is the domain specific, adaptive character of the local or global mode of representation as an equilibrium. The equilibrium is usually thought of as a fixed point, but in the case of reversible figures it is to be regarded as a periodic orbit or a "strange" attractor. This equilibrium is reached by a trade off between two tendencies in the system. The first is the tendency from local precedence to global preference just discussed. The second follows directly from a necessary assumption for local precedence to occur. This is the tendency of an information processing unit to reset itself and make a fresh start (producing initially first-order information only). The frequency with which the second tendency is actuated gives rise to the **reset rate**. Reset rate is assumed to be a control variable of an information process. The reset rate sets an asymptote to the amount of globality reached within an information process. The higher the reset rate, the more local the representations used in that process.

It is an empirical issue which factors influence reset rate. Reset rates are assumed to depend at least partly on the incoming stimulation. There is reason to assume that in preparation for incoming information, there is perturbation in the system that reduces the order of information (Freeman,

1990). That the order of information in the neural signal is reduced in **anticipation** of a stimulus rather than at its onset, is in accordance with the view that a strategic factor like reset rate is responsible. Reset rate may have become adapted to the sensoric representation, for instance because the task of extracting information from a changing, retinotopic patchwork is optimally performed with relatively high reset rates. As a consequence, the representation of a visual pattern will be more locally organized in the presence of a stimulus than in its absence. This principle is illustrated when we look at Figure 8 below (adapted from Kanizsa, 1970). We perceive this figure as a mosaic of a black cross lying in a checkerboard pattern (an interpretation stressing local regularities, as the cross dominates the checkerboard) rather than as a checkerboard pattern with one square made black (an interpretation stressing global regularities, as the checkerboard dominates). However, if we try to imagine or recall the figure, we prefer the global interpretation. In absence of the stimulus, reset rate might therefore be lower (see also Wulff, 1922).

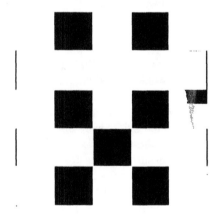

Fig. 8 Kanizsa figure: A cross instead of a square occludes the checkerboard

The reset rate in **perceptual** domains seems mostly not to adapt in accordance to what we remember. Otherwise, one would expect that the global interpretation would become dominant for the Kanizsa figure, mainly by looking longer[1]. Likewise, certain perceptual illusions do not disappear under the influence of our beliefs or strategies. As "information encapsulation" of perception, this last fact played a role as an argument for the modularity of mind (Fodor, 1983). Fodor concluded that there had to be fixed, structural constraints on context information availability in early perceptual processes. A

---

[1] Note, however, that some individuals actually report seeing a white patch covered by a black square (the global representation) after looking long at the stimulus. The trick might be to focus attention on the white diagonals.

fast reset rate could provide an alternative explanation for information encapsulation. Reset rate is not a fixed constraint, but a control **variable**. Such an explanation would take into account that the encapsulation is strategic, and could disappear, for instance by an exploration leading to a change in vantage point (Kennedy & Portal, 1990). That strategic factors should account for the piecemeal perception in Peterson and Hochberg's studies is suggested by our studies. Reset rate could change (and thus the characteristics of piecemeal perception disappear) if circumstances invite this.

10 Reset rate as a psychophysiological hypothesis

The reset rate is a frequency common to a certain information process. This yields psychophysiological predictions that could be tested by investigating the spectrum of electrocortical brain potentials during the execution of a task. The reset rate should be the dominant frequency in the power spectrum within the appropriate range. The same stimulus material could be processed context dependently or independently, and the dominant frequency should vary in accordance with that: Context-dependent processing should yield lower dominant frequency.

A simple, preliminary test is possible under the assumption that reset rate has a subject-specific component. Then it is possible to predict differences between individual subjects in the peak-alpha frequency of their background EEG. A correlation with the subject's average context dependency is predicted. Figure 9 illustrates the prediction. Background EEGs were obtained.[2] The criterion was the average influence of the surrounding context on the interpretation of the ambiguous Necker cube, as measured in the task of van Leeuwen, Weitenberg and Smit (1993). Over 15 subjects, a correlation of .615 between peak-alpha frequency and the context dependency criterion was reached (Smit, Kohl, Heslenfeld, Kenemans & van Leeuwen, 1993). We know of no other study showing such a large correlation with EEG peak-alpha frequency, so our data seem to support the interpretation of EEG alpha peaks as reset rates. By this, we may also have found a theoretical basis for the personality trait concept of cognitive style (Witkin, Moore, Goodenough & Cox, 1977).

---

[2] Average power spectra were calculated for one hundred 5.5 s periods of sampling over 2 occipital and 1 parietal electrode.

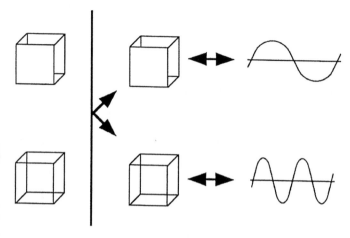

Fig. 9 Results of Smit et al. (1993): r = 0.615; N = 15

11 Summary

A theory was presented, predicting local precedence/global preference in perception and cognition. The theory could further be used to explain task- and subject-specific components in the usage of context information. I discussed applications of this model to the experimental results obtained in recent investigations on ambiguous stimuli, such as the Necker cube. The cube could be perceived locally or globally, dependent on the strategy. Which strategy is used determines which information is used to disambiguate the cube. Some progress was made in showing that the only parameter of this general computational theory, the reset rate that determines the dependency on global context in information processing, has a psychophysiological basis.

I conclude that there is some preliminary evidence for the principles of NET. The computational principles of the theory had explictly been stated before we started to build neural network models. Because theories without a model are, in principle as bad as models without theories, we have recently implemented some computer simulations of the NET principles. Further developments along these lines have to show whether these principles could be useful in the usual domains of neural network applications, like pattern recognition and categorization.

References

Anderson, J.R. (1990): Cognitive Psychology and its Implications. New York: Freeman.
de Graef, P., Christiaens, D. & d'Ydewalle, G. (1990): Perceptual effects of scene context on object identification. Psychological Research 52, 317 – 329.
Fodor, J.A. (1983): The Modularity of Mind. Cambridge, MA: MIT Press.
Freeman, W.J. (1990): On the problem of anomalous dispersion in chaoto-chaotic phase transitions of neural masses, and its significance for perceptual information in the brain. In: H. Haken & M. Stadler (Eds.), Synergetics of Cognition, pp. 126 – 143. Berlin: Springer.
Gibson, J.J. (1979): The Ecological Approach to Visual Perception. Boston: Houghton Mifflin.
Hogeboom, M. & van Leeuwen, C. (1993): Complexity effects in local and global visual search strategies. (in press).
Hogeboom, M., van Leeuwen, C. & Bakker, L. (1993): Perceptual organization, stimulus complexity and sameness detection strategy. (in press).
Kanizsa, G. (1970): Amodale Ergänzung und Erwartungsfehler des Gestaltpsychologen. Psychologische Forschung 33, 325 – 344.
Kennedy, J.M. & Portal, A. (1990): Illusions: Can change of vantage point and invariant impressions remove deception? Ecological Psychology 2, 37 – 53.
Koffka, K. (1935): Principles of Gestalt Psychology. New York, NY: Harcourt Brace.
Peterson, M.A. & Gibson, B.S. (1991): Directing spatial attention within an object: Altering the functional equivalence of shape description. Journal of Experimental Psychology: Human Perception and Performance 9, 183 – 193.
Peterson, M.A. & Hochberg, J. (1983): Opposed-set measurement procedure: A quantitative analysis of the role of local cues and intention in form perception. Journal of Experimental Psychology: Human Perception and Performance 9, 183 – 193.
Nattkemper, D. & Prinz, W. (1984): Costs and benefits of redundancy in visual search. In: A.G. Gale & F. Johnson (Eds.), Theoretical and Applied Aspects of Eye Movement Research, pp. 343 – 351. Amsterdam, NL: North Holland.
Navon, D. (1977): Forest before trees: The precedence of global features in visual perception. Cognitive Psychology 9, 353 – 383.
Sekuler, A.B. & Palmer, S.E. (1992): Perception of partly occluded objects: A microgenetic analysis. Journal of Experimental Psychology: General 121, 95 – 111.
Smit, D., Kohl, K., Heslenfeld, D., Kenemans L., & van Leeuwen, C. (1993): Event related potentials and background EEG predictors for context dependency in visual ambiguity resolution. (Manuscript in preparation).
Swinney, D.A. (1979): Lexical access during sentence comprehension: (Re)consideration of context effects. Journal of Verbal Learning and Verbal Behavior 18, 645 – 659.
Stins, J. & van Leeuwen, C. (1993): Context influence on the perception of figures as conditional upon perceptual organization strategies. Perception & Psychophysics 53, 34 – 42.
van Leeuwen, C. (1993): Network Ecology Theory and Local Precedence as a Principle of Processing in Perception and Cognition. (in press).
van Leeuwen, C., Buffart, H. & van der Vegt, J. (1988): Sequence influence on the organization of meaningless serial stimuli: Economy after all. Journal of Experimental Psychology: Human Perception and Performance 14, 481 – 502.
van Leeuwen, C., Weitenberg, N. & Smit, D. (1993): Context influences on the perception of the Necker cube. (in press).

van Leeuwen, L., Smitsman, A. & van Leeuwen, C. (1993): Tool-use in infancy; perception of a higher-order relationship. Journal of Experimental Psychology: Human Perception and Performance. (in press).

Witkin, H.A., Moore, C.A., Goodenough, D.R. & Cox, P.W. (1977): Field-dependent and field-independent cognitive styles and their educational implications. Review of Educational Research 47, 1 – 64.

Wulff, F. (1922): Über die Veränderung von Vorstellungen (Gedächtnis und Gestalt). Psychologische Forschung 1, 333 – 373.

Multistability –
More than just a Freak Phenomenon

A. C. Zimmer

Institute of Psychology, University of Regensburg, D-93053 Regensburg, Germany

Abstract: A classification of forms of multistability reveals that multistability is not confined to experimental settings but a pervasive, albeit inconspicuous phenomenon also in complex scenes and events.

Analyzing the conditions of multistability and its relation to spatial perception results in the notion of parallel, cooperative as well as competitive processes contributing to the emergence of a stable spatial percept.

The epistemological implications of such an approach are discussed and termed 'Interactive Realism;' according to which the processes underlying the 'constructive mind' have evolved under the constraints of the transphenomenal (physical) world.

1 What is multistability about:
The real world or the laboratory?

If one looks at the astounding examples of multistability, e.g. Jastrow's duck/rabbit (Fig. 1) which has influenced Wittgenstein in his Philosophical Investigations or the looking glass/skull (Fig. 2), or if one experiences the involuntary switches in the perception of the Necker cube and compares these with the apparent stability of everyday perception one tends to write off these phenomena as freak phenomena albeit interesting, perhaps telling as much about the processes of perception as the Manierist anamorphic pictures (see e.g. Leeman, Elffers & Schnyt, 1976) tell about Renaissance art.

This point has been made in the framework of Ecological perception according to which either illusions happen only if the information in the provided stimulus is underdetermined (Michaels & Carello, 1981, p. 180, 181) or 'the perception is not in error' but the researcher who erroneously equates measures of physical energy with reality (Michaels & Carello, 1981, p. 71).

Regarding their venerable status in the history of perception calling phenomena like Rubin's vase, Jastrow's duck/rabbit or Necker's cube freak phenomena might ring of facetiousness but if one takes into account the fact

Fig. 1 Jastrow's duck/rabbit form

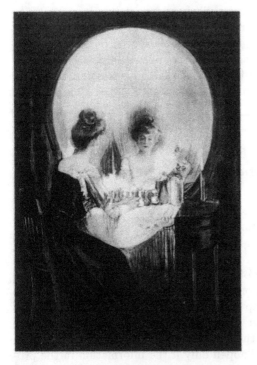

Fig. 2 The looking-glass/skull picture

that these phenomena exhibit multistability only under very specific conditions or in the case of the wire cube or wire staircase seem to lack ecological validity, one is tempted to doubt their importance for more general rules of perception. However, the field of multistability is diverse and therefore before starting a detailed analysis, a classification of types of multistability might be helpful.

The distinctions I have made are not as clear cut as it appears, for instance, the major characteristic of the geometrical category, namely the induction of 3-dimensionality in 2-D displays, can also be found, albeit to a lesser degree, in the other categories: In the looking glass/skull example as well as in the 'View of a City' the semantic and/or the Fig.-ground disambiguation induce depth. However, as Brandimonte & Gerbino (1993) have demonstrated, semantic priming can be used to classify types of multistability because only for semantically ambiguous forms does priming cause the perception to be uniquely specified. That a combination of different types of multistability can achieve striking effects, demonstrate Bradley & Petry (1977) in their Necker cube viewed through peep-holes in a white surface before a black background or in front of 8 black circles on top a white background; if the second view is taken, virtual contours appear as in Kanizsa (1955).

Table 1 Classification of types of multistability

Category	Example (only those examples which are depicted in this article)
semantic	rabbit/duck (Fig. 1), looking glass/skull (Fig. 2)
Fig.-ground	"View of a City" after Metzger (1975) (Fig. 3)
illusory contours	Bradley and Petry's view of a Necker cube (Fig. 4)
geometrical multistability (a) depth (b) concave-convex (c) classification	the Necker cube (Fig. 4) Japanese patterns (Fig. 5) the cupola of S. Giovanni degli Eremiti (Fig. 6) Roman-Byzantine tilings (Fig. 7) non-periodic Penrose tilings (Fig. 8)

My goal is to show that by the observable regularities in the domain of geometrical multistability we gain insights into the tuning of perception for picking up objects and spatial relationships between these objects and the perceiver. The multistability of the Necker cube will serve to define conditions for spatial impressions from flat graphical designs.

That multistability is more than a negligible fringe effect in visual perception becomes obvious if one analyzes the amount to which the information giving rise to a stable percept really specifies the seen objects and their spatial arrangement in regard to each other and to the viewer. Figure 9 gives

Fig. 3 "The View of a City" after Metzger, 1975

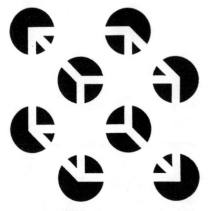

Fig. 4 A Necker cube seen through eight peepholes or before eight black circles (after Bradley & Petry, 1977)

the view of a cathedral together with 3 different groundplans (out of many other possible once) compatible with this view.

However, this 'real-world' multistability in a complex situation remains unnoticed because only the interior with a rectangular groundplan is 'seen'. This can be interpreted as the result of two complementary processes in perception, that of stability and its tendency towards simple forms (Wertheimer, 1912) and that of singularity which specifies uniquely the position of the viewer relative to the perceived objects. Kanizsa and Luccio (1986) were the first to point out that the term 'Prägnanz' in Gestalt theory really is ambiguous; Zimmer (1991) in commenting them has shown that the relation of singularity and stability is that of a complementarity: "The second aspect of the complementarity, namely that of interaction between local vs. global optimization, can best be exemplified in the field of spatial perception where the forked effect of local optimization (stability leading to the transformation of ellipses into circles. of arbitrary rectangles into squares etc.) and the uniqueness (singularity) of the point of view give rise to a stable

Fig. 5 Japanese patterns inducing a depth effect

Fig. 6 The cupola of S. Giovanni degli Eremiti in Palermo exhibiting a concave-convex switch in the upper pendentives of the main cupola

image of the surrounding world despite the fact that any given projection can originate from a multitude of spatial arrangements. LaGournerie in the last century (see Fig. 9) and Ames since about 1935 have shown this and how local orthogonalizations or symmetrizations together with the singularity of the viewing point result in unique spatial impressions even if they contradict 'known facts' as e.g. the relative height of people in the Ames-room. One prediction from this assumption that it is the tension between a global tendency towards stability and the sensitivity to local disturbances which generates the impression of space, is that this impression should be strongest if the forked

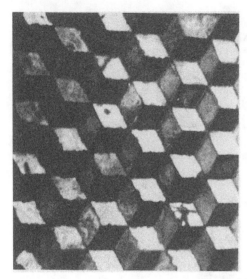

Fig. 7 A Roman-Byzantine mosaic (about 500 AD)

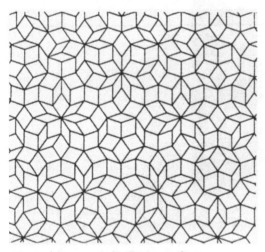

Fig. 8 A non-periodic Penrose tiling

effects are about equal. Since the effects cannot be measured directly, this hypothesis can be tested indirectly by showing that in 2-dimensional drawings the strength of the spatial impression is not maximal for drawings that obey perfectly the rules of perspective but for those which form a compromise between stability of partial forms and perspective distortions" (Zimmer, 1991, pp. 277 – 278).

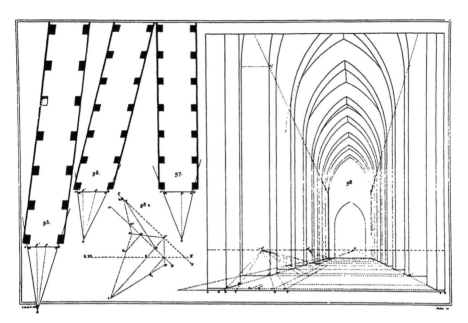

Fig. 9 A view of a cathedral with 3 compatible groundplans (after LaGournerie, 1859)

That the relationship between stability and singularity is a complementarity can be seen from the fact that in the perception of simple stimuli, for instance angles or triangles, even minute deviations from orthogonality or triangularity are detected immediately, which Goldmeier (1982) attributes to the effect of singularity, but that in remembering the same stimuli (e.g. triangles) there is a strong tendency for nearly all stimuli to be identified with the equilateral or orthogonal ones (Zimmer, 1991), an effect which can be modelled according to the laws of gravitation, that is, a process with a tendency towards most stable configuration. This implies that with any potential landscape characterized by sinks, which specify the control parameters in the synergetic approach, there is associated an inverse landscape where these points of stability become singularities (Fig 10 a, b).

One implication of such a model is that the traditional separation of perception and memory should be abandoned; instead of it a unitary if complex coping of organisms with their environment is postulated.

The duality of perceiving objects as they are and at the same time perceiving the spatial relations between the perceiver and the perceived seems to be fundamental for the process of perception. Nevertheless, in experimental research this duality is usually dissolved: Either object and/or feature recognition or distance, time-to-contact or spatial perception are analyzed. My hunch is that this dissolution is partially due to the fact that it is difficult to construct stimuli which are complex and exactly analyzable and

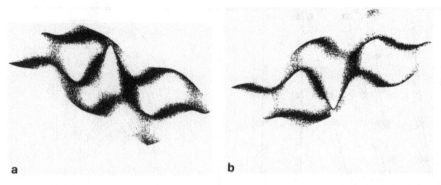

Fig. 10 Two complementary potential landscapes: (a) after Haken (1991); (b) its inverse

controllable at the same time. This is one of the reasons why I think that inherently complex stimuli such as multistable forms and perspective drawings offer an opportunity for an experimental analysis of space perception without dissolving this fundamental duality.

The epistemological approach most in line with this line of argumentation might be termed Interactive Realism, because it accounts well for the apparent order in the perceived world.

2 Perception and epistemology: Do we pick up what is already there or do we construct it out of our heads or both?

Before delving into the questions concerning the order in nature of in our perception of it, some words about Realism seem to be necessary. In the plainest sense, Realism is the epistemological position which claims that the objects we see, feel, remember and talk about have a correspondence 'out there' in the physical realm, that is, the mind is intentional. In this sense the entire pre-Socratic philosophy plus Aristotle and even Plato adhered to a realist position. The extreme counterposition of solipsism is so contrary to the world of everyday experiences that even Idealists tended to subscribe to a primitive Realism as a kind of epistemological truism.

What advances the Realist position toward more specific epistemological theories are the questions about natural categories (e.g. humans), about natural laws (e.g. can we say something about gravity that refers to something as real as an individual stone?), about the relation between mental representations of objects and the physical objects and about persistence over time. If one answers all these questions in the affirmative, one arrives at the position of Direct (or derogatorily called Naive) Realism as championed by J.J. Gibson:

"It seems to me that these hypotheses (namely (1) the existence of stimulus information, (2) the fact of invariance over time, (3) the process of extracting invariants over time, and (4) the continuity of perception with memory and thought make reasonable the common sense position that has been called by philosophers direct or naive realism" (Gibson, 1967, p. 168).

Gibson discards alternative Realist positions, for instance Holt's, as circular in their support for this common sense position because they regard perception as sensation-based; in contrast, his main 'new reason for Realism' is the notion of the information-based perceptual system. Appealing as this common sense position is, one should nevertheless ask oneself why this position has come to be questioned at all. I want to do this in a paradigmatic fashion in regard to the Realism-Nominalism debate in Scholastic philosophy because it can be shown that this old debate is far from outdated.

The core question of this debate is about universals: Are universals real, independent of the mind, as William of Champaux (1070 – 1121) claimed, or are they nomina, nothing but an exhalation of the voice (flatus vocis), as Roscelin (1050 – 1125) objected? It is apparent that William of Champaux' claim cannot be generally true, there exist without doubt classes of objects made up by a specific mind or being the result of a cultural convention. But on the other hand to deny the reality of natural categories implies that no planned interaction with nature (including science) is possible because then only individuals at specific times and places are real and any predictions about classes of individuals or about individuals over time have no basis in reality.

This problem is stated most succintly by Lichtenberg, a German physicist and philosopher of Enlightenment:

"Everything that we as humans recognize as being real, is real for humans. Otherwise, if it were no longer possible to be able to judge every natural constraint as reality, it would be impossible to think even in terms of concrete principles." (Georg Christoph Lichtenberg from "Bemerkungen" in the years 1766 and 1799).

Modern versions of Roscelin's Nominalism underly Whorf's thesis of cultural relativism and, for instance, Quine's postulate that classes are generally abstract and not real entities, is genuinely nominalistic. Furthermore the 'language of thought' (Fodor, 1974) or the computation approach (Pylyshyn, 1984) to human information processing are in a way nominalistic as well as van Fraassen's or Putnam's antirealism. A side effect of the debate shows in modern logic where the universal quantifier (as in "all men are mortal") is paraphrased (as in "for any x, if x is a man, x is mortal").

Already Abelard (1079 – 1144), William's pupil, and later especially Roger Bacon (1220 – 1292) and Ockham (1285 – 1349) have concentrated on questions like 'what constitutes natural categories' and 'what is the relation between the concept of a class (the representation in the mind) and its counterpart in the physical world?' Abelard approached the problem from an ontological Platonic angle: There is no common essence in a class but a common idea present when God created it, that is, from God's point of view, universals are 'ante rem' but from the human perspective, they are 'post rem'. In contrast, Roger Bacon and Ockham searched for empirical criteria and arrived in the end at a position of conceptual empirism: Bacon's "Deficiente sensu deficit scientia" postulates that the concepts of science (as pure cases of natural categories) have to be based on observables. Ockham gave this position an even more radical twist when he denounced the theories of mediated perception (especially that of Duns Scotus) and claimed that perception is direct, making him a predecessor of Gibson and modern Direct Realism.

A further attempt to save Realism from the Nominalist attack is concerned about the relation between the representations in the mind and their objects in the physical world leading to the Representational Realism of J. Locke; the shortcomings of which were resolved in Santayana's 'Critical Realism' (1920) distinguishing the phenomenological world and its transphenomenal counterpart.

An early attempt to define non-mentalistic processes inducing a self-organisation of the phenomenal world under the same constraints as present in the physical world was W. Köhler's field theory (1920) which set the frame for the Gestaltist theory of perception.

For terminological clarity, I want to point out that I use the term 'physical world' (not 'world of physics') equivalent with the term 'transphenomenal world' as introduced by Santayana (1920) in his treatise on 'Critical Realism'. In this position the Nominalist fallacy is refuted according to which the physical world (res) and the cognition operations of the human mind upon it (Roscelin's 'flatus') can be neatly separated. The tradition of that line of argumentation can be found in Hume's theory of perception, refined by Kant in his theory of schemata which link the 'images' of the physical world and the categories of thinking. However, Kants interactionist position on perception has not been very influential for the development of psychological theories of perception because starting with Helmholtz' hierarchy of perceptual processes: Sensations, unconscious inferences, conscious cognitions, the Nominalist tradition of separating sensations as directly influenced by the physical world and perceptions as the cognitively processed sense informations is very alive albeit disguised in modern theories of perception, for instance, in Marr's (1982) or Pylyshyn's computational approach (1984).

The problems about the relation between our mental representations and the world 'out there' have not lost their importance and some of the riddles of Realism are far from solved (e.g. how is the information produced that constitutes Gibson's ecological perception which allegedly solves these problems?) but also a radical Nominalism is untenable, or at least implausible and impractical for science. Therefore further investigations into Realism are necessary. Multistability can be regarded as a paradigmatic approach to the question how the order in the phenomenal world is related to that of the physical world and by doing this, one arrives at a position which might be termed Interactive Realism; a position claiming that for the perceiver the order of the world which is in the mind is a constituting element of this world itself. Contrary, however, to the position of Radical Constructivism (Maturana, 1982) it is claimed that the meaning producing mechanisms of the mind as – for instance – perception have evolved under the constraints of the physical world and in interaction with it.

How the external world might impose constraints on perceptual mechanisms has been modelled by Rechenberg's evolutionary algorithms (1990) and by means of cellular automata (e.g. Li, Packard & Langton 1990 analyze the behavior of cellular automata dependent on the parameter λ.); one consequence of this kind of modelling is that the resulting solutions are very often thus that a single, albeit complex, algorithm cannot be found which describes the entire function of the mechanism correctly but that two or more competetive, complementary, or cooperative algorithms in parallel are necessary. In some cases where the coupling of the external world and the behavior of an organism might even be described by a single algorithm, as for instance the τ-algorithm for the time-to-contact control of action, it turns out that the observable behavior relies on more than one but simpler algorithms; Borst and Egelhaaf (1992) have shown this to be the case in the landing behavior of house flies.

Apparently such a coupling of external constraints and perceptual processes is more stable against perturbations and at the same time more sensitive to singularities, than entirely internally or externally driven can be. The prototypical example of the evolution of perceptual mechanisms under the constraints of the external world (physical reality) is the sensitivity of the eye (in humans as well as in most animals) to those electromagnetic wavelength with especially high energy in the sunlight which reaches the surface of the earth (see Land & Fernald, 1992, for an overview). Any sense organ tuned to other wavelengths would need a much higher sensitivity to result in a comparable discriminative power (acuity). Interestingly, the distribution of sensitivity for colours mirrors not the distribution of wavelengths in sunlight but the distribution in reflectance of objects in the environment. Shepard (1990) has sketched an evolutionary process for trichromatic vision pointing out that it achieves a maximum for the robustness-informativeness trade-off (Zimmer, 1982) in perception. It seems plausible to assume that this trade-off plays a central role in coordinating the physical world (reality) and the mental representations of it.

What I have termed Interactive Realism can already be traced back to Aristotle's theory of perception. In his Metaphysics, he stresses the point that 'seeing' means 'seeing something', that is, it is an intentional act not a mere registering of what is phenomenologically external. However, in the same context he mentions the importance of the viewer: "[seeing is always] the seeing of the person whose seeing it is."

Following this line of argumentation, one would agree with the Radical Constructivist (Maturana, 1982) position according to which percepts are not projections from the physical world but constructions of the perceiving organism reflecting its intentional position as well as its mechanisms of perception. However, and here I think the Realist stance cannot be parted with, the perceptual mechanisms also have evolved under the constraints of the physical world. In such a way already Koffka (1935) has answered to the question "Why do things look as they do?" (p. 76) namely because in perception "...the processes organize themselves under the prevailing dynamic and constraining conditions" (p. 105) and only such processes survive which optimize the interaction of the organism with its environment.

The position of Interactive Realism in regard to space perception has been circumscribed by Koenderink (1990) as follows: "... homunculi or God-given local signs are not required in a formal scientific understanding of the geometrical expertise of brains, that is, the apparent ability to generate efficacious potential action based on optical exposure. The brain can organize *itself* through information obtained via interactions with the physical world into an embodiment of geometry, it becomes a veritable *geometry engine*. Whether space is in the head, or the head is in space, remains undecided here. In the final analysis the distinction is scientifically meaningless anyway in view of the inherent circular nature of vital processes including optically guided behavior." (p. 8). Investigations into the multistability of wire cubes or related forms and into the bag of tricks used in visual arts to induce depth in flat displays will help to clarify the relations between the world of things (reality) and that of actions including perception (actuality).

3 The wire cube as a paradigm for space perception

In Fig. 11 six skeleton cubes are presented which differ in regard to their strength in inducing 3-dimensionality (apparent depth) and to the degree of multistability: In both regards, Fig. 11b induces the strongest effects because it has two equally strong semi-stable views which correspond to alternate spatial organisations.

Only slighly weaker is the effect in 11 a because here a third very transient state can be observed in which a 2-D pattern with a 45° axis of symmetry appears, this effect can be enhanced by rotating the axis of symmetry into a vertical position (see Zimmer, 1986). From an Empiricist point of view only 11d–f should give the impression of depth because only they correspond to real perspective projections with one, two, or three vanishing points; in contrast, however, 11a is a perspectively impossible view of a wire cube and

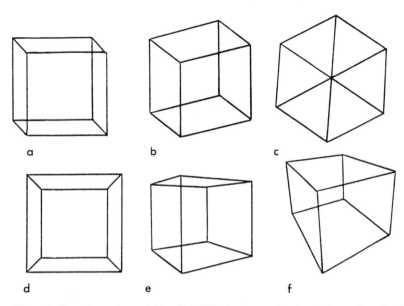

Fig. 11 Six wire cubes which all fulfill the general criteria for cubes (8 threefold vertices plus quadrupelwise parallel connecting lines) but differ according to the degree in which they obey the laws of projective geometry

11b as well as 11c are parallel projections without a vanishing point or one infinitely far off, that is, they are not possible in a finite environment. On the other hand, because of their unique perspective specification 11d–f should not induce multistability. As it turns out, they induce a much weaker effect regarding apparent depth than 11a and b – apparent from the fact that minor perturbations, namely, removing the vertices, destroy the 3-D effect - but they nevertheless give rise to bistability albeit with a bias towards the cube. This bistability seems to defy the Gestalt theoretic minimum principle (Köhler, 1920), according to which deformations of forms exhibiting 'Prägnanz' (singularity) are avoided by perceptual mechanisms (Koffka, 1935). However, practically none of the closed geometrical forms in 11e or f shows the high degree of symmetry typical for stable 2-D forms (Wulf, 1922); the cube is a 3D form with multiple axes of symmetry and therefore highly stable but none of these characteristic features is preserved in the perspective drawings: The lines are not parallel and of equal length, the angles are not orthogonal. At least partially, these features are preserved in 11a–d, where this preservation plus the breaking of symmetry induces the strong 3-D effect especially in 11 a and b. If, in contrast, the optimal symmetry of a 2-D projection of a cube exhibits maximal symmetry (6 axes of symmetry as in c and 8 in d), no or only a very weak and transient depth effect is induced.

But back to the startling effect of multistability in the unique projections of 11e and f: If one assumes that the forks and arrows at the vertices with their

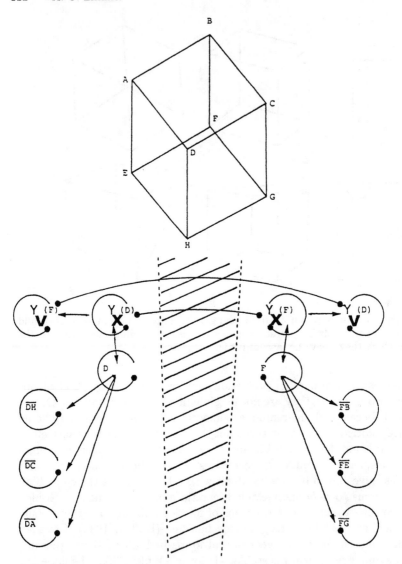

Fig. 12 Zimmer's (1989) network model for the distribution of attention during the inspection of a wire cube with excitatory and auto-inhibitory connections. Arrows indicate excitatory relations, dots inhibitory ones. Y_v and Y_x indicate fork junctions which are either concave (V) *or* convex (x). The edges indicated by DH and the vertices (capital letters) are always in front. Since only these vertices and edges are attended to, the others are regarded as inactive. The shaded area indicates the region where D and F cannot be discriminated and no depth effect and perspective reversal happen (Fig. 11c)

ambiguity of being either convex or concave can undergo selective satiation or, if modelled as a network, have auto-inhibition (Zimmer, 1989), the switching of only one node immediately triggers the switching of all other nodes into their alternate states (Fig. 12). What results, however, from this structural change is a form which fulfills Attveave's (1981) model of a generalized soap bubble, not as perfect as a cube but nevertheless convex.

Among the wire cubes of Fig. 11 there still remains one puzzling exemplar, namely, 11d which preserves many features of the cube *and* is a possible projection but nevertheless appears flat or as a picture frame but not as a cube. Perkins (1968) has used this instance to derive laws for the induction of 3-dimensionality in pictures:

(i) A fork juncture is perceived as the vertex of a cube if and only if the measure of each of the three angles is greater than 90 degrees.

(ii) An arrow juncture is perceived as the vertex of a cube if and only if the measure of each of the two angles is less than 90 degrees and the sum of their measures is greater than 90 degrees.

These laws predict correctly that Fig. 11d is seen as flat. However, if one breaks the symmetries as in Fig. 13, forms are generated which defy Perkins' laws, especially a and c.

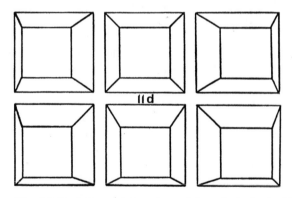

Fig. 13 Variations of the view of the wire cube in Fig. 11d with different degrees of symmetry breaking

The wire cubes in Fig. 13 cannot only induce a depth effect, they even exhibit multistability between a cube seen from the inside and a cut-off pyramid seen from outside. There is no clear cut preference for any of these perspective orientations because the subjects are about evenly split between an initial identification as a cube and a pyramid; afterwards subjects tend to stick to their initial percept. This is probably due to the fact that the biases for the convex form (pyramid) and for the most regular form (cube) are about equally strong. A similar 'real-world' example is the view of the cupola of S. Giovanni degli Eremiti in Palermo where the cupola is usually seen as concave but the pendentives tend to induce a convex orientation in spite of the knowledge about the constructive rules for building cupolas over square groundplans (see Fig. 6 above). A couple of lessons about multistability and its role in depth perception can be derived from this analysis of different projections of wire cubes:

(i) If the necessary conditions for perceiving a wire cube are given, namely 8 vertices with 3 connections each, plus convexity, apparent depth is induced if and only if the symmetry of the display is broken; an orientation of 30° relative to the fronto-parallel plane seems to be optimal.

(ii) The depth effect as well as the symmetry of the multistability are strongest if as many as possible features of the cube are preserved in its projection.

(iii) Experience in the sense that the subjects might have seen such a projection or in the sense that the rules of projective geometry are known and applied seem to play a negligible role.

These results agree very well with the results of a factor analysis of 'reversible-perspective drawings of spatial objects' by Hochberg and Brooks (1960). According to that study the apparent depth of a drawing depends on three factors: (i) simplicity vs. complexity (measured by the number of angles), (ii) good continuation vs. segmentation (measured by the number of line segments), and (iii) symmetry vs. asymmetry (measured by the relative number of different angles). What this factor-analytic approach implies, however, is the additivity of these components: "The greater the *complexity*, the *asymmetry*, and the *discontinuity* of the projection of a given tridimensional object in two-dimensions, the more three-dimensional it will appear. We may, in reality, be dealing with only one dimension – 'figural goodness'" (Hochberg & Brooks, 1960, p. 354). This runs counter to the comparison of the induced depth effects in Fig. 11: The line drawings of cubes in a and b have a much stronger effect than those in e and f. Which implies that the interaction between the components identified by Hochberg and Brooks (1960) is not additive but that they have to be modelled as competetive processes producing the maximum joint effect if they all are of comparable magnitude. That this interpretation is not restricted to the exemplars in Fig. 11 can be illustrated by the analysis of generalizations of wire-cube drawings: (i) Repetitions and glide symmetries can be used to produce tilings with the Necker cube as constituting elements (Fig. 14), (ii) the dimensionality of the generating spatial object can be increased from 3 to 4; a 4-dimensional hyper-cube is defined as consisting of 32 edges and 16 four-fold vertices (see figures 15 & 16), or (iii) the cube as one exemplar of the Platonic body can be exchanged against a more complex one, the dodecahedron (Fig. 17).

In all cases, the factor-analytic criteria plus the convexity criterion of the generalized soap bubble would predict that the spatial effect should even be stronger than that in Fig. 11b. By inspection, it can be seen that this is

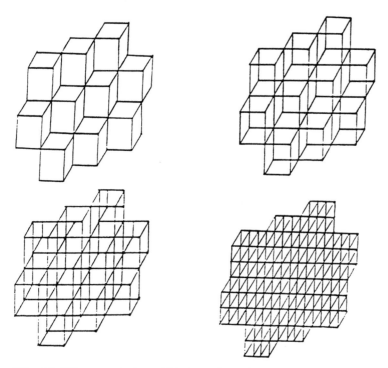

Fig. 14 Tilings consisting of Necker cubes which induce a depth effect and multistability only on a low level of complexity

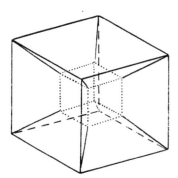

Fig. 15 A hypercube projected into 2 dimensions with 16 fourfold vertices

not the case and experimental results concerning the depth effect after an occlusion of the vertices support this.

In regard to the tilings in Fig. 14 one could argue that due to multistability different perspective orientations cancel out the over-all depth effect, however, such different perspective orientations at the same time do not occur: If one segment switches, all other become enslaved at such a speed that

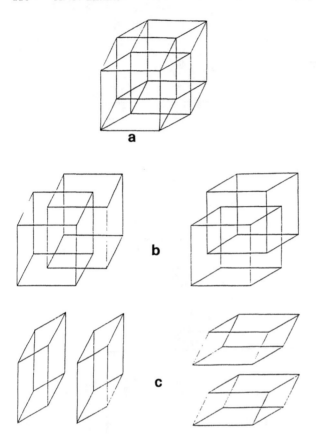

Fig. 16 A different partial view of the 4-dimensional Necker cube which does not exhibit multistability and induces only a weak depth effect. However, if disected into constituting parts (parts b and c show two alternate ways to do it), the parts exhibit multistability in depth

selfobservation cannot determine the exact and detailed time course of the process or the subprocesses of enslavement.

However, the theory of tilings offers one alternative interpretation: The patterns in Fig. 14 are all periodic, that is, there exists a subpattern which is repeated over and over resulting in glide symmetries; if the periodicity is broken, as in the non-periodic tiling of Fig. 8, multistability is given in spite of the complexity in the pattern. My conjecture is that aperiodicy, convexity, and symmetry breaking are the decisive factors for the depth effect in drawings.

In the following part I want to show how artists have made use of these factors in producing strikingly realistic pictures which at closer inspection deviate from a direct projective transformation as implied by the devices for perspective drawing which Alberti and Dürer invented. These deviations

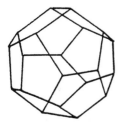 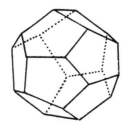 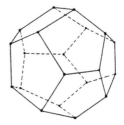

Fig. 17 Dodecahedrons which induce only a strong depth effect – but no multistability – if the lines in front and in the back are different

accentuate especially the factors of symmetry breaking and convexity whereas non-periodicity is usually already given by the chosen subject.

4 What is really perceived: Neither perspective nor an arrangement of immutable objects

In his analysis of art and geometry, Ivins (1946/1964) observes that the perception of space can be visual as well as tactile: "Tactually, things exist in a series of *heres* in space, but where there are no things, space, even though "empty," continues to exist, because the exploring hand knows that it is in space even when it is in contact with nothing. The eye, contrariwise, can only see *things*, and where there are no things there is nothing, not even empty space, for that cannot be seen. There is no sense of contact in vision, but tactile awareness exists only as conscious contact." (p. 5). If the representation what the "hand" perceives, there results absolute size and form constancy; if, however it is founded on the perception by the eye, the transformations of affine geometry, deformation and foreshortening, result. For Ivins "...intuitively the Greeks were tactile minded, and when ever they were given the choice between a tactile or visual way of thought they instuctively chose the tactile one." (p. 6). He explains this as being due to the fact that even in the "great period" the Greeks were primitive people, implicitly equating geometric competence with primitiveness.

Ivins equates the progress in visual art with the observation of perspective. However, as simple as this equating suggests, the role of perspective in art is far from straightforward. The inventions of Alberti's and Dürer's window have been heralded as at the same time the foundation of projective geometry and as the breakthrough in art for the naturalistic representation of the world. The first proposition might be true, the second is patently not as can be seen by comparing the projections of Dürer's window (Fig. 18 a,b) with his drawings (Fig. 18c).

In a way, in his drawings Dürer has disambiguated what in proper projections becomes ambiguous by adhering to form constancy, as can be seen

Fig. 18a The apparatus of the Dürer window (from Dürer's Unterweyssung der Messung, 1538)

Fig. 18b What a projection would look like

in Fig. 20 where different forms induce the same projection on the retina of the eye; in contrast Dürer has kept similar forms as similar as possible even if the projections are different and vice versa.

The comparison of the reclining body of Christ by Dürer (1505) and by Mantegna (after 1501) makes apparent how much closer Mantegna adheres to what is seen in a device like the Alberti window but even he 'corrects' the projection towards a greater form and size constancy (Fig. 20a,b).

How the combined manipulation of size and form constancy induces striking spatial effects has been well known in Italian Renaissance architecture, a

Fig. 18c What Dürer has produced under comparable perspective conditions (Das Meerwunder, 1498)

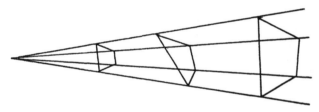

Fig. 19 Different quadrangular forms giving rise to the same projection on the retina

good example being the interior of the transept tower of S. Sidone in Torino (Fig. 21).

How this effect is diminished when different forms are used can be seen in the picture of the cupola of S. Carlo alle quattro fontane in Rome (Fig. 22) supporting the result of the preceding section according to which complexity reduces the spatial effect. This in turn sheds light on the limitations of space perception due to the processing demands.

The position implicit in Ivins' treatise of 'Art and Geometry' is that visual art in obeying the laws of projective geometry produces pictures which give the veridical account of the physical world from a specific point of view. The comparison of the projections in the Dürer window with the actual drawings and paintings of Dürer, however, makes apparent that only by means of systematic deviations from the results of projective geometry can the artist induce a veridical representation in the perceiver. How extreme these corrections sometimes have to be can be shown in Raphael's 'School of Athens' (Fig. 23a) where Euclid and another person carry perfect spheres which, as

Fig. 20 Two views of the reclining body of Christ a) by Dürer (1505), b) by Mantegna (after 1501)

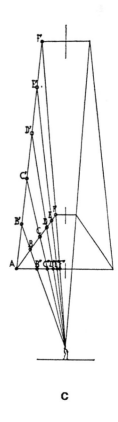

Fig. 21 The tower of S. Sidone, Torino a) view of the tower from inside b) view from outside c) constructive principles

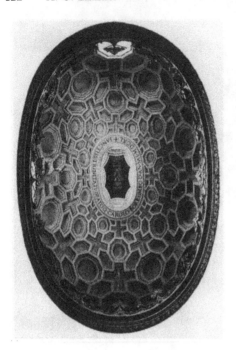

Fig. 22 The cupola of S. Carlo alle quattro fontane, Rome

LaGournerie (1859, p. 170) has demonstrated, should have the appearance given in Fig. 23b.

LaGournerie had the circles changed into the perspectively correct ellipses and presented them to colleagues who found them quite unacceptable. "They did not at all look like spherical objects – whereas Raphael's do so" (Pirenne, 1970, p. 122). The spheres and all the other figures are not drawn according to the rules for central projections. Each is drawn from one specific subsidiary center of projection just in front of the figure. Pirenne (1970, p. 122 – 123) comments on this for an Empirist puzzling result as follows: "Thus it appears that the spectator looking at Raphael's picture of the spheres must make a complicated intuitive compensation. On account of natural perspective, the circles appear foreshortened to him. They do not form in his eyes the retinal images which would be formed by actual spheres. But, on the basis of his knowledge of the shape and position of the surface of the painting, he recognizes them as circles drawn on a flat surface. Since real spheres always look circular, he concludes that these circles represent spheres. It will be noted that all this, which must somehow occur unconsciously, can be done as well when the spectator uses both eyes, and is in the wrong position. To most spectators, the *School of Athens*, in which the perspective is in parts inaccurate, appears as an outstanding example of the use of perspective."
However, the 'complicated intuitive compensation' on behalf of the spectator

Multistability – More than just a Freak Phenomenon   123

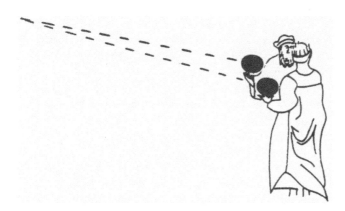

Fig. 23 (a) Raphael's School of Athens; (b) Euclid and companion with perspectively correct spheres

is effortless and as fast as rejecting the ellipses as out of place; therefore it indicates automatic, not analytic processes. The way spatial depth and constancy of form are at the same time present in the School of Athens implies two theoretical results: (i) perception is top-down and bottom up at the same time, and (iii) the constituents of complex scenes are processed in a piecemeal fashion, glance for glance with every glance having its owen frame of reference.

This conception of perception as the automatic integration of glances has been proposed by Hochberg (1962); it accounts not only well for the form constancies as in Raphael's School of Athens, furthermore it makes plausible why these regions which are especially attended to are perceived enlarged in comparison to those not, or less, attended to, and finally it explains why the patent inconsistencies in forms like the Penrose triangle are not perceived – not even by scrutiny – but have to be analyzed.

The distribution of attention while perceiving a scene or a picture is determined by bottom-up as well as by top-down processes: Before the information in the visual pathway has been analyzed in the visual cortex and in the parietal lobes it already triggers and directs the saccades in the superior colliculus by which the next glance is sampled; superimposed on these saccades are the voluntary eye movements being either the result of the analysis of the information in the visual pathway, the conscious percept, or of the intentional state the perceiver is in. As an aside it should be remarked here that Wundt (1874) had already found that there is no one-to-one relation between eye movements and the focus of attention but the correlation is high. The result of a differential sampling of glances in a percept is illustrated in Fig. 24 a–c.

Figure 24c gives only a rough approximation of the percept: Inside of each glance linearity, orthogonality, and form constancy are preserved, therefore the curvilinearity in the Escher drawing is not what is normally perceived. That the percept really is made up by glances, Hochberg (1962) has shown in his analysis of Escher prints of impossible scenes; another example is R.N. Shepard's 'Doric Dilemma' (Fig. 25) where it is obvious that the bases and the capitals are not sampled in the same glances.

Something additional about this notion of perception-frame-glances can be derived from the perception of this Fig.: The puzzlement induced by this drawing is due to the lack of an explicit boundary between the conflicting upper and lower part of the temple. The conceptual distinction between a square base and discrepant upper part with fitting capitals does not induce a perceivable edge – a result very much in accordance with Kanizsa's (1955) analysis of illusory contours which are neither due to top-down nor to bottom-up processes but the result of the interactions (or the resultant forces) in the perceptual field.

For the central perspective in Raphael's most famous paintings the piecewise corrections for form and also for size are mandatory – making use of and being necessitated by the glance-to-percept characteristic of human vision – however, this is not the cause for the impression that the depth appears flatter than intended by Raphael in contrast to the immediate and striking impression of depth one finds in the Illusionist paintings of the Baroque period, most famous Pozzo's painting on the ceiling of S. Ignazio in Rome (see Fig. 26 a & b).

Fig. 24a The undistorted view

Whereas the Classicist ideal in the Renaissance implied the central perspective, in the Baroque period artists liberated themselves from that iconographic convention and made use of all possibilities of projective geometry (e.g. 3 vanishing points); this progress is obvious when one compares scene designs of the respective periods (Figs. 27a,b).

The design by Serlio resembles in its effect more perspective tilings (see Fig. 16) than a scene in depth, but a nearly irresistible trompe d'oeil effect is only achieved if as in Fig. 28b the symmetry is broken, a principle already to be found in the foregoing analysis of the Necker cube. A combination of these techniques, symmetry breaking, form and size constancy is characteristic for the Veduti di Roma by Piranesi.

The phenomenon of mixed perspectives, that is, using for the different parts in a picture different vanishing points, and that of parallel projection have been interpreted as indicators for a lack of mastery in techniques of perspective and projective geometry in general. However, if perception is governed by the complementary tendencies towards stability and singularity, no singular point of view in a picture can induce an impression which mimicks perception in a natural scene. What distinguishes the comparatively crude

Fig. 24b A possible distribution of glances (circles)

16th century view of Amberg's castle (Fig. 28) from Piranesi's Veduti di Roma (Fig. 29) is not the veridicality of the depiction as measured by the strict adherence to the laws of projective geometry but the excellent technique of Piranesi in making the switching from one vanishing-point view to the other or from one size frame to the other smooth and imperceptible.

Figure 29 shows the Forum Romanum (Campo Vaccino) by Piranesi. Levit (1976) has taken a photograph from the same window from where Piranesi had made a sketch with a comparable angular aperture (Fig. 29).

Figure 30 shows where Piranesi deviates systematically from projective geometry: He accentuates the size of objects which can serve as points of orientation (e.g. the Collosseo), he shifts the orientation of the arch making it more stable, and he induces the impression of singularity by shifting and enlarging the ruins of the temple in the foreground. In a conversation with his friend Hubert Robert, Piranesi remarked that the sketches of the city served him only as memorizing device, the real sketch he had in his head. The comparison of the Campo de Vacco as depicted by Piranesi with Levit's photograph who took the same **ventage point** from the Capitol and the same angular aperture shows 3 specific **systematic** distortions Piranesi applies:

(i) *increasing the size* of objects which serve as orienting points for the viewer,
(ii) *shifting and rotating objects to* optimize the spatial effect, and
(iii) *changing the height/width relation.*

Fig. 24c The resulting percept (after Escher)

Fig. 25 R.N. Shepard: The Doric Dilemma

Fig. 26a Pozzo's painting

Fig. 26b The architecture implied in Pozzo's painting (after Uttal)

Multistability – More than just a Freak Phenomenon 129

Fig. 27a Scenic design by Serlio (1551)

Fig. 27b Bibiena (1740)

Fig. 28 Part of a Renaissance aerial view of the city of Amberg (1598) showing the castle

Fig. 29 G.B. Piranesi 'Veduta di Campo Vaccino' (after 1748) from 'Vedute di Roma'

Fig. 30 Levit's (1976) photograph of the Forum Romanum showing the same scene as in Fig. 29

In order to pinpoint these techniques I have made comparable line drawings from the Piranesi etching as well as from the Levit photograph. Where the excavation since the last century have changed the location I have taken from Piranesi only the still extant objects and have added to them what is now visible. Thus I have produced two drawings which differ only in the systematic distortions used by Piranesi (Fig. 31a,b).

Showing these drawings to tourists which had visited Rome (n=32) with the question which drawing depicted most correctly the real scenery, resulted in a majority (23) for the Piranesi-type drawing, 7 being undecided and only 2 voted for the perspectively correct drawing. This makes clear that Piranesi's distortions are not ideosyncratic whims but that 'in his head' the same accentuating processes took place as in the normal perceiver. This combination of exact observation and 'reconstruction out of his head' allows Piranesi even to show his scenes from that point of view which induces best the intended effect even if it is physically not accessible.

The comparison of the views depicting 3 Napolitan baroque churches of comparable structural complexity makes apparent how important the chosen point of view is for the induction of a spatial impression: (i) The fronto-parallel view of S. Anna a Porta Capuana (Fig. 32a) leaves the symmetries in

Fig. 31 (a) Line drawing after Fig. 29; (b) line drawing after Fig. 30

the architectural design unchanged, the induced spatial effect is negligable, if existent at all. (ii) The view from below of S. Maria della Sanità (Fig. 32b) tries to mimic the view as given when entering the church from the side porch; that there is an induced depth effect can be seen from the fact that in the cupola and in the apse a concave-convex bi-stability occurs; this, however, impedes the formation of a global 3-D effect. (iii) The view from above of Concezione a Monte Calvario (Fig. 32c) induces a strong and globally consistent 3-D effect.

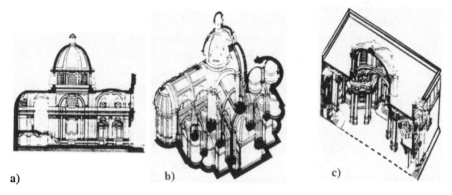

Fig. 32 Line drawings of three Napolitan Baroque churches; (a) S. Anna a Porta Capuana, (b) S. Maria della Sanita, (c) Concezione a Monte Calvario

All three views have in common that they are parallel projections, a fact which obviously does not impede a 3-D effect if the dominant symmetries are broken; this together with the strong global spatial impression of Fig. 32c which is taken from an inaccessable and for normal surface-bound perception unusual vantage point implies that the 3-D effect in pictures is not due to conventions of perceiving acquired through experience with the rules of depiction, but to strong local processes of perception which interact, sometimes in a competetive way as in Fig. 32. Apparently, Piranesi in "the perfect marriage of hard fact with an incomparable poetic eloquence" (Levit, 1976 p. X.) knew about the sometimes imagery vantage point producing the strongest 3-D effect because he usually depicts the major objects from the side ($15 - 30°$) and slightly from above ($7 - 15°$).

The analysis of multistability so far supports the theoretical position that in space perception bottom-up and top-down processes interact. One could, however, argue that in the choices of examples already lies a kind of question begging, namely that in relying on specific forms experience as a top-down process necessarily plays a role in constructing depth in 2-D displays. That form perception in itself is at least partially inborn has been shown by Fantz (1961), but even for the prototypical example of a bottom-up processing of information, namely the perception of colour, it can be shown that only in the case of experimenting with colour chips in isolation pure bottom-up processing can be assumed; if the complexity of the perceptual task is increased, top-down influences can be observed. It is interesting to note that in the history of art (Ruskin) and in the theory of art (Mondrian) radical positions of bottom- up processing in colour vision have been proposed, developed independently from psychological theories of perception and very much earlier than Marr's influential treatise 'Vision' (1982).

J. Ruskin (1857) in writing about William Turner presents his bottom-up theory as follows: "The perception of solid form is entirely a matter of experience. We see nothing but flat colours; and it is only by a series of experiments that we find out that a stain of black or grey indicates the dark side of a solid substance, or that a faint hue indicates that the object in which it appears is far away. The whole technical power of painting depends on our recovery of what may be called the *innocence of the eye* that is to say, of a sort of childish perception of these flat stains of colour, merely as such, without consciousness of what they signify, – as a blind man would see them if suddenly gifted with sight." (XV, p. 27).

Ruskin's speculations are – at least partially – supported by clinical data: Gregory and Wallace (1963) report the case of S.B, born blind in 1906 and given a corned graft in 1959 which allowed him to see: "The difficulty with pictorial representations was general, and it persisted; he did not respond to the common geometric illusions, and perspective seems not to have suggested depth to him. He never learned to read properly" (Morgan, 1977, p. 182).

In 1958 Mondrian has described his programmatic stance from a comparable Empiricist position:

"In cubism abstraction has not been developed until its final goal: The expression of pure reality I felt that this reality could only be achieved by pure construction. In its utmost expression pure construction is independent from subjective emotions and imaginations. It took a long time until I realized that the specific form and the natural colour induce subjective mental states which obscure pure reality. If one wants to construct the Gestalt of pure reality, it is necessary to reduce natural forms to the constant elements of form and the natural colours to the elementary colours" (1958, p. 9 translation by the author)

However, that this is not the only possible interpretation becomes clear when we turn to Turner on whose paintings Ruskin founded his cited position. Turner regarded his painting as an explicit application of Goethe's theory of colour which he commented in writing and painting, e. g. §808 of 'Zur Farbenlehre' (1810) "Is the totality of colour like an object brought to the eye from the outside, then it is agreable to it because the sum of its own activity approaches it as reality" is commented by Turner in the following way - "this is the object of Paintg (painting)". His own paintings defy Ruskin's concept of the "innocence of the eye" which perceives colours and not meaning, especially his illustrative sketches for Goethe's theory of colour.

Turner was the first to use colour not to enhance the geometrical perspective induced by foreshortening but as a means in itself to give a 2D display the impression of a third dimension; prototypical examples are his 1819 water colour picture "Colour Beginning" and his "Light and Colour (Goethe's Theory)" of 1843. Goethe himself in §155 of his "Zur Farbenlehre" (1810) has described the spatial impressions related to the colour blue: "...In the mountains during the day the sky appears to show a Prussian blue because only a very faint mist have before the infinite dark space; when descending, the blue becomes lighter until it changes entirely into a whitish blue in certain regions when the density of the mist increases." Especially Turner's "Colour Beginning" shows that the spatial impression due to colour is not confined to different densities of mist. As for instance Wolfgang Metzger (1975) has observed any colour induces a typical distance with blue being the farthest away. But Turner systematizes the spatical impression of colour further: He is the first to analyze systematically the conditions of transparency as shown in his water colours where he depicts overlapping colours. Insofar there is a direct tradition of research starting with Goethe's theory of colours, taken up again by Metelli (1975), and analyzed in formal approaches by Beck et al. (1984) and Gerbino (1988). This tradition highlights the fact that contrary to the positions taken by Ruskin or Mondrian and by Marr's bottom-up theory (1982), perception it is not a sequential process which starts with colours, subsequently derives forms, and finally constructs space but a parallel process tuned for 3-dimensionality.

5 Conclusion: By exploring the role of multistability we realize that reality and actuality meet

The demonstrated phenomena of multistability illuminate one important aspect about the relation between reality (the transphenomenal world of physical objects) and actuality (the mentally represented world which directs actions and interprets the results of them): It is not only many-to-one (the classification of objects and events into equivalence classes, Aristotles' problem) but also one-to-many (identical stimuli can give rise to different percepts). This indeterminacy in the relation between the physical world and the mental representation with its intentionality does not imply an anythinggoes stance because in the cases of multistability usually only small number of attractors exist what, however, is principally unpredictable is the exact basin in which stability will be reached. That is, the physical world constrains but does not determine the actuality of the mind.

The notion of competing processes in perception which reach points of stability, that is, an equilibrium where the effective forces are balanced, appears to bridge the gap between the concept of a perceptual field, central for Gestalt psychologists as Köhler (1920) and Koffka (1935), and that of parallel distributed processes as developed by Rumelhart and McClelland (1986). At the same time it helps to clarify the notion of a complementarity between stability and singularity in perceptual and cognitive processes: The combination of effective forces characterizing a stable percept is achieved by fixing each of the competing processes at a point which is singular and therefore not a stable basin. The standard example for this is the Necker cube which exhibits its spatial impression only if neither the overall symmetry nor the similarity with the geometric characteristics of the cube are maximal. Insofar, in the perception of even moderately complex scenes (e. g. the Necker cube, the Schröder stairs, or the Thiery Fig.) I regard the role of the so-called Gestalt laws of perception as processes pitched against each other not as adding up in their effects. This is even to a higher degree characteristic for the more complex scenes we come across in everyday-day life; that in these cases the multistability inherent in competing processes is not experienced can be attributed to the high complexity regarding the number of processes involved and of local multistability patterns possible.

My conjecture about the relation between structural or geometric multistability and spatial perception is that space perception under normal conditions, that is, during and by perambulation, means construction and disambiguation of spatial relations at the rate prescribed by the speed of perambulation. The spatial invariants enabling the perceiver to build up a stable spatial frame of reference therefore consists of the affine transformations and not of the objects and their relative distances. This process is mimicked by multistability. However, the closer analysis of the conditions for multistability reveals that the extraction of affine transformation is in turn constrained by temporal and/or spatial frequencies and the complementarity of symmetries and complexities of scenes and objects. As demonstrations can serve the tilings in Fig. 15 and the 2-D projection of a 4-D Necker cubes in Fig. 16.

Shepard (1987) has summarized the epistemological stance I have called Interactive Realism very succintly in the following way: "I have argued that to the extent that the principles of the mind are not merely arbitrary, their most likely ultimate sources are the abiding regularities in the world.... Among such external regularities, the most abiding are the ones that in the long run should have the greatest opportunity to become internalized – however abstract those regularities may be:.....[t]he facts that space is three-dimensional, that objects have six degrees of freedom of global motion, that light and darkness alternate with a fixed period, and that sets of objects having the same significant consequences tend to form a compact region in an appropriate parameter space ... (p. 269)."

References

Attveave, F. (1981): Prägnanz and soap bubble systems: A theoretical exploration. In: J. Beck (Ed.), Organization and Representation in Perception. Hillsdale, NJ: Lawrence Erlbaum Ass.

Beck, J., Prazdny, K. & Ivry, R. (1984): The perception of transparency with achromatic colours. Perception and Psychophysics 35, 407 – 422.

Borst, A. & Egelhaaf, M. (1992): Im Cockpit der Fliege. MPG Spiegel 3, 14 – 16.

Bradley, D.R. & Petry, H.M. (1977): Organizational determinants of subjective contour: The subjective Necker cube. American Journal of Psychology 90, 253 – 262.

Brandimonte, M.A. & Gerbino, W. (1993): Mental image reversal and verbal recoding: When ducks become rabbits. Memory and Cognition 21, 23 – 33.

Fantz, R.L. (1961): The origin of form perception. Scientific American 204, 66 – 72.

Fodor, J.A. (1974): The Language of Thought. New York: Crowell.

Gerbino, W. (1988): Models of achromatic transparency. Gestalt Theory 10, 5 – 20.

Gibson, J.J. (1967): New reasons for realism. Synthese 17, 162 – 172.

Goethe von, J.W. (1810): Zur Farbenlehre. Tübingen: Cotta.

Goldmeier, E. (1982): The Memory Trace: Its Transformation and its Fate. Hillsdale, NJ: Erlbaum.

Gregory, R.L. & Wallace, J.G. (1963): Recovery from Early Blindness. EPS Monograph, No. 2. Cambridge: Heffer.

Haken, H. (1991): Synergetic Computers and Cognition. Berlin: Springer.

Hochberg, J. (1962): The psychophysics of pictorial perception. Audio-visual Communications Review 10 (5), 22 – 54.

Hochberg, J. & Brooks, V. (1960): The psychophysics of form: Reversible-perspective drawings of spatial objects. American Journal of Psychology 73, 337 – 354.

Ivins, W.M. jr. (1946): Art and Geometry. A Study in Space Intuitions. New York: Dover Publications.

Kanizsa, G. (1955): Condizioni ed effetti della transparenza fenomenica. Rivista di Psicologia 49, 3 – 19.

Kanizsa, G. & Luccio, R. (1986): Die Doppeldeutigkeiten der Prägnanz (transl. A. Hüppe & M. Stadler). Gestalt Theory 8, 99 – 135.

Köhler, W. (1920): Die physischen Gestalten in Ruhe und im stationären Zustand. Erlangen: Verlag der Philosophischen Akademie.

Koenderink, J.J. (1990): The brain a geometric engine. Report Nr. 19/1990 Research Group on Mind & Brain, ZiF Bielefeld.

Koffka, K. (1935): Principles of Gestalt Psychology. New York: Harcourt Brace.

Land, M.F. & Fernald, R.D. (1992): The evolution of eyes. Annual Review of Neuroscience 15, 1 – 29.

LaGournerie, J. de (1859): Traité de perspective linéaire contenant les tracés pour les tableaux plans et courbes, les bas-reliefs et les décorations théatrales, avec une théorie des effets de perspective, vol.1 and atlas of plates. Paris: Dalmont & Dunod; Mallet-Bachelier.

Leeman, F., Elffers, J. & Schuyt, M. (1976): Hidden Images. New York: Harry N. Abrams.

Levit, H. (1976): Views of Rome Then and Now. New York: Dover Publications.

Li, W., Packard, N.H. & Langton, Ch.G. (1990): Transition phenomena in cellular automata rule space. In: H. Gutowitz (Ed.), Cellular Automata, pp. 77 – 94. Amsterdam: North-Holland.

Marr, D. (1982): Vision. New York: Freeman

Maturana, H.R. (1982): Erkennen: Die Organisation und Verkörperung von Wirklichkeit. Braunschweig: Vieweg.

Metelli, F. (1975): The perception of transparency. In: G.B. Flores d'Arcais (Ed.), Studies in Perception: Festschrift for Fabio Metelli, pp. 445 – 487. Milano: Martello-Giunti.
Metzger, W. (1975): Gesetze des Sehens. Frankfurt: Kramer (3rd ed.).
Michaels, C.F. & Carello, C. (1981): Direct Perception. Englewood Cliffs, NJ: Prentice Hall.
Mondrian, P. (1958): Lebenserinnerungen und Gedanken über die "Neue Gestaltung". Das Kunstwerk 11, 9 – 10.
Morgan, M.J. (1977): Molyneux's Question. Cambridge: Cambridge University Press.
Perkins, D.N. (1968): Cubic corners. Quarterly Progress Report 89, M.I.T. Research Laboratory of Electronics, pp. 207 – 214 (Reprinted in Harvard Project Zero Technical Report No. 5, 1971).
Pirenne, M.H. (1970): Optics, Painting and Photography. Cambridge, MA: University Press.
Pylyshyn, Z. (1984): Computation and Cognition. Cambridge, MA: MIT Press.
Rechenberg, I. (1990): Evolutionsstrategie – Optimierung nach Prinzipien der biologischen Evolution. In: J. Albertz (Ed.), Evolution und Evolutionsstrategien in Biologie, Technik und Gesellschaft. Schriftenreihe der Freien Akademie, Bd. 9. Hofheim: Hofheimer Druck- und Verlagsanstalt (2nd ed.).
Roth, G. (1992): Das konstruktive Gehirn: Neurobiologische Grundlagen von Wahrnehmung und Erkenntnis. In: S.J. Schmidt (Ed.): Kognition und Gesellschaft, pp. 277 – 336. Frankfurt: Suhrkamp.
Rumelhart, D.E. & McClelland, J.L. (1986): Parallel Distributed Processing: Explorations in the Microstructure of Cognition, vol. 1: Foundations. Cambridge, MA: MIT Press.
Ruskin, J. (1857/1987): The Art Criticism. New York: Da Capo Press.
Santayana, G. (1920): Three proofs of realism. In: D. Drake, A.O. Lovejoy, J.B. Pratt, A.K. Rogers, G. Santayana, R.W. Sellars & S.A. Strong (Eds.) (1968), Essays in Critical Realism. New York: Gordian Press.
Shepard, R.N. (1987): Evolution of a mesh between principles of the mind and regularities of the world. In: J. Dupré (Ed.), The Latest on the Best: Essays on Evolution and Optimality, pp. 251 – 275. Cambridge, MA: MIT Press.
Shepard, R.N. (1990): A possible evolutionary basis for trichromacy. Paper presented at the SPSE/SPIE Symposium on Electronic Imaging, Santa Clara, CA.
Wertheimer, M. (1912): Über das Denken der Naturvölker. Zeitschrift für Psychologie 60, 321 – 378.
Wulf, F. (1922): Über die Veränderung von Vorstellungen (Gedächtnis und Gestalt). Psychologische Forschung 1, 333 – 373..
Wundt, W. (1874): Grundzüge der Physiologischen Psychologie. Leipzig: Engelmann.
Zimmer, A.C. (1982): Criteria of optimality for retrieval schemes: Informativeness vs. Robustness. In: G. Luer (Ed.), Bericht über den 33. Kongreß der Deutschen Gesellschaft für Psychologie. Göttingen: Hogrefe.
Zimmer, A.C. (1982): Are some triangles heavier than others? A graviational model of form perception. Psychologische Beiträge 24, 167 – 180.
Zimmer, A.C. (1986): The economy principle, perceptual mechanisms, and automatic cognitive processes. Gestalt Theory 8, 174 – 185.
Zimmer, A.C. (1989): Gestaltpsychologische Texte - Lektüre für eine aktuelle Psychologie? Gestalt Theory 11, 95 – 121.
Zimmer, A.C. (1991): The complementarity of singularity and stability. A comment on Kanizsa & Luccio's "Analysis of the Concept of 'Prägnanz' ". Gestalt Theory 13, 276 – 282.

Recognition of Dynamic Patterns by a Synergetic Computer

R. Haas[1], A. Fuchs[1], H. Haken[1,2], E. Horvath[2], A.S. Pandya[2] and J.A.S. Kelso[2]

[1] Institute for Theoretical Physics and Synergetics, University of Stuttgart, D-70569 Stuttgart, Germany

[2] Program in Complex Systems and Brain Sciences, Center for Complex Systems, Florida Atlantic University, Boca Raton, FL 33431, USA

Abstract: We show how a synergetic computer recovers patterned structure from motion, in this case point light displays created by complex human movements (after Johansson). A procedure is introduced to 'connect the dots', that is, it determines which points belong to the same extremity. By using order parameters or collective variables that characterize time-independent patterns, the network identifies and discriminates complicated dynamic patterns even when embedded in noise.

1 Introduction

This contribution extends previous work in which a synergetic neural architecture was presented that recovers the structure of moving objects from several single camera views of point lights (Haken, Kelso, Fuchs & Pandya, 1990). The specific case treated in that research emerged from studies by (Kelso, Wallace & Buchanan, 1989) of movement patterns produced by a human arm. Infrared light emitting diodes were placed on a subject's shoulder, elbow, wrist and hand generating "moving dots" over time. The synergetic algorithm was able to discriminate and categorize the resulting patterns of coordinated motion, even when the patterns changed abruptly. The algorithm uses identified collective variables or order parameters that characterize the macroscopic behavior of this multidegree of freedom system. Here we elaborate the approach to computer generated point light displays of human gait. We show that the algorithm works even when gait patterns are embedded in a moving field of random dots (random dot kinematograms).

As we mentioned in our previous paper, classical studies by Johansson (1973) have revealed the importance of motion to the recovery of biological structure. Light bulbs fixed on the joints of a person reveal no information to a human observer if the display is static. Once motion is introduced, however, perceivers can readily determine what the, otherwise invisible, person is doing. Further extensions of these original observations have occurred, including recognition of gender (Cutting & Proffitt, 1981) and accurate judgement of the amount of weight lifted by a person (Runeson & Frykholm, 1981). All of these observed phenomena are based only on the constellation of point lights changing in time which constitute, in some sense, a dynamic signature. Recent psychophysical experiments (Horvath, Kelso & Pandya, 1993), have demonstrated the remarkable ability of naïve human observers to identify gait patterns (e.g. whether a person is walking, running, limping, hopping) composed of moving point lights, even when the points are embedded in random dot kinematograms. We show here that our pattern recognition algorithm can also identify and categorize such dynamic patterns of point lights even in noise.

2 The synergetic strategy to multidegree of freedom motion

Although biological systems are composed of very many degrees of freedom, a most fundamental feature is the high degree of harmony among the parts. Viewing the human body simply as an aggregate of hinge joints, around 100 mechanical degrees of freedom may be involved in typical daily activities. The state space, in other words, is high dimensional. For a number of years, through a close theory-experiment relation, we have sought to identify relevant macroscopic variables that characterize the cooperative activity of such complex systems. In this synergetic strategy, we use phase transitions or bifurcations – qualitative changes in a system's behavior – as a way to identify collective, macroscopic variables for different patterns and their nonlinear dynamics (multistability, loss of stability and so forth). Around phase transitions or bifurcations, phenomenological description turns to prediction: The essential processes governing a pattern's stability, change and even its selection can be uncovered. The control parameter(s) that promote(s) instabilities can also be found (for many recent examples, see Haken & Stadler, 1990).

As noted elsewhere (Haken, 1987a) the discovery of phase transitions in rhythmical bimanual coordination was a first step toward the goal of identifying order parameters and control parameters in complex, biological systems. In that case, the relative phase, ϕ, was identified as a collective variable capturing the ordering relations among the individual components. Multistability and transitions among phase-locked states were observed at critical values of a control parameter, in this case the movement frequency. En route to these transitions, predicted features of synergetic systems including enhancement of order parameter fluctuations, critical slowing down and switching time of the order parameter were observed experimentally (for reviews, see Haken, 1988, ch. 11 and e.g. Kelso et al., 1987; Schöner & Kelso, 1989). Since then, numerous examples and theoretical modeling of phase transitions (a fundamental form of self-organization) have emerged in the literature (Kelso, 1990);

(see Turvey, Carello & Nam-Gyoon Kim, 1990 for reviews). In all these cases, an attraction to a limited set of relative phases among the interacting components captures the coordinative structure of the motion.

Why might this research on the generation and formation of coordinated behavioral patterns be relevant to dynamic pattern recognition by computer? The reason, hinted at earlier, is that our approach uses the order parameters that characterize multidegree of freedom biological motion as the relevant macroscopic variables upon which to base the encoding of so-called prototypes. In the present case of recognizing point light displays of human walkers, we suppose that meaningful information (for human observers and our synergetic computer) lies in the collective variable relative phase dynamics. Put another way, we consider pattern recognition as a dual process to pattern formation (Haken, 1987b) in synergetic systems that undergo nonequilibrium phase transitions.

3 Synthetic generation of realistic point light displays

The basic technique used to generate human point light walkers derives from Johansson, who attached light bulbs to the articulating joints of subjects and filmed them in the dark so as to eliminate all cues except motion. Typical patterns of light points are shown in Fig. 1.

Fig. 1 Graphical depiction of one sample of the light spots of movement pattern 1 shown to the observer

Obviously it is virtually impossible for a human observer to make any sense of these static configurations of points. Once put in motion however, the spatiotemporal patterns create a compelling impression of human motion (see also Hoenkamp, 1978).

Computer simulations of dot patterns (with all other cues eliminated) are based on the following considerations. The internal dynamics of the human gait figure are comprised of two constituents, relative displacement, and relative topography among the parts. These constituents are, respectively, dynamic and static relational invariants of the figures. The internal dynamics of the figure consist of the motion of the points relative to one another. The relative topography of the parts is the length of the components (upper arm, lower arm, etc.). The second constituent of the figure is the center of moment which is the point in space about which the internal dynamics of the figure is specified. The center of moment is the reference point for the description of the motion and locations of the parts of the figure (Cutting, 1978).

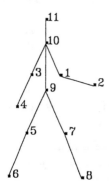

Fig. 2 This figure shows the eleven points of the walking man

The display with eleven points denoting the head (11), shoulder (10), hips (9), right elbow (1), right wrist (2), left elbow (3), left wrist (4), right knee (5), right ankle (6), left knee (7) and left ankle (8) is shown in Fig. 2. The elbow moves with a sinusoidal motion while the wrist movement is sinusoidal relative to the elbow motion. The shoulder acts as a pivot for this double pendular type of motion between the elbow and wrist. The knee motion is also sinusoidal, while the ankle motion is relative to the knee and is defined by a sawtooth function,

$$-(a - \theta_{\text{low}})/(360.0 * bk) \qquad (1)$$

and

$$(a - \theta_{\text{low}})(360.0(1 - bk)) \qquad (2)$$

where a is a constant.

The break point bk, and the θ_{low} value may be varied in order to generate different kinds of gaits, e.g. walk, run, skate. Figure 3 shows the angles that determine how the limbs move. Angle 1 and 3 are sinusoids, which define, respectively, the motion of the right arm and left knee and the motion of the left arm and right knee. Angles 2 and 4 generate the sawtooth function for

Recognition of Dynamic Patterns by a Synergetic Computer 143

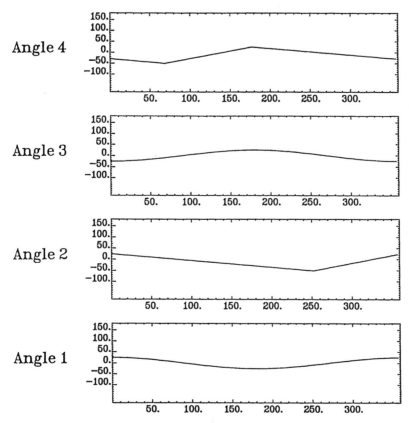

Fig. 3 These four angles determine how the limbs move. Angle 1 and 3 are sinusoids. They define the motion of the right arm and left knee and the motion of the left arm and right knee, where angle 2 and 4 generate the sawtooth function for the motion of the two ankles. The figure shows the temporal behavior of the angles over one cycle

the ankle motion for the right and left ankle for parameter values $bk = 0.7$ and $\theta_{low} = -50.0$.

The relative position of the points on the computer display is specified as follows: During the first stage of motion, when $ang2 > bk * 360.0$, the right ankle is fixed, and thus acts as the center of movement. The right knee moves forward along with the hip in a way that straightens the leg back from the position where it is extended for a step. At the same time the left leg is brought forward by the movement of the hip. A second stage of motion evolves, when $ang2 \leq bk * 360.0$ and $ang1 < 180.0$ in which the hip remains fixed and the left leg continues to kick forward as the right leg kicks back. Once the left leg is fully extended forward, a third stage begins in which the left ankle now behaves as the pivot moving the hip forward just as the right

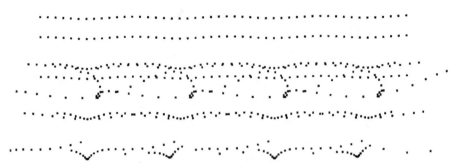

Fig. 4 Stroboscopic motion of the computer-generated man over two cycles

ankle did in stage one. At each stage the shoulder and head move in the same manner as the hip. Fig. 4 shows the point light stroboscopic motion of the computer generated man over two cycles. This pattern agrees with the motion of real human gait as shown in work by Johansson.

4 Solving the connectedness problem

Before our pattern recognition algorithm can be applied to the point displays shown in Fig. 1, it is necessary to determine the phase angles. Thus we need a procedure to determine which points are connected to the same extremity. The connections are determined by finding pairs of dots whose position relative to each other remains time independent. By denoting each of the points P by coordinates $x_i(t), y_i(t)$, the procedure sorts out the pairs j, k for which the condition

$$\Omega_x + \Omega_y < \Omega, \tag{3}$$

where $\Omega_x = |(x_i - x_j)(\dot{x}_i - \dot{x}_j)|$ and $\Omega_y = |(y_i - y_j)(\dot{y}_i - \dot{y}_j)|$ is met at least τ (threshold) number of times in one cycle. The resulting stick figure when the interconnections between points is made is shown in Fig 5.

Once these interconnections have been determined the phase angles between the limbs may be found by the following trigonometric method, since the vectors $(\mathbf{p}_j - \mathbf{p}_k)$ and $(\mathbf{p}_k - \mathbf{p}_i)$ are known.

$$\Phi_{kl} = \arccos(\frac{(\mathbf{p}_j - \mathbf{p}_k)(\mathbf{p}_k - \mathbf{p}_i)}{||(\mathbf{p}_j - \mathbf{p}_k)(\mathbf{p}_k - \mathbf{p}_i)||}) \tag{4}$$

The phase angles as determined by the algorithm are shown in figure 6. In a next step the computer plots the time dependence of the phase angles as shown in Figs. 7 and 8. Now we are in the same position as in our previous paper, where we took the time dependence of the angles $\phi_k(t)$ at the joints as a starting point of our analysis. In complete analogy to our previous work we then Fourier analyze the phase angles

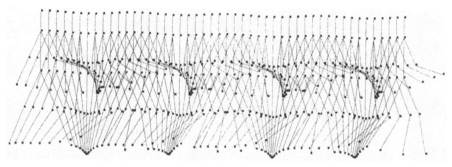

Fig. 5 Shows the movement pattern 1 after the reconstruction of the underlying figure due to eq. (3)

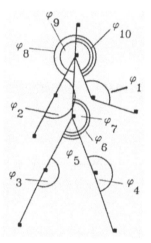

Fig. 6 Enumeration of the angles between constant limbs

$$\psi_k(t) = a_{k_0} + \sum_{l=1}^{\infty} A_{k_l} \cos(\omega_l t + \phi_{k_l}). \tag{5}$$

The left hand side of fig. 9a shows the absolute value of the Fourier coefficients, the right hand side the corresponding phases. The abscissa denotes the index of the Fourier component l = 0, 1, 2 etc. Figure 9a is continued on Fig. 9b.

Figures 10a and 10b represent the corresponding results but for the second movement pattern. The two types of movement patterns are encoded by means of the following vectors. (Haken, Kelso, Fuchs & Pandya, 1990):

$$\mathbf{v}^{(u)} = (a_{k_0}^{(u)}, A_{k_l}^{(u)}, \phi_{k_l}^{(u)} - l\phi_{1_l}^{(u)}, c^{(u)}) \tag{6}$$

146 R. Haas et al.

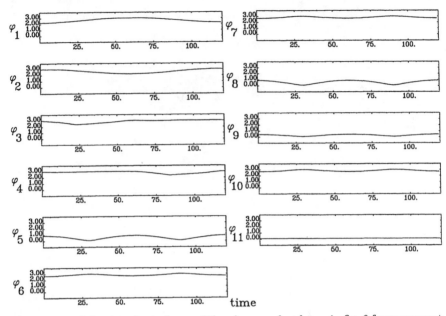

Fig. 7 Plot of the time dependence of the phase angles shown in fig. 6 for movement pattern 1, a walking man

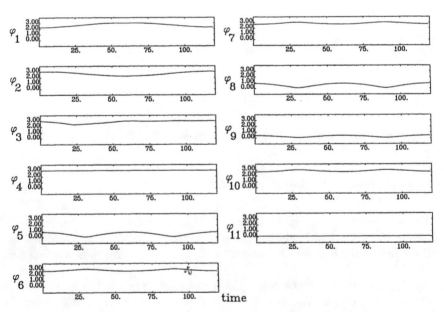

Fig. 8 Plot of the time dependence of the phase angles for movement pattern 2, a limping man

Recognition of Dynamic Patterns by a Synergetic Computer 147

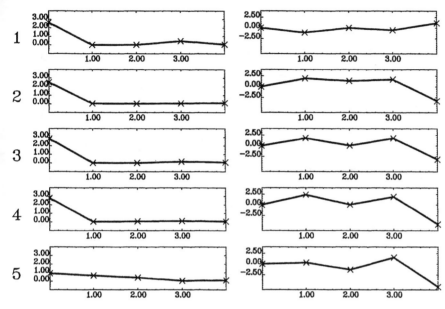

Fig. 9a The left-hand side of this picture shows the absolute value of the Fourier components denoted on the abscissa. The right-hand side shows the corresponding relative phases of pattern 1

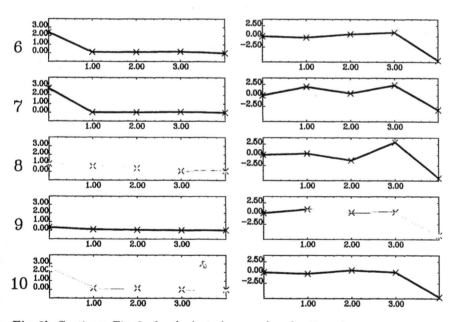

Fig. 9b Continues Fig. 9a for the last phase angles of pattern 1

148 R. Haas et al.

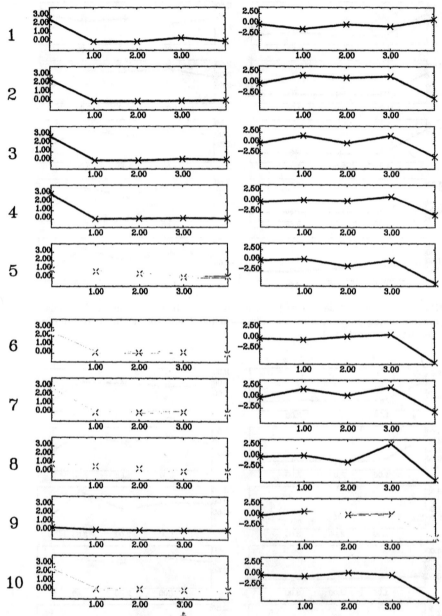

Fig. 10a,b The same as in Fig. 9 but for the second pattern

Where $\mathbf{v}^u$ for u=1,2,...,n are prototype vectors. In order to distinguish the two prototype patterns from each other we have added the labels $c^{(u)}, c^{(1)} = 0.5$ and $c^{(2)} = -0.5$. For practical applications we have kept the terms of $l = 1...4$, $k = 1...11$ corresponding to the prototype patterns 1 or 2. To decide whether a test pattern vector $\mathbf{q}$ is more similar to $\mathbf{v}^{(1)}$ or $\mathbf{v}^{(2)}$, we apply the concept of the synergetic computer. (Haken, 1987 b; Fuchs & Haken 1988). We introduce the adjoint vectors by means of the requirements:

$$(\mathbf{v}^{(u)+} \mathbf{v}^{(u')}) = \delta_{uu'}, \qquad \mathbf{v}^{(u)+} = \sum_{u'=1}^{M} g_{uu'} \mathbf{v}^{(u')} \qquad (7)$$

We then form $\xi_u = (\mathbf{v}^{(u)+} \mathbf{q})$ and subject ξ_u to the dynamics as follows

$$\dot{\xi}_u = \xi_u(\lambda - 2D + \xi_u^2), \qquad D = \sum_{u'=1}^{M} \xi_{u'}^2 \qquad (8)$$

Figures 11 and 12 show the temporal evolution of the ξ's according to eq. 8 above. In figure 11, ξ_1 increases to a saturation value of 1 whereas the other pattern ξ_2 decays to zero. The label $c^{(1)} = 0.5$ runs to the appropriate value for the ξ_1 pattern vector. The opposite case occurs in figure 12 which correctly classifies the ξ_2 pattern. Of course it is possible to generate as many prototypes as necessary for a given application (e.g. limping, running etc.)

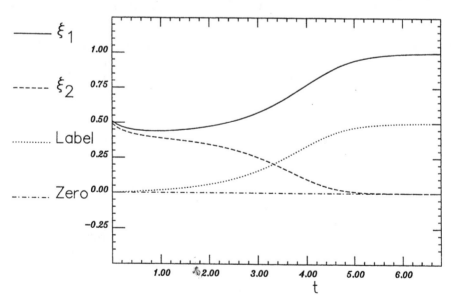

Fig. 11 Simulation of the pattern recognition dynamics. As initial value of $\mathbf{q}$ pattern 1 was chosen but with label $c = 0.0$. The figure shows the time evolution of the scalar product $\xi_u = (\mathbf{v}^{(u)+} \mathbf{q})$. After a short time the dynamics reaches its stationary point and the label is determined with $c = 0.5$ according to the recognized initial value

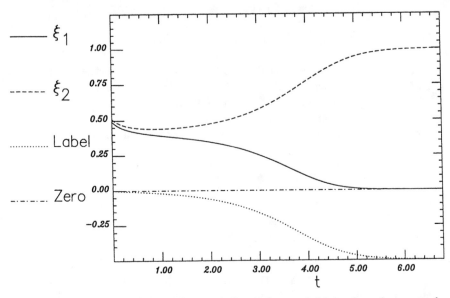

Fig. 12 Shows the same dynamics as in fig. 11 but as initial value of q we took $\mathbf{v}^{(2)}$ with label $c = 0.0$. Movement pattern 2 could be recognized by the final label of $c = -0.5$

5 Further results: Detecting dynamic patterns in noise

The type of human gait pattern can be identified even when the pattern is embedded in noise (random dot kinematograms). In order to generate the random dot kinematogram frame into which the point light display of the human walker is placed, we initially create 68 random dots. The spatial density of the dots is 1.7 per 100 (pixel^2). Each dot then takes an independent two dimensional random walk with a step size equal to ten times the displacement in the x,y direction of the figure's shoulder (other trials have the dots move with a constant step size). The direction in which any dot moves is independent of its own previous displacement and the displacements of other dots. The special case of static perception of the human gait figure, when positioned in a field of random dots, provides only information about the relative topology of all the points. Once motion is added, the human point light walker cannot immediately be detected by an inexperienced observer. After a cycle or two, however, naïve observers always perceive the figure (Horvath et al., 1990), demonstrating the ability of the human visual system to detect local correlations in the presence of noise. A similar mechanism may allow humans to detect Moiré effects from random dots, since this also is related to the detection of local correlations and the combination of local correlations from different regions to form a percept (Glass, 1969). Recently, Williams, Phillips and Sekuler (1986) had subjects view a field of dots in which some dots moved with an inherent direction of motion within a random dot kinematogram. The viewer's task was to extract the mean direction of the dots from the field, similar to judging the average direction of the ocean's surf, despite the fact that individual waves move in different paths (Williams et al., 1986). The human point light walker locomoting in a random dot field is another example of such a scene. The cooperative network that allows one to perceive the non-random vectors corresponding to the point light walker may be the same as the one that allows one to perceive the mean direction of motion in other displays. In order to show that the algorithm given in equation (3) can also determine the connections between the points of the human gait pattern when it is embedded in noise, we generate a field of random points. The points' positions are updated every two frames. The number of times condition (3) is met over the cycle is again computed. If this value is less than some threshold, τ, the points are connected. The only modification of the algorithm is to not compute the connections between the random points which are fixed for two frames. Naturally, these points remain in the same relative position for 180 times during a cycle, and this value exceeds the threshold τ. The identification of the human gait figure within noise is shown in Fig. 13.

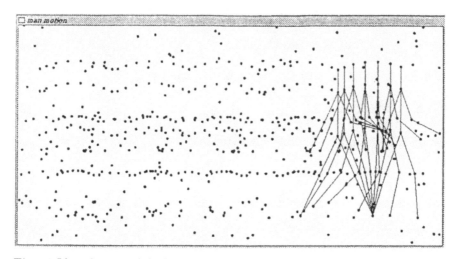

Fig. 13 Identification of the human gait figure within noise

6 Generalization of the approach

We have so far assumed that the movements are in a plane perpendicular to a viewer or camera. We assume also that the angle α between the plane on which the point light walker moves and the plane of the screen on to which the coordinates are projected is time-independent. Here we extend our approach to take into account the third dimension by assuming rigid motion. We also treat the case where the angle α is time-dependent. First we assume there is a finite distance l between the plane in which the center of the point light walker moves and the plane of limb movement. The relation between x, y (the coordinates of the screen of the observer) and x', y' (those of the moving objects in its plane of movement) is given by

$$x = x' \cos \alpha \tag{9}$$

$$y = y' \tag{10}$$

which then generalizes to

$$x = l \sin \alpha + x' \cos \alpha \tag{11}$$

Before making α time-dependent we have to check the relation

$$\frac{(\Delta x_{ij})^2}{\cos^2 \alpha} + (\Delta y_{ij})^2 = \text{const?} \tag{12}$$

Differentiation of eq. (12) within respect to time yields

$$\frac{\Delta x_{ij} \Delta \dot{x}_{ij}}{\cos^2 \alpha} + 2 \frac{(\Delta x_{ij})^2}{\cos^3 \alpha} \sin \alpha \dot{\alpha} + \Delta y_{ij} \Delta \dot{y}_{ij} = 0 \tag{13}$$

or when we introduce suitable abbreviations

$$K(t) + L(t) \frac{\sin \alpha}{\cos \alpha} \dot{\alpha} + \cos^2 \alpha M(t) = 0. \tag{14}$$

We consider eq. (14) as a differetial equation for α. Making the substitution

$$\eta = -\ln \cos \alpha \tag{15}$$

from which

$$\dot{\eta} = \frac{\sin \alpha}{\cos \alpha} \dot{\alpha} \tag{16}$$

and

$$\ln \cos \alpha = -\eta \tag{17}$$

result, eq. (14) can be cast in the form

$$\dot{\eta} + e^{-2\eta} \tilde{M}_{ij}(t) = \tilde{K}_{ij}(t). \tag{18}$$

In addition we consider a different pair kl which is assumed to rotate in the same plane as the pair ij again with the fixed distance l. We then obtain

$$\dot{\eta} + e^{-2\eta}\tilde{M}_{kl}(t) = \tilde{K}_{kl}(t). \tag{19}$$

Using the abbreviations

$$\tilde{K} = -\frac{K}{L} \tag{20}$$

$$\tilde{M} = \frac{M}{L} \tag{21}$$

In order to remove the time derivative $\dot{\eta}$, we take the difference between eqs.(18) and (19) to obtain

$$e^{-2\eta}(\tilde{M}_{ij}(t) - \tilde{M}_{kl}(t)) = \tilde{K}_{ij} - \tilde{K}_{kl} \tag{22}$$

We may resolve eq. (22) with respect to $e^{-2\eta}$

$$e^{-2\eta} = \frac{\Delta \tilde{K}}{\Delta \tilde{M}}. \tag{23}$$

This possesses the solution

$$-2\eta = \ln \frac{\Delta \tilde{K}}{\Delta \tilde{M}}. \tag{24}$$

The right-hand side of eq. (24) must obey eqs. (18) and (19). A different self consistency check is obtained by establishing eq. (23) for a further pair $i'j'$ so that the following relation

$$e^{-2\eta} = \frac{\Delta \tilde{K}'}{\Delta \tilde{M}'}. \tag{25}$$

holds. Dividing (23) by (25) we find the relation

$$\frac{\Delta \tilde{K}'}{\Delta \tilde{M}'} = \frac{\Delta \tilde{K}}{\Delta \tilde{M}}. \tag{26}$$

Making these different self-consistency checks for the different pairs, the synergetic computer may identify those pairs for which the relation

$$|\mathbf{x}_i(t) - \mathbf{x}_j(t)|^2 = \text{constant} \tag{27}$$

holds, where the constant is time-independent. We now know the motion of the limbs and proceed as before (see Section 4). In general, to resolve any ambiguities due to rotation of plane, it is necessary to incorporate further points into the point light displays. For example, to resolve perspectival changes, points on both shoulders and hips representing fixed anatomical distances can be included.

7 Summary and conclusion

The human visual system is remarkably adapt at picking up local correlations in the presence of noise and can combine them from different regions in the visual field in such a way that a global percept arises. We have recently found in psychophysical experiments that Johansson-type point light walkers can also be identified by naïve observers even in the presence of a moving random dot field. Here we present a pattern recognition algorithm, based on concepts of synergetics, that successfully identifies and discriminates synthetically generated patterns of moving points, such as human gait patterns. Our algorithm essentially looks for local correlations by finding pairs of points whose position relative to each other remains time-independent. Once the points are connected, the phase angles among the extremities are computed along with their Fourier coefficients. Test pattern vectors are then created and the order parameters determined. Any number of prototype gait patterns are identified and classified. Our algorithm has been tested on several sets of computer generated point light patterns and distinguishes different kinds of gaits even when they are embedded in random dot kinematograms. Further generalizations of the approach to include rotation on the plane and time-dependence are shown to be possible and are the subject of current numerical and experimental research. Although neural cooperative processes underlying global percept formation are difficult to study, the present work which blends theoretical concepts (the duality of pattern formation and pattern recognition process in synergetic systems), computational methods (pattern recognition by computer) and psychophysical experiments appears to represent a promising approach. In particular, we show that both man and machine are capable of detecting patterned structure on the basis of macroscopic order parameters.

Acknowledgement

Financial support was provided by the Volkswagenwerk Foundation within the project on Synergetics. NIMH (Neurosciences Research Branch) Grant MH 42900, contract N00014-88-J-1191 from the U.S. Office of Naval Research and the Florida High Technology and Industrial Council.

References

Cutting, J.E. (1978): A program to generate synthetic walkers as dynamic point-light displays. Behavior Research Methods and Instrumentation vol. 10, 91 – 94.

Cutting, J.E. & Proffitt, D.R. (1981): Gait perception as an example of how we perceive events. In: R.D. Walk & H.L. Pick (Eds.), Intersensory Perception and Sensory Integration. Ney York: Plenum.

Fuchs, A. & Haken, H. (1988): Pattern recognition and associative memory as dynamical processes in a synergetic system I + II, Erratum. Biological Cybernetics 60, 11 – 22, 107 – 109, 476.

Glass, L. (1969): Moiré effect of random dots. Nature 223, 578 – 580.

Haken, H. (1983): Synergetics. An Introduction. Berlin: Springer (3rd ed.).

Haken, H. (1987a): Information compression in biological systems. Biological Cybernetics 56, 11 – 17.

Haken, H. (1987b): Synergetic computers for pattern recognition and associative memory. In: H. Haken (Ed.), Computational Systems - Natural and Artificial. Berlin: Springer.

Haken, H. (1988): Information and Selforganization. Berlin: Springer.

Haken, H., Kelso, J.A.S., Fuchs, A. & Pandya, A.S. (1990): Dynamic pattern recognition of coordinated biological motion. Neural Networks 3, 395 – 401.

Haken, H. & M. Stadler (Eds.) (1990): Synergetics of Cognition. Berlin: Springer.

Hoenkamp, E. (1978): Perceptual cues that determine the labeling of human gait. Journal of Human Movement Studies 4, 59 – 69.

Horvath, E.I., Kelso, J.A.S. & Pandya, A.S. (1993): Detection of Human Gait Pattern within Random Dot Kinematograms. (in press).

Johansson G. (1973): Visual perception of biological motion and a model for its analysis. Perception and Psychophysics 14, 201 – 211.

Kelso, J.A.S.(1990): Phase transitions: Foundations of behavior. In: H. Haken & M. Stadler (Eds.), Synergetics of Cognition, pp. 249 – 268. Berlin: Springer.

Kelso, J.A.S., Schöner, G., Scholz, J.P. & Haken, H. (1987): Phase locked modes, phase transitions and the component oscillations in biological motion. Physica Scripta 31-1, 79 – 87.

Kelso, J.A.S., Wallace, S. & Buchanan, J. (1989): Phase transitions and trajectory formation in single multijoint limb patterns. Psychology of Motor Behavior and Sports, Kent State University, June 1-4, 33.

Runeson, S. & Frykholm, G. (1981): Visual perception of lifted weights. Journal of Experimental Psychology: Human Perception and Performance 7, 733 – 740.

Schöner, G. & Kelso, J.A.S. (1989): Dynamic patterns in behavioral and neural systems. Science 239, 1513 – 1520.

Turvey, M.T., Carello, C. & Nam-Gyoon, Kim (1990): In: H. Haken & M. Stadler (Eds.), Synergetics of Cognition, pp. 269 – 295. Berlin: Springer.

Williams, D., Phillips, G. & Sekuler, R. (1986): Hysteresis in the perception of motion as evidence of neural cooperativity. Nature 324, 253 – 255.

Part III

Attractors in Behavior

Multistability and Metastability in Perceptual and Brain Dynamics

J.A.S. Kelso, P. Case, T. Holroyd, E. Horvath, J. Rączaszek,
B. Tuller and M. Ding

Center for Complex Systems, Florida Atlantic University,
Boca Raton, FL 33431, USA

Abstract: In this paper we demonstrate that vision, speech and language may exhibit self-organizing dynamic properties including transitions between perceptual states, multistability, instability, and hysteresis. Theses features illustrate a crucial characteristic of perpecptual organization, namely, the ability to function coherently yet retain some degree of flexibility. We propose the generic dynamical mechanism of intermittency as a way to flexibly enter and exit perceptual states, and suggest that this mechanism is exploited in coordinated perceptual and neural behavior.

1 Introduction

The 'what' of perception as exemplified by information processing and ecological approaches has received by far the most attention from experimental psychology and neurophysiology. The 'how' of perception – the determination of perception by influences intrinsic to the perceiver – is only now receiving the attention it deserves. Yet the contents of perception and the intrinsic dynamics of perception are intimately related. Perceptual states, by definition, are low-dimensional, self-organized entities formed by the nervous system. They exist, not as fragmentary pieces of information gleaned from stimuli, but as integrated wholes. Phenomena such as multistability, transitions between perceptual states, hysteresis and so forth attest to the existence of an underlying perceptual dynamics. In the view presented here, there is no need to dichotomize theories that deal with the contents of perception from theories that deal with the dynamics of perception. Both aspects may be brought under the aegis of theoretical concepts of pattern formation in nonequilibrium systems, i.e. synergetics (e.g. Haken, 1990) and the corresponding tools of nonlinear dynamics.

In this paper, we shall try to accomplish three things. First, we draw attention to the shared dynamical principles between coordinated action and perception. We believe this goes far beyond mere analogy but rather suggests that the neural representations underlying both take the form of self-organized dynamical systems (see also Haken & Fuchs, 1988). In the second part of this paper, we disambiguate the neural representations of purely physical aspects of a signal from those processes in the brain associated with percept formation and change. For both these aims, the concepts of attractors, multistability, instability and so forth play a central role. In the third part, we suggest a rather more flexible or fluid view of brain and perception in which the underlying dynamics are intrinsically *metastable*, hence intermittent. Rather than residing in attractors of a neural network, the system dwells for varying times near instability where it can switch flexibly and quickly. The brain under these less well-defined conditions may be viewed as a "twinkling" system living on the brink of instability.

1.1 Multistable Glass Patterns

Random dot Moiré patterns constitute one of the most eye-catching phenomena in visual psychophysics. Glass (1969) created these patterns by taking a field of random dots and superimposing it on itself with a slight transformation. For example, with a small rotation between the two fields, a circular pattern emerges that dissolves to randomness if the rotation is further increased. The physical simplicity of Moiré patterns and the near immediacy of their detection belies their psychological and neurophysiological complexity. It has been proposed by Glass and others (e.g. Julesz, 1971) that perception of Moiré fringes is based on the ability of the human visual system to extract information about local correlations from different regions in the visual field when forming a global percept.

The study of the formation or destruction of Glass patterns has been limited to static displays. Yet Glass patterns afford a wonderful opportunity to expose the underlying intrinsic dynamics of percept formation and change. For example, the evolution from a random pattern to an ordered pattern can be generated by allowing the following pair of differential equations to evolve in time.

$dx/dt = ax + cy$

$dy/dt = dx + by$

Glass patterns possess the same morphologies as the trajectories of a phase-plane representation of two coupled autonomous linear differential equations undergoing an infinitesimal transformation (Glass & Perez, 1973; Leftshetz, 1963). Depending on the specified parameters (see Table 1), four basic archetypes may be generated as shown in Fig. 1. By systematically varying a parameter, or by simply allowing the equations to evolve in time, we may obtain insights into perceptual transitions between disorder and order, multistability (different perceptual effects for the same physical stimulus configuration), switching from one percept to another, hysteresis (persistence

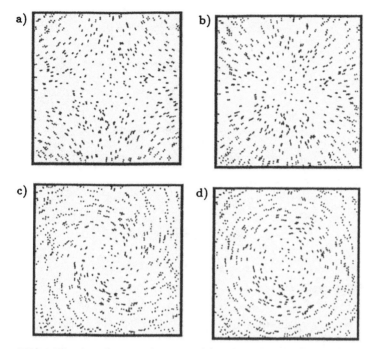

Fig. 1 The four basic archetypes: (a) saddle; (b) degenerate node; (c) spiral; (d) vortex

Table 1 Parameters specified to generate four basic archetypes

a	b	c	d	archetype
1.0	−1.0	0.0	0.0	Saddle
1.0	1.0	0.0	0.0	Degenerate Node
1.0	1.0	−2.0	2.0	Spiral
0.0	0.0	−2.0	2.0	Vortex

of a percept despite parameter changes that favor an alternative), as well as other perceptual effects.

We (Horvath & Kelso) have studied Glass patterns under three main experimental conditions: Two involved sequential presentations of patterns that change from disorder to order and from order to disorder and one involved the random presentation of patterns. Each pattern was displayed for 500 milliseconds, followed by an interval of 2 seconds during which the subject had to decide (by clicking a mouse button) whether the pattern was random or ordered. Then the pattern was advanced one frame (referred to as position

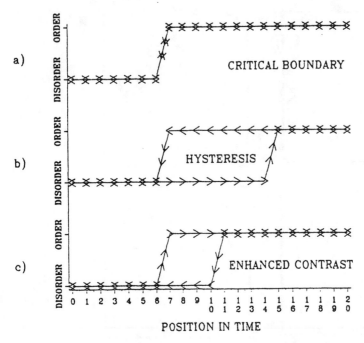

Fig. 2 Three distinct perceptual patterns for sequential presentation of Glass patterns. (a) critical boundary; (b) hysteresis; (c) enhanced contrast (see text for details)

in time) and the perceptual judgment task repeated. Three distinct effects were observed for the sequentially ordered patterns (See Fig. 2).

The first is *critical boundary:* This shows that the percept depends only on the pattern displayed, not the order of display. The second perceptual effect is *hysteresis*, a larger position in time is necessary to perceive a change from disorder to order than from order to disorder. The presence of hysteresis, namely the dependence on past input and the direction of parameter change constitutes strong evidence for nonlinearity and multistability in visual perception. The third perceptual effect indicates that a switch from disorder to order occurs at a smaller position in time than when the patterns evolve from order to disorder. We refer to this phenomenon as *enhanced contrast*, implying that the boundary between patterns shifts to enhance differences between percepts. Later on we will describe a theoretical model that accommodates all of these effects.

Figure 3 shows the results of one of our experiments. Except for a single subject (OB), all subjects exhibit a hysteresis effect of varying magnitude. Hysteresis is by far the most dominant perceptual effect occurring on 72 % of the trials compared to 21 % and 7 % for enhanced contrast and critical boundary respectively.

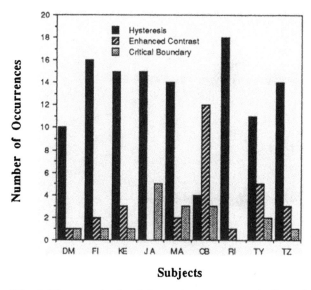

Fig. 3 The magnitude of the various perceptual effects for nine subjects

In contrast to classical studies of hysteresis and multistability in visual perception, such as Fisher's (1967) man-girl reversible figure in which many aspects of the stimulus configuration change from one configuration to the next, so-called Glass patterns allow us to manipulate the vector field of a dynamical system systematically and study corresponding perceptual effects. The results of Figure 2 clearly demonstrate: a) spontaneous formation of an ordered percept from a disordered stimulus array at a critical parameter value (a given position in time corresponding to a systematic deformation of the vector field); b) a transition from one perceptual organization to another; c) multistability, i.e., the same stimulus configuration can give rise to two different perceptual effects; and d) hysteresis, i.e., the percept that is formed persists and depends upon the direction of parameter change. Such experiments reveal the intrinsic organizational tendencies of visual perception and open up exciting possibilities for detecting brain events corresponding to the formation of percepts and for a deeper understanding of the neural basis of percept transitions (see below).

1.2 Categorical Changes in Meaning

The effects displayed in Figure 2 are by no means unique to visual perception but may also be observed in studies of ambiguous sentences. In real life, the inherent ambiguity of sentences is resolved by the context within which they are perceived but it is also possible to perform experiments in which local acoustic cues are varied as putative control parameter(s) and study consequent effects on sentence interpretation. Is it possible to change

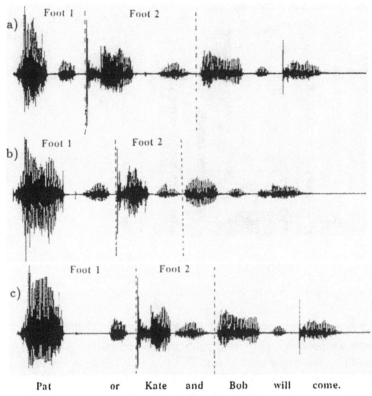

Fig. 4 (a) The original version of '(Pat or Kate) and Bob will come.' (b) The original version of 'Pat or (Kate and Bob) will come.' (c) One altered version of (a) shifted toward the alternative meaning

interpretation from one meaning to the other by such methods, and if so, what is the dynamics of such change? Figure 4 shows how the sentence
"Pat or Kate and Bob will come"
was systematically manipulated in an experimental design similar to the previously mentioned study of visual perception (Rączaszek, Case, Tuller & Kelso). But here two prosodic cues marking a syntactic boundary were systematically manipulated, namely the duration of the last stressed vowel before the boundary and the pause duration at the boundary itself. These cues contribute to a "foot" duration in the sentence.

The putative control parameter in this case effectively corresponds to a "foot ratio" as shown in Figure 4. An original sentence was digitized and re-synthesized with the foot ratio systematically varying in equal logarithmic steps until the sentence approached the foot ratio observed in the alternative interpretation. Of course, there are many other cues that could be manipulated in such settings that we did not attempt to control. Any cues not manipulated are, in a sense, "enslaved" even if they are in disagreement with the altered meaning. Once the meaning is perceived as the alternative, such cues may be said to be incorporated in the new pattern. Figure 5a shows a typical response function for two individual runs of sequentially ordered presentations.

In the top half of Figure 5a, the original meaning of the sentence was 'Pat or (Kate and Bcb) will come,' labeled as P(KB). The x-axis indicates the sequential alteration in foot ratio and the arrows indicate the direction of the alteration. Follow the solid line beginning at the left side of the figure. The subject switched to the alternative meaning, '(Pat or Kate) and Bob will come,' labeled as (PK)B, at the eleventh step. As the foot ratio was altered in the opposite direction, indicated by the dashed line, the subject initially heard the meaning, '(Pat or Kate) and Bob will come,' and did not switch back to the original meaning until the fifth step. That is, the subject exhibited hysteresis. Hysteresis is also seen for the trial shown in the bottom half of Figure 5a, in which the subject first heard the meaning '(Pat or Kate) and Bob will come.' Of the ten subjects, seven showed hysteresis, two enhanced contrast and one a critical boundary. The conditional probabilities in the random condition show the same pattern (Figure 5b). The probability of interpreting a sentence as having a particular meaning was higher when the subject perceived the same meaning immediately before and was lower when a different meaning was perceived before.

These results suggest that dynamical methods are useful to study the problem of sentence ambiguity and categorical processes in meaning. Although the evidence is preliminary it suggests that the same principles underlie pattern formation processes not only in perceptual systems but also at rather high levels of cognitive functioning. Further work is necessary to verify the interpretation of these experiments in terms of the dynamics of meaning as opposed to attention to local cues. Interesting questions pertain to the minimum amount of information necessary to switch meaning (a kind of catalytic linguistics, cf. Iberall, Soodak & Hassler, 1978).

1.3 Speech Categorization

The basic issue of how we sort a continuously changing signal into the appropriate category is still incompletely understood, whether the categories concern speech, objects, emotions, individuals, etc. (see, for example, Harnad, 1987). Here the question is whether the perceptual stability of speech sounds may also be characterized as a pattern formation process, i.e. in terms of nonlinear dynamics. In typical studies listeners judge stimuli as belonging to the same speech category despite large variations in the manipulated acoustic parameter. Only when the value of the manipulated parameter reaches a critical boundary does the percept change and then it typically does so abruptly and discontinuously. Usually the stimulus set is presented to listeners in random order with the aim of eliminating short-term sequential effects known to modify category boundaries. In our work (Tuller, Case, Ding &

166 J.A.S. Kelso et al.

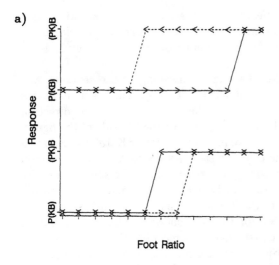

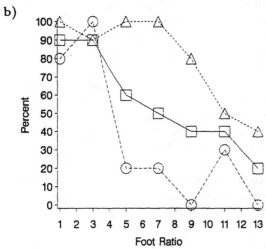

Fig. 5 (a) Patterns of response for two individual runs (see text for details). (b) Conditional probabilities associated with the random identification function (□ =random, △ = same, and ○ = different). See text for details

Kelso, 1993), rather than employ procedures designed to eliminate possible dynamical effects, we studied the latter in their own right because we believe them to reflect intrinsic properties of the speech perception system. We employed a stimulus continuum for which categorical perception has often been demonstrated, but varied the acoustic parameter sequentially, i.e., as a control parameter, in order to explore the temporal evolution of phonemic categorization and change. The stimuli chosen comprise a 'say-stay' continuum, with sequential variation of the silent gap duration after the initial fricative. The subject's task was to identify each stimulus as either 'stay' or 'say' and to indicate the response on an answer sheet during a two second inter stimulus interval. Stimuli were presented binaurally through headphones at a comfortable listening level, approximately 70 dB. The sequentially ordered patterns involved: 1) silent gap duration increasing in 4 ms. increments from 0 ms. to 76 ms., then decreasing back to 0 ms; and 2) silent gap decreasing in 4 ms. steps from 76 ms. to 0 ms., then increasing back to 76 ms. Once again, three distinct patterns of results were observed. Critical boundary occurred only 17 % of the time across subjects and conditions. Hysteresis was far more common than critical boundary, occurring in 41 % of runs across subjects and conditions. Enhanced contrast, namely boundary shifts that act to enhance the contrast between speech categories, occurred on 42 % of runs across subjects and conditions. In the following section, we propose a theoretical model that captures all three patterns of category change within a unified dynamical account. The idea is that this theoretical description handles the dynamics of category change regardless of modality or level of description. The general principles behind the various effects observed in vision, speech perception and sentence interpretation are deemed to be model-independent.

1.4 Theoretical Modelling of Categorical Effects

Determination of a dynamical model is based on mapping stable perceptual categories onto attractors of a dynamical model. In this we follow essentially the same strategy as employed by Haken, Kelso & Bunz (1985) in their theoretical treatment of coordinated movement. For a single perceptual category, a local model containing a fixed point is adequate. However, if several stable percepts coexist a nonlinear dynamical model must be found. Task variables (e.g. deformation of a vector field in vision, foot ratio in sentences, acoustic parameters in speech) are presumed to act as parameters on the underlying dynamics. Figure 6 shows the attractor layout for the following dynamical system:

$$\mathbf{V}(x) = kx - x^2/2 + x^4/4 \qquad (1)$$

where x is the perceptual form and k is the control parameter specifying the direction and degree of tilt for the potential $\mathbf{V}(x)$. The equation of motion from which the potential function is derived is

$$dx/dt = -d\mathbf{V}(x)/dx \qquad (2)$$

in which the long term behavior of x is captured by stable fixed points (the minima shown in Figure 6). For ease of visualization, Figure 6 shows the potential for several values of k.

With $k = -1$, only one stable fixed point exists corresponding to a single percept. As k increases, the potential landscape tilts but otherwise remains unchanged in terms of the composition of attractor states. However, when

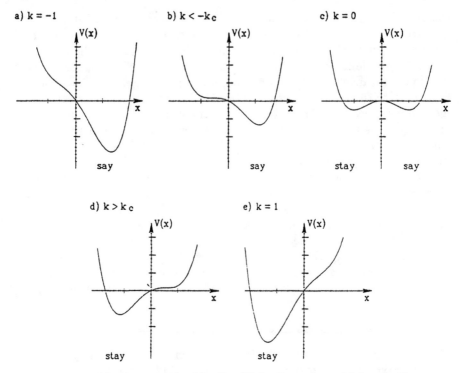

Fig. 6 Potential landscape defined by Eq. (1) for five values of k (see text)

k reaches a critical point, an additional attractor appears by a saddle-node bifurcation in which a point attractor (with $x < 0$) and a point repeller (maxima in Fig. 6) are simultaneously created. The co-existence of two percepts continues until $k = k_c$ where the attractor corresponding to one percept ceases to exist via a reverse saddle-node bifurcation, leaving only one stable fixed point in the system. Further increases in k only serve to deepen the potential minimum corresponding to the alternative percept. As stressed again and again by synergetics, any accurate portrayal of a real world problem must take into account the influence of random disturbances, sources of which in a perception experiment may correspond to factors such as fatigue, attention, boredom, etc. Mathematically, spontaneous switches among attractive states occur as a result of these fluctuations, modelled as random noise. For a given point attractor, the degree of resistance to the influence of random noise is related to its *stability*, which, in general, depends on the depth and width of the potential well (basin of attraction). As k is increased successively in Figure 6, the stability of the attractor corresponding to the initial percept decreases (the potential well becoming shallower and flatter) leading to an increase in the likelihood of switching to the alternative percept. This implies that perceptual switching is more likely with repeated presentations of a stimulus near the transition point than with repetition of a stimulus far from the transition point. Such predictions have, in fact, been confirmed in experiments (see Tuller, Case, Ding & Kelso, 1993).

In order to account for the three response patterns observed i.e., a critical boundary, hysteresis, and enhanced contrast, it is necessary to examine in more detail the dependence of the parameter k on the experimentally manipulated parameter of gap duration. Without going into details, the tilt of the function (equation 1) may be shown to depend on the initial conditions, a parameter λ that is linearly proportional to the gap duration, n which is the number of perceived stimuli in a given run and ϵ, a parameter that represents cognitive factors such as learning, linguistic experience and attention. Note that the introduction of criteria stemming from cognitive processes is not without precedent, for example, attention and previous experience play a large role in synergetic modelling of perception of ambiguous visual figures (Haken, 1990; Ditzinger & Haken, 1989, 1990) and contribute to factors that determine adaptation level in Helson's work (Helson, 1964).

When n and/or ϵ are sufficiently small the tilt of the potential is only dependent on gap duration and the initial configuration. Figure 7a & b shows three regions corresponding to different states of the system in the $\epsilon-\lambda$ plane, in the first half of each run when n is small (Fig. 7a) and in the second half of each run when n is large (Fig. 7b). White regions indicate the set of parameter values for which a stimulus has but a single possible categorization in the represented portion of the run. Shaded regions indicate the set of parameter values for which a stimulus may be categorized as either one form or the other (the bistable region) and thus represent the condition from $-k_c$ (the lower border of each shaded region) to k_c (the upper border of each shaded region). Consider the initial condition with $k_0 = -1$ and the parameter λ increasing. As λ increases, the stimuli are categorized as "say" for any value of ϵ so long as the $\epsilon - \lambda$ coordinate remains below the shaded region. Within the shaded region, the stimuli are categorized as "say" despite the percept becoming progressively less stable. As λ continues to increase the percept switches from "say" to "stay" at the upper boundary of the shaded region (the heavy line) after which "say" is not a possible percept. Note that for different values of ϵ the switch to a new percept occurs at different durations of silent gap.

In the second half of the run, λ is decreasing and the resulting division of the $\epsilon - \lambda$ plane looks somewhat different. This portion of the figure should be read from top to bottom, from large gap durations to small ones. As λ decreases, the stimuli are categorized as "stay" for any value of ϵ so long as the $\epsilon - \lambda$ coordinate remains above the shaded region. Again assuming the absence of a perturbing force, the subject continues to categorize the stimuli as "stay" within the bistable (shaded regions) despite the percept becoming less stable. As λ continues to decrease, the *lower* boundary of the shaded region, (the heavy curve) marks the switching from "stay" back to "say". In

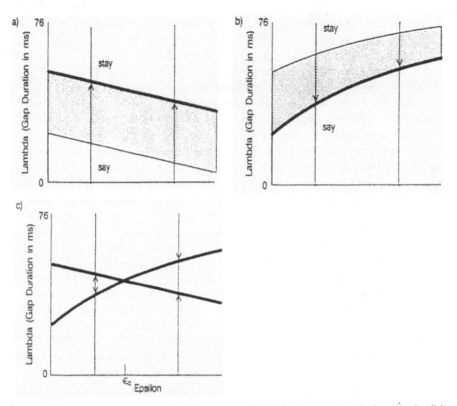

Fig. 7 (a) Perceptual states in the $\epsilon - \lambda$ plane for increasing λ (small n); (b) perceptual states in the $\epsilon - \lambda$ plane for decreasing λ (large n); (c) superposition of reversed saddle node bifurcation points in a and b. See text for details. [From Tuller, Case, Ding & Kelso, 1993]

Figure 7c, we have simply superimposed the boundaries at which switching occurs in Figures 7a and 7b because each describes a different segment of a run. The line with negative slope represents the category switch in the first half of the run and the curve with positive slope represents the switch back to the initial category. Their intersection yields a critical value of ϵ, ϵ_c for which critical boundary would be observed in the "say-stay" continuum. For $\epsilon > \epsilon_c$, (e.g. the vertical line to the right of ϵ_c), the system exhibits the enhanced contrast effect. In that region, λ is smaller for the negatively-sloped line than for the positively-sloped curve. For $\epsilon < \epsilon_c$, (e.g. the vertical line to the left of ϵ_c), the classical hysteresis phenomenon is obtained: λ is larger for the line than the curve. The main qualitative features of the observed data are thus captured in the model. Moreover, the parameter plane looks essentially the same when the effect of random fluctuations is explicitly considered. The effect of such fluctuations is to enhance the likelihood of observing switches to the other category as the boundary is approached rather than crossed.

The model accommodates all three perceptual patterns observed and spawns further experiments. The potential landscape is seen to deform with variations in a parameter with the rate of deformation depending on the number of perceived repetitions, as well as an "experience-factor" with the stimuli. For example, near ϵ_c small fluctuations in ϵ are more likely to lead to a change in pattern than when the system is far from ϵ_c (see Fig. 7c). The observable consequence is that if a subject shows both hysteresis and enhanced contrast, then the overlap and "underlap" regions should be small. On the other hand, if the subject shows only hysteresis or only enhanced contrast then the overlap or underlap region should be large. Simply ranking the size of the overlap or underlap confirmed this prediction. Subjects who showed only hysteresis or only enhanced contrast had the two largest overlap and underlap regions respectively.

A second prediction concerns the influence of cognitive factors, ϵ, as the experiment proceeds in time. If a subject shows at least two of the three possible response patterns the model predicts a smaller probability of hysteresis in the second half of the experiment compared to the first half. This prediction was confirmed in five of seven subjects. A third prediction is that spontaneous changes in percept should occur only in the bistable region. One way to test this is to repeat the last stimulus in a run, thereby maximizing the opportunity to see spontaneous changes. In principle, as a category loses stability the shallower minimum in the potential has a less restraining influence on the fluctuations and changes in category should occur earlier with repetition than with single stimulus presentation. This prediction was confirmed experimentally. Regardless of whether an individual's response patterns displayed hysteresis, enhanced contrast or a critical boundary, stimulus repetition maximized the probability of category change near the boundary.

In summary, the traditional approach to understanding cognitive and perceptual processes is to emphasize static representational structures that minimize time dependencies and ignore the potentially dynamic nature of such representations. Increasingly this approach is being challenged by researchers in a wide variety of fields who believe that synergetic concepts provide a more appropriate framework for the study of cognition than symbolic computation, and have built dynamics directly into their theories and models (see e.g. contributions in Haken & Stadler, 1990). The present experiments – ranging from vision through speech perception and sentence interpretation – are a striking testament to the basic notions of synergetics, such as control parameters, transitions, attractors, multistability, instability and hysteresis. Of course, we have made no effort to distinguish particular cues or specific mechanisms involved in these processes, instead we have focused on uncovering the dynamical nature of categorization and change. Our assumption is that the same general principles apply across particular cues or categories, thus serving to unify a broad set of data and providing a coherent basis for predicting contextual effects.

2 Brain dynamics: Exploiting hysteresis

The presence of hysteresis creates a unique situation for studying the brain dynamics underlying perception. For example, in the case of speech the subject may hear two different words for the identical acoustic stimulus, depending on the direction of change in the gap duration parameter. So we can *use* or *exploit* the nonlinear effect of hysteresis to dissociate brain processes pertaining to the physical (acoustic) *stimulus*, from brain activity patterns specific to phonetic *perception*.

In what follows we present some recent results using a multi-sensor SQUID array to monitor neural activity patterns in the human brain (for details of placement and recording set-up, see Kelso et al. 1992). SQUID stands for Superconducting Quantum Interference Device, sensors that are capable of detecting signals that are less than one billionth of the earth's magnetic field. SQUIDs pick up signals generated by intracellular dendritic currents in the brain directly, without having to stick electrodes in the brain or on the skull. Because the skull and scalp are transparent to magnetic fields generated inside the brain and because the array is large enough to cover a substantial portion of human neocortex, this new research tool opens a (noninvasive) window into the brain's dynamic patterns and their relation to cognition.

The SQUID array was placed over the left auditory evoked potential field before the start of the experiment. Figure 8 shows the location of the montage with respect to the subject's head and the cortex itself which was reconstructed from MRI slices. The speech categorization paradigm was very similar to that described in the previous section. A "say-stay" continuum (4ms–48ms gap duration, changing in 4ms steps) was presented with gap duration sequentially increasing and decreasing. During the interstimulus interval of 1.5 s, the subject's task was to press one of a pair of microswitches to indicate which word was heard. Analysis of responses showed that at a gap duration of 32ms., 50 % of the stimuli were perceived as 'say' and 50 % as 'stay.' For each of the 37 sensors, we averaged the raw signal across tokens with the same response then calculated the root mean square amplitude of the averaged waveforms. Figure 9 shows a sequence of topographic distributions of the field amplitude corresponding to the percepts 'say' and 'stay.'

We remind the reader that the physical stimulus is identical for the two percepts. The numbers correspond to specific acoustic events: 1) onset of /s/ frication; 2) offset of /s/ frication; 3) onset of vowel periodicity; 4) offset of vowel. It seems quite clear that differences in brain activity between the two percepts are minimal up to the offset of the vowel. However, a few milliseconds following vowel offset (5), a large difference in field amplitude between the two percepts is observed in the posterior part of the array. By the end of the epoch (6) the fields are again identical.

These results are encouraging and appear to open up a wide range of exciting possibilities, not just in speech but in other modalities as well. The

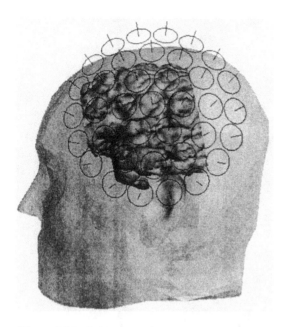

Fig. 8 This figure is a phantom view of the subject's head showing the cortical surface as reconstructed from MRI data, and the SQUID sensor array's position relative to skull and cortex. Not all the MRI slices were converted; this explains the vertical cutoff on the posterior portion of the cortex. The temporal lobe and Sylvian fissure are clearly visible, however. The circles floating above the head show the positions of the lower SQUID loop of a pair forming a gradiometer. The short line segments originating from the centers of the SQUID loops show the orientation of the gradiometer

methodology introduced here allows a clear separation between the neural analogues of the acoustic (or any other) signal, and what the brain is doing when people perceive. Although preliminary, to our knowledge this is the first clear evidence of brain activity patterns that are specific to changes in phonetic percept.

3 Mind switching: From multistability to metastability

The mind cannot fix long on one invariable idea. (John Locke)
or *Try not to think of an elephant for the next 15 seconds.*[1]

Ambiguous stimuli are those that can be perceived in two or more possible configurations and whose perceived organization – without any physical

---

[1] The quote from John Locke is taken from Kawamoto & Anderson (1985). The quote following is a favorite of Arnold Mandell's.

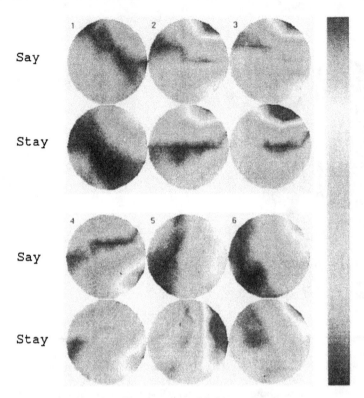

Fig. 9 Topographic distributions of field amplitude

change – may shift spontaneously from one stable configuration to another (e.g. Attneave, 1971). Once a particular configuration is perceived the assumption is that the percept remains stable for a period of time before spontaneously changing. For example, the Necker cube, which is a wire-frame cube drawn in two dimensions (usually in an orthographic projection), can be perceived in one of two orientations, front face up and front face down. A well known fact is that during continuous observation of the same unstable figure, the figure switches back and forth between the different percepts (see Ditzinger & Haken, this volume for data description and synergetic modelling).

In order to investigate this switching process, we (Holroyd & Kelso) had subjects view a Necker cube displayed on a computer screen in one of eight randomly presented orientations. Each orientation was displayed for approximately 1 or 5 minutes with a brief rest interval between viewing periods. Subjects were instructed to press a mouse button if their percept changed, thus generating a sequence of switching times.

The orientation of the cube was determined as follows: The cube was assumed to be three-dimensional. The vertical direction was fixed at 30 deg.,

Perceptual and Brain Dynamics   175

and the figure was rotated along the polar (or θ) direction in increments of 10 deg. ranging from 10 to 80 deg. Zero and 90 deg. rotations result in a 2-dimensional square and so were excluded.

In Figure 10 (top) are shown the experimental conditions. From left to right the orientation shifts from 10 deg. to 80 deg. One can see that at 40 deg. (top right) the figure is almost, but not quite, a flat symmetric 2D hexagon. Likewise at 80 deg. the figure is nearly a flat square.

Also shown in Figure 10 is a time series for a single (40 deg.) Necker cube orientation illustrating the pattern of switching time (and conversely, the duration of a percept). The time series begins on the lower left of the box and continues to the upper right. One can see that bursts of switching are interspersed with prolonged periods during which no perceptual change takes place. There does not seem to be any consistent pattern in the time series data. For example, switching rate does not appear to asymptote with viewing time, a percept held a long time is not necessarily followed by briefly persisting states, and so forth.

In Figure 11, we show a statistical analysis of the switching time data for 3 subjects (total number of switches $\approx$ 8000) as a function of experimental condition. It is quite clear that the mean switching time is only modestly affected by changing orientation, with slight increases for the most symmetric conditions (40 deg. and 80 deg.). All the distributions are positively skewed, i.e. asymmetric. By far the most sensitive parameter dependence emerges in the variance (a measure of dispersion) and kurtosis (a measure of peakedness). As the figure approaches symmetry (40deg. and 80deg.) both the second and fourth moments of the switching time distribution increase quite remarkably.

In Figure 12, we show histograms for a single subject's switching times when exposed to three different orientations of the Necker cube. Thirty deg. and 50 deg. conditions flank the 40 deg. condition, the nearly symmetric 2-D figure. All frequency distributions have a single hump of varying height, but the one for the 40 deg. condition has an extended tail, meaning that occasionally a given orientation is perceived for a long time without switching. One can see this also in the time series data shown in Figure 10. Borsellino and colleagues (1972) have performed extensive switching time studies of ambiguous figures and remark that it is "well-known that subjects get blocked." Because of this they discard switching times (percept durations) greater than 3 standard deviations around the mean. Although the present distributions are similar to the gamma distribution fit by Borsellino, i.e. a nonuniform, exponentially distributed mass characteristic of a Poisson process, the parameter dependence we have found (Figures 11 and 12) may be, in part, due to *not* discarding outliers.

What kind of underlying dynamics might generate such distributions? Elsewhere (Kelso, DeGuzman & Holroyd, 1991a; 1991b; Kelso & DeGuzman, 1991) we have demonstrated, both empirically and theoretically, that the neural coordination dynamics is seldom (except under highly prepared con-

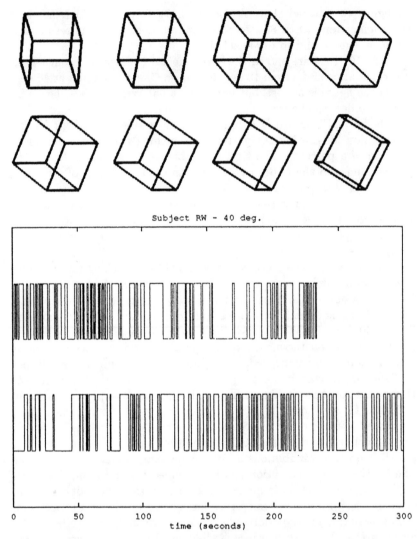

Fig. 10 Necker cube experimental conditions and time series of switching behavior (see text)

ditions) frequency and phase-locked in an asymptotic sense. Rather, to retain flexibility and stability the CNS lives near, but not in mode-locked states. This view contrasts strongly with claims regarding the significance of frequency and phase-synchronization in the nervous system for "binding" local stimulus features into global percepts (e.g. Crick and Koch, 1990; see also Crick & Koch, 1992). Our model is a discrete version of a coupled nonlinear oscillator model derived by Haken, Kelso and Bunz (1985). The "neurons"

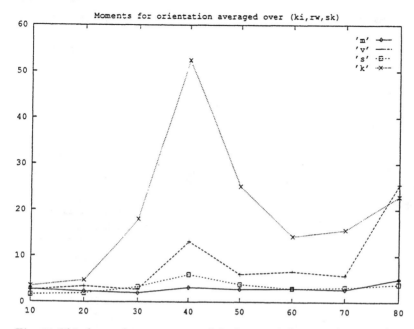

Fig. 11 This figure shows moments of the inter-switch interval averaged over three subjects, for different orientations. The x-axis shows rotation of the Necker cube about the θ axis. Notice the increase in both variance ('v') and kurtosis ('k') for orientations 40 and 80 deg; these orientations correspond to nearly symmetrical views of the Necker cube, and the distributions of inter-switch intervals for these orientations are long-tailed, indicating the presence of a slow mode in the dynamics

in this case are modelled as nonlinear oscillators coupled together by a phase relation, ϕ.

In such a description, the phase relation carries the essential information, i.e. as an order parameter relating the individual units, whose intrinsic properties may exhibit wide spatial and temporal variation. Here we study the relative phase variable and its time evolution as representative of the behavior of the coupled neural dynamics.

The essential idea is that the distance from a fixed point of this nonlinear dynamical system (corresponding to a frequency and phase-locked state) varies directly with stimulus parameters (here the orientation of a Necker cube). Figure 13(a-c), shows the parameter dependence of the model. The boxes on the top of each figure plot the function ϕ_{n+1} versus ϕ_n as the coupling parameter, k is changed by a small amount. Note that this function never quite crosses the 45 deg. line, i.e. the local slope near the fixed point is never less than 1. Iterates of the function run through the corridor between the curve and the line; how fast they move through the gap depends on the width of this corridor.

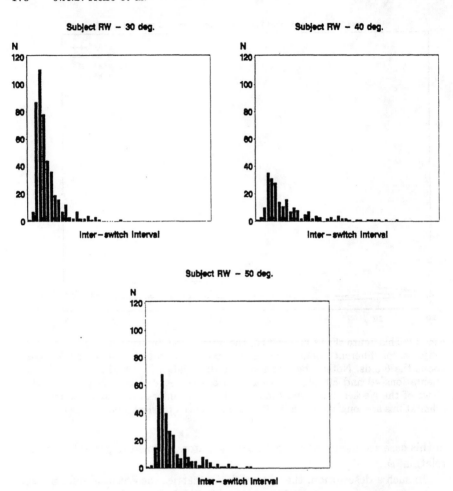

Fig. 12 Histograms of switching times for RW at 3 different orientations

Figures 13a–c(bottom) show the distribution of 'dwell times' near a 1:1 mode-locked state. The dwell time refers to how long the system spends in the narrow channel before exiting. Here, dwell-time is equivalent to the persistence of a given percept before switching occurs. In Figure 13a the distribution has a sharp peak and a rapid fall-off (compare with Figure 12). The time series in the inset of the figure shows the evolution of the phase variable. Note the rather short pauses (seen as plateaus in the time series, where the variable is trapped in the channel) and the rapid escape (corresponding to a switch) followed by reinjection.

As the parameter moves the system closer (Figure 13b) and closer (Figure 13c) to the tangent line, the system resides longer and longer near the fixed point. In Figure 13c, we see that the system stays in the neighborhood of such

Perceptual and Brain Dynamics 179

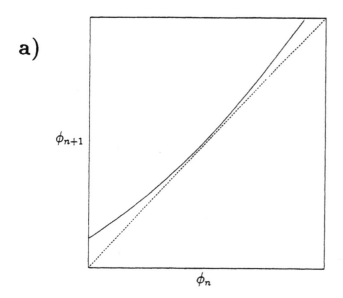

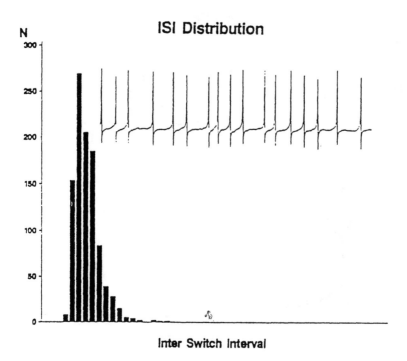

K = 0.498, Omega = 0.0796

Fig. 13a Model simulation (see text)

b)

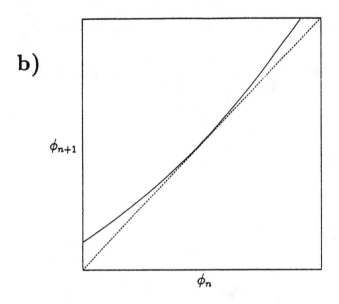

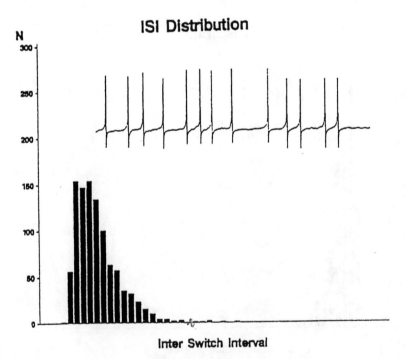

K = 0.499, Omega = 0.0796

Fig. 13b Model simulation (see text)

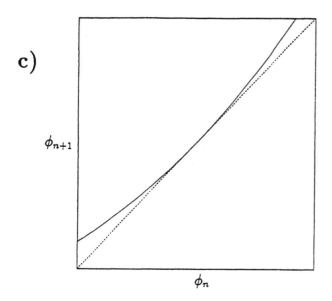

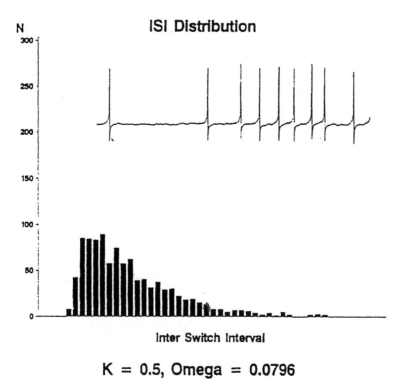

K = 0.5, Omega = 0.0796

Fig. 13c Model simulation (see text)

a point for a variable amount of time exhibiting occasional long dwell times and more frequent short dwell times. The comparison with the experimental data (Figure 12) is quite compelling. Obviously the closer the system is to a mode-locked state (here corresponding to a single, unambiguous perceptual state) the less likely it is to switch.

The dynamics shown in Fig. 13 are called *type 1 intermittency* (Pomeau & Manneville, 1980) which occurs near *tangent* or *saddle-node* bifurcations. We suggest that the observed switching time behavior is generated by a coupled, nonlinear neural dynamics residing in this intermittent régime. Intermittency means that the perceptual system is intrinsically *metastable*, living at the edge of instability where it can switch spontaneously among perceptual states. Indeed, perceptual states themselves may be metastable, as opposed to corresponding to stable, fixed point attractors. In the intermittency régime, there is *attractiveness* but, strictly speaking, no attractors. We remark at the formal resemblance between perceptual intermittency here and our previous analysis of relative coordination (von Holst, 1973) in terms of intermittent dynamics (e.g. Kelso, DeGuzman & Holroyd, 1991).

4 Conclusions

In this paper we have demonstrated that the basic theoretical concepts of synergetics and the mathematical tools of nonlinear dynamics are adequate to capture a wide range of perceptual phenomena. The existence of attractive states, multistability, instability, transitions, hysteresis and theoretical modelling of these complex behaviors suggest that the underlying neural representation takes the form of a self-organized dynamical system. A new aspect is that the dynamics of perceptual (and motor) self-organization is intrinsically metastable. Basically, models of nervous function are based on coupled, nonlinear dynamical systems (whether expressed as a neural network or not). Here we have shown that observed switching time distributions for ambiguous stimuli may be generated by an underlying neural dynamics that is intermittent. Such a generic mechanism may be very important for flexibly entering and exiting coherent patterns, and avoiding resonant mode-locked states (which are asymptotically stable). Rather than requiring active processes to destabilize and switch from one stable state to another (e.g. through change in parameter(s), increase of fluctuations), here intermittency is built in to the dynamics. At least when faced with ambiguous stimuli the perceptual system is marginally stable or metastáble operating *close* to stable fixed points. As we remarked in previous papers in the context of coordination, the intuition is that mind, brain and behavior exploit this metastability feature – the *tendency* to phase- and frequency synchronize – as a means of retaining a vital mix of coherent yet flexible function.

Finally, studying the brain (e.g. with large SQUID arrays, multichannel EEG, functional magnetic resonance imaging and other methods) under conditions where the input is constant but the percept switches, appears to be a powerful paradigm for understanding the neural basis of perceptual organization (see Section 3). Such a paradigm shifts the focus to intrinsic (dynamical) determinants of perception and goes beyond static approaches based on the processing of stimulus features.

Acknowledgements

This work was supported by NIMH (Neurosciences Research Branch) Grant MH 42900, BRS Grant RR07258, NIDCD Grant DC001411 and NSF Grant DBS-9213995.

References

Attneave, F. (1971): Multistability in perception. Scientific American 225, 62 – 71.
Borsellino, A., DeMarco, A., Allazetta, A., Rinesi, S. & Bartolini, B. (1972): Reversal time distribution in the perception of visual ambiguous stimuli. Kybernetik 10, 139 – 144.
Crick, F. & Koch, C. (1990): Towards a neurobiological theory of consciousness. Seminars in the Neurosciences 2, 263 – 275.
Crick, F. & Koch, C. (1992): The problem of consciousness. Scientific American 267, 152 – 160.
Ditzinger, T. & Haken, H. (1993): A synergetic model of multistability in perception. In this volume.
Ditzinger, T. & Haken, H. (1989): Oscillations in the perception of ambiguous patterns: A model based on synergetics. Biological Cybernetics 61, 279 – 287.
Ditzinger, T. & Haken, H. (1990): The impact of fluctuations on the recognition of ambiguous patterns. Biological Cybernetics 63, 453 – 456.
Fisher, G.H. (1967): Measuring ambiguity. American Journal of Psychology 80, 541 – 547.
Glass, L. (1969): Moiré effect from random dots. Nature 223, 578 – 580.
Glass, L. & Perez, R. (1973): Perception of random dot interference patterns. Nature 246, 360 – 362.
Haken, H. (1990): Synergetics as a tool for the conceptualization and mathematization of cognition and behavior - How far can we go? In: H. Haken & M. Stadler (Eds.), Synergetics of Cognition, pp. 2 – 31. Berlin: Springer.
Haken, H. & Fuchs, A. (1988): Pattern recognition and pattern formation as dual processes. In: J.A.S. Kelso, A.J. Mandell & M.F. Shlesinger (Eds.), Dynamic Patterns in Complex Systems, pp. 33 – 41. New Jersey: World Scientific.
Haken, H., Kelso, J.A.S. & Bunz, H. (1985): A theoretical model of phase transitions in human hand movements. Biological Cybernetics 51, 347 – 356.
Haken, H. & Stadler, M. (Eds.) (1990): Synergetics of cognition. Berlin: Springer.
Harnad, S. (1987): Categorical Perception: The Groundwork of Cognition. Cambridge: Cambridge University Press.
Held, R. & Richards, W. (1976): The organization of perceptual systems. In: Perception: Mechanisms and Models. San Francisco: Freeman.
Helson, H. (1964): Adaptation Level Theory: An Experimental and Systematic Approach to Behavior. New York: Harper and Row.

Iberall, A.S., Soodak, H. & Hassler, F. (1978); A field and circuit thermodynamics for integrative biology. II Power and communicational spectroscopy in biology. American Journal of Physiology 234 (1), Re – R19.

Julesz, B. (1971): Foundations of Cyclopean Perception. Chicago: University of Chicago Press.

Kawamoto, A.H. & Anderson, J.A. (1985): A neural network model of multistable perception. Acta Psychologica 59, 35 – 65.

Kelso, J.A.S., Bressler, S.L., Buchanan, S., DeGuzman, G.C., Ding, M., Fuchs, A. & Holroyd, T. (1992): A phase transition in human brain and behavior. Physics Letters A 169, 134 – 144.

Kelso, J.A.S. & DeGuzman, G.C. (1991): An intermittency mechanism for coherent and flexible brain and behavioral function. In: J. Requin & G.E. Stelmach (Eds.), Tutorials in Motor Neuroscience, pp. 305 – 310. Dordrecht: Kluwer.

Kelso, J.A.S., DeGuzman, G.C. & Holroyd, T. (1991a): The self-organized phase attractive dynamics of coordination. In: A. Babloyantz (Ed.), Self-organization, Emerging Properties and Learning, Series B: vol 260, pp. 41 – 62. New York: Plenum.

Kelso, J.A.S., DeGuzman, G.C. & Holroyd, T. (1991b): Synergetic dynamics of biological coordination with special reference to phase attraction and intermittency. In: H. Haken & H.P. Köepchen (Eds.), Rhythms in Physiological Systems, pp. 195 – 213. Berlin: Springer.

Leftshetz, S. (1963): Differential Equations: Geometric Theory. New York: Interscience. (2nd ed.).

Pomeau, Y. & Manneville, P. (1980): Intermittent transitions to turbulence in dissipative dynamical systems. Communications in Mathematical Physics 74, 189.

Tuller, B., Case, P., Ding, M. & Kelso, J.A.S. (1993): The nonlinear dynamics of speech categorization. Journal of Experimental Psychology: Human Perception and Performance. (in press).

von Holst, E. (1939/1973): Relative coordination as a phenomenon and as a method of analysis of central nervous function. In: R. Martin (Ed.), The Collected Papers of Erich von Holst, pp. 33 – 135. Coral Gables, FL: University of Miami.

Self-Organizational Processes in Animal Cognition

G. Vetter

Institute of Psychology and Cognition Research, University of Bremen,
D-28334 Bremen, Germany

Abstract: There is increasing evidence that the brain functions as a self-organizing system. Nonlinear phase transitions most succinctly reveal the autonomous order formation in such systems. Nevertheless they are only occasionally reported in psychological research, and if so, mostly referring to human and not to animal cognition. The aim of the following contribution is to demonstrate, that synergetic effects can be shown there, too. For this matter, results of stimulus generalization experiments are reanalyzed, since there we have a control parameter which releases the non-equilibrium phase transitions from one to another stable state, when continuously enhanced. The article presents examples of phase transitions as well as of critical fluctuations, critical slowing down, and hysteresis. Together this allows the categorization of the phenomenon as consequence of a self-organization process.

1 Introduction

A constructivistic theory of cognition postulates a cognitive system which is an open system with respect to energy and a closed system regarding information; i.e. physical stimulation does not transmit meaning, but only sets the unspecific initial and limiting conditions for the system, which influence the energetic states of its chaotic–deterministic processes (Schmidt, 1987). Order formation in such an autonomous system develops out of the free inner dynamics of its own elementary units. Studying the innersystemic principles of such order formation is subject of an interdisciplinary theory of self-organization of dynamic systems (Haken, 1977; Prigogine, 1985; Maturana, 1985; Köhler, 1958).

In the field of psychology it was Gestalt theory which from its very beginning studied such autonomous order formation in cognitive systems. As a result, Gestalt theory formulated many fundamental principles of such a self-organization. The best known and the most important of these is the concept of goodness or Prägnanz. This concept has a twofold meaning. (1) a tendency to achieve the utmost order, consistency and regularity which can be found in all cognitive areas, like perception, learning, thinking etc. (Rausch, 1966); and (2) the tendency of such structures towards a maximum of stability and persistence to change (Kanizsa & Luccio, 1990). The concept of *Prägnanz* corresponds to the general self-organizational principle of the energetic minimum of a field-like distribution of forces and is a principle optimizing all of the single Gestalt factors.

Self-organization theory especially studies such processes in all areas of nature where a continuous flow of the energetic state of a system (so-called control parameter) first leads to increasing fluctuations and thereafter to non-equilibrium transitions into new ordered states (order parameter). The aforementioned concept of *Prägnanz* can be viewed as such an order parameter. The spontaneous appearance and disappearance and the autonomous restructuring of macroscopic order out of the elementary dynamics of the system is called a phase transition. To change from one stable state to another or to approach a stable state, the system has to pass through or to be in a phase of instability. In self-organization theory intra-systemic instability is of major importance for the process of autonomous order formation. During instability various subtle and far-reaching interactions between the elements of the system occur which enable the spontaneous order formation. Haken (1977) describes the dynamic behaviour of a self-organizing system by the overdamped movement of a particle in a potential field. This mathematical description may be visualized by a ball moving in a landscape.

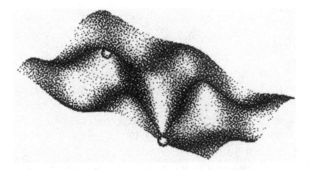

Fig. 1 Potential landscape with a ball (see text)

If the system is in a stable state, the ball is in one valley and it's own movements are not strong enough to leave it. A transition from one stable state to another (*phase transition*) is only possible when the shape of the landscape changes or when the movement of the ball is increased up to a critical degree (*critical fluctuations*). When the ball moves from one valley to another, it has to pass a point of complete instability at the top of the hill. This situation is called *symmetry breaking* because it has to be decided in which direction the ball will move, whence the symmetry of possible stable states will be broken. In such situations of symmetry breaking the behaviour of the system is open to very subtle influences. Even a force of otherwise negligible influence is able to determine which stable state the system will reach. In the phase of maximal instability little causes have great effect.

Research on the self-organization of biological and cognitive processes understandably focuses on such areas of instability, because non-linear phase transitions most succinctly reveal the dynamics of the system which underlies autonomous order formation. In the result of such analysis, several synergetic effects in cognition and behaviour have so far been reported for human cognitive systems (Haken & Stadler, 1990; Stadler & Kruse, 1991, 1992; Kelso, 1990; Bischof, 1990). The question of whether such effects are also found in animal cognition, however, is widely unanswered as yet, especially on the phenomenological level. The following article is the result of searching the relevant experimental literature on animal psychology as to whether there, too, is evidence for such self-organization processes.

2 The stimulus generalization paradigm

Searching for autonomous processes in cognitive systems demands – as already mentioned – that a new role is assigned to the stimulus situation. From a constructivistic point of view external stimuli function as unspecific initial conditions for autonomous order formation, and cognitive systems react extremely sensitively in the range of phase transitions and so-called bifurcations. Here, variation of the stimulus situation produces non-linear effects. Instances where increasing or decreasing variation of a stimulus parameter results in non-linear behavioural covariation, in sudden changes on the phenomenological side are therefore of great importance for a theory of self-organization; because this corresponds to the paradigmatic situation for synergetics, that continuous variation of a control parameter results in nonequilibrium phase transitions of an order parameter.

And it was Gestalt theory again which first led attention to this kind of system behaviour. As early as 1923, Wertheimer pointed out that in perception gradual variation of an objective stimulus does by no means always lead to comparable phenomenological covariation, but often exhibits sudden jumps and bends. And it was found that some elements in the sequence of the phenomonological counterparts of this gradual variation are experienced as better than others: The so-called stages of goodness (Prägnanzstufen). Increasing variation of the wave length of light from 400 nm to 700 nm for example leads through four such stages of goodness: The main colours blue, green, yellow and red (Metzger, 1954). And the transitions between stages of goodness are often experienced as rather abrupt, or – as Wertheimer put it – the course of phenomenological change exhibits sharp bends, *Knicke* (Wertheimer, 1923).

Now, gradual variation of a given stimulus value and registering the co-varying behavioural change is a paradigm, which – under the name of stimulus generalization – is well known in the psychology of learning and quite frequently studied especially with animals (Mostofsky, 1965; Honig & Urcuioli, 1981).

In a typical variant of these paradigm, animals are trained to show some kind of behaviour in the presence of one stimulus, S^+, and not to show it in the presence of another stimulus, S^-. In the following phase of generalization testing, S^+ is varied gradually and the concomitant change in the amount of conditioned behaviour registered. This constitutes the so-called generalization gradient. If processes of self-organization occur in animal cognition, than it might be expected, that in at least some of these studies one should find cases of non-linear behavioural covariation; i.e. gradients of stimulus generalization with sudden jumps and bends, indicating abrupt and sudden behavioural changes. This indeed is the case and, as a first instance, results of our own experiments shall be reported, which were conducted in another context some time ago.

3 Behavioural evidence for restructuring during gradual variation of a stimulus

Following Vetter & Hearst (1968), a stimulus generalization experiment with pigeons was run, where rhombuses in different rotational positions were used as the stimulus dimension (see Figure 2).

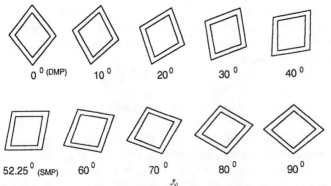

Fig. 2 The rhombus in its ten different rotational positions. DMP = Diagonal Main Position; SMP = Sides Main Position. (See text)

As can be seen, two figures of this sequence stand out: 0° and 52.25°. Although the rhombus is only rotated, i.e. all the figures are objectively the same, these two variants phenomenologically appear as different figures, have a different structure. The rhombus in the 0°-position appears more pointed and slim, with an accentuation of the corners. In the 52.25°-position it seems more rectangular. Not the corners, but the parallel sides are prominent in this case. The reason for this is, that in the 0°-position the figural extensions are parallel to the diagonals. Here the rhombus is, as Rausch (1952) – who studied this very thoroughly for human subjects – puts it, diagonal parallel (d.p.) structured. In the other case, the rhombus is side S^- parallel (s.p.) structured, the figural extensions are parallel to the figures sides. Now, exactly these two rotational positions are positions of goodness, *Prägnanzlagen*. Rausch speaks of *Diagonal Main Position* (DMP) and *Sides Main Positions* (SMP). They are distinct, because either one (in SMP) or both figural extensions (in DMP) coincide with the actual horizontal or vertical.

Around the respective positions of goodness there finally is a range of rotational positions, the *Prägnanzbereich*, where the respective structure, d.p. or s.p., remains unchanged. Rotating the rhombus over the limits of one range into the other results, then, in a *restructuring* of the figure.

If this is also true with animals, than at this point we could expect a phase-transition-like, sudden and abrupt change in the gradient of responses. To test that hypothesis, pigeons were trained to discriminate between two values of the rhombus rotational dimension. These rhombuses could be projected on the rear of a transparent response key located on one wall inside an usual Skinner box. According to the respective experimental conditions, the animals were reinforced with food for pecking this key in the presence of one kind of rhombus (S^+); whereas this behaviour was extinguished in the presence of another kind of the figure (S^-). The day after mastering this discrimination, the generalization test was run. Now all the different rotational positions of the rhombus successively were presented, and this in random order - 9 times. The results are shown in Figures 3 and 4. It can be seen there, that with increasing rotation of a rhombus the S^+-behaviour indeed covaries discontinuously.

If the pigeons had learned to peck the key in the presence of the rhombus in 0°-orientation (DMP), S^+, and not to peck in the presence of either the 40°-, 52.25° (SMP)-, 60°-, 70°- or 80°-orientation, S^-, than during the generalization test S^+-behaviour was also emitted in the presence of the 10°-, 20°-, 30°- and 90°-orientations; whereas S^-- behaviour occurred in the presence of the 40° - through 80°-orientations. If, on the other hand, the 0°-orientation of the rhombus was not S^+ but S^- and the 52.25°-orientations not S^- but S^+, than the registered generalization test behaviour changed in a mirror image like fashion (see Figure 4). The pigeons showed the go-behaviour while being presented the 40°- to 80°-orientations of the rhombus, and suppressed pecking not only in the presence of S^- (0°), but also with the 10°-, 20°-, 40°- and 90°-orientations. As Figs. 3 and 4 show, there is a small range of test values, between 30° and 40°, where a sharply declining response rate gradient is found.

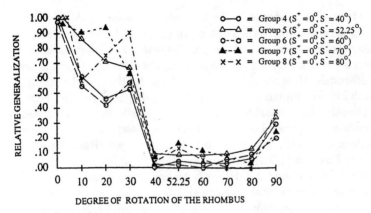

Fig. 3 The average generalization gradients of experimental groups 4 - 8. The values in paranthesis in the upper right stand for the different orientations of S^+ and S^- in the respective experimental groups during training

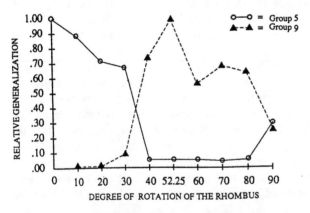

Fig. 4 The average generalization gradients of experimental groups 5 (S^+ = rhombus in 0°-, S^- = rhombus in 52.25°-rotational position) and 9 (S^+ = rhombus in 52.25°-, S^- = rhombus in 0°-rotational position)

Although these findings are generally in agreement with what one would have expected from the point of view of self-organization theory, rival explanations cannot be excluded only on this basis. Thus, one could argue that the reported result could very well fit into a linear function. It therefore seems necessary to look for supporting evidence.

4 System fluctuations, symmetry breaking, and generalization behaviour

If it is true that the reported sharp and abrupt decline of the response rate in the course of successive variation of the stimulus parameter is due to a phase transition from one stable state to another in the sense of self-organization theory, than indeed one would expect still other synergetic effects to occur in stimulus generalization data. The prerequisite and motor of phase transitions are system fluctuations. In the instable phase the system exhibits fluctuating behaviour, which increases up to critical fluctuation, leading to symmetry breaking and thereafter to a new stable state. When instability is at its maximum, behaviour becomes unpredictable. The more so, the sharper the alternative conditions are separated from each other. From this it follows: If in a stimulus generalization experiment of the kind just described the animal had been trained to discriminate S^+ from S^-; i.e. exhibits behaviour A in presence of S^+ and behaviour B in the presence of S^-, then it could be hypothesized, that – given the adequate stimulus continuum – successive variation of S^+ to S not only leads to an abrupt decline in generalization behaviour. Rather one should expect, that inside the range of this bend there are test stimulus values, in the presence of which the animals alternately show behaviour A the one time and behaviour B the other - and this to one and the same test value. In this range of stimuli the system is assumed to be in a phase of instable equilibrium between the alternatives A and B. The animal can solve this conflict only by means of kind of an act of decision which breaks the symmetry and chooses one alternative. Since here a small hint or impulse of minimal energy can suffice to solve the conflict and to direct the behaviour in one of either ways, it should be intra- and interindividually unpredictable, in which direction the decision goes and to what test stimulus this bifurcation point corresponds. Such kind of symmetry breaking in behaviour was repeatedly be referred to particularly trying to analyze the Lewinian conflict situations on the basis of such synergetic consideration (Stadler & Kruse, 1986; Haken, 1984).

Since the possible significance of such data was not expected at the time the above-reported experiment was conducted, they were not recorded. Evidence which could prove this assumption must therefore be looked for elsewhere. This causes a little problem, though. Usually in stimulus generalization experiments animals are taught a *go/no-go behaviour*. For instance, pigeons learn pecking a response key in the presence of one stimulus, S^+, and not to peck in the presence of another stimulus, S^-. Under these conditions, the described fluctuating behaviour is difficult to demonstrate, because alternating go and no-go behaviour, if it occurs in the presence of one test stimulus, is averaged out over the total presentation time. All that is reported as a lower response rate. Whether such averaging is responsible for the diminished response rate to distinct test stimuli was studied in an experiment by Migler & Millenson (1969). The results of this experiment are very interesting, because again they not only document the abrupt and sudden decline in generalization behaviour, but the mentioned fluctuation in the range of this abrupt transition from behaviour A to behaviour B as well.

Two albino rats were used as subjects and they were run in one of the usual operant conditioning chambers for rodents. A Gerbrands pigeon key was mounted on the rear wall 1.5 in. from the floor (at nose height) and could be illuminated from behind. Two response levers were mounted on the front wall, 6 in. apart centre-to-centre. A pellet hopper was located between these levers near the floor. Each animal was trained to emit the following response sequence: Pushing the illuminated nose key on the rear wall turned off illumination and turned on a clicking noise, either of low frequency (2.6 clicks/sec.) or of high frequency (25 clicks/sec.). This defined the beginning of a trial. During the low frequency clicks (S_L) only pressing on the left lever (R_L) was reinforced, whereas during high frequency clicking (S_R) only pressing on the right lever (R_R) was reinforced. Using appropriate reinforcing schedules further guaranteed that the animals finally had learned to press the left lever (R_L) with a high response frequency in the presence of S_L and to press the right level (R_R) with a low frequency in the presence of S_R. Following stabilization of this discrimination, the generalization test was run. It consisted of inserting probe trials during the last 32 experimental sessions. Now, pushing the nose key produced a click frequency, responding however remained unreinforced for one minute. Each of such probe trials were followed by 10 regular training trials, etc. During these stimulus generalization test probes, pushing the illuminated nose key produced a click rate having one of the following eight values: 0.4; 1.6; 2.5 (= S_L); 9.5; 15.6; 20; 25 (= S_R); 55 clicks/sec. This constituted the stimulus continuum. There were eight sessions required to complete a test across the entire continuum. Four such complete replications were carried out.

Figure 5 shows the average total lever-response rate during probe trials for each of the four replications of generalization testing. The dependent variable represents the average rate of lever pressing, disregarding lever position, during the various test click frequencies.

Looking from left to right, click frequencies up to 10 clicks/sec. are associated aproximately with the same relative high rate that is found at the original 2.5 click/sec. training point. Then a sharply declining gradient is found. Thus, we again see a discontinuous behavioural covariation associated with continuous variation of the stimulus parameter. Comparably nonlinear is the picture, when the percentage of time is registered, which the animals spend "on" the right lever R_R, during the various probe stimuli. These data are presented in Figure 6. The figure reveals, that up to 10 clicks/sec. practically no time was spent "on" the right lever. Then there follows a rather abrupt and steep transition: Now almost 100 % of the time was spent on this lever.

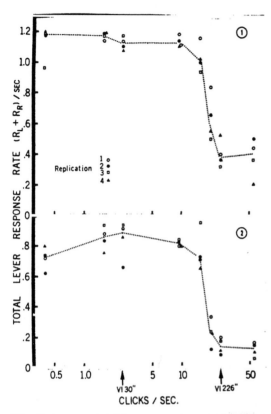

Fig. 5 Average total lever response rate in probe trials, for each rat, for each of the four replications. Each point represents the data for all the probe trials in one session and is calculated from cumulative $R_L + R_R$ frequencies over cumulative probe times for that session. Dotted lines connect medians (from Migler & Millenson, 1969)

Of especial importance, however, is the fact, that in the range of this jump-like transition, viz. between 16.5 and 20 clicks/sec., this clear association breaks down. In the presence of one and the same stimulus value, the animals instead chose the left and the right lever as well. And this occurred – look at the distribution of points in this area – with a remarkable inter- and intraindividual variation.

Such fluctuating behaviour can be seen even more clearly in Figure 7, where running response rates on each lever during the probe stimuli are presented. The figure shows, especially in the upper half, that in this range of transition (between 16.5 and 20 clicks/sec.) the otherwise clear association between stimulus value and learned response breaks down. Here the two curves overlap: The animals alternately show low frequency and high frequency responding in the presence of identical probe values. Migler & Millenson note in this context, that in these cases cumulative response records (not presented)

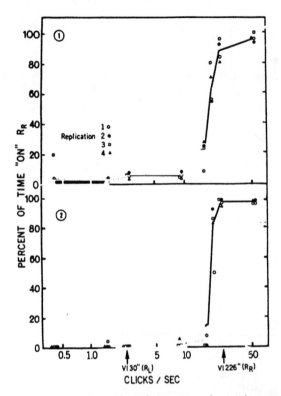

Fig. 6 Percentage of total time in probe trials spent on the right lever, for each rat, for each of the four replications. Each point represents the data for all the probe trials in one session, and is calculated from total time on R_R divided by the total time on both levers. Lines connect medians (from Migler & Millenson, 1969)

show that the animal select one lever on one probe trial and the other on a later probe trial. Switching between levers occured rarely during a probe test trial.

These results seem to be in full accordance with the outlined assumptions of self-organization theory. Not only is there again is a jump-like, discontinuous behaviour covariation associated with continuous variation of a stimulus parameter, but we also find the postulated fluctuation in the range of transition from behaviour R_L to behaviour R_R. Thus, the interpretation seems to be confirmed, that this is a case of phase transition from one stable state to another, whereby the system passes through a phase of instability including situations of symmetry breaking which result in sudden and unpredictable behaviour changes.

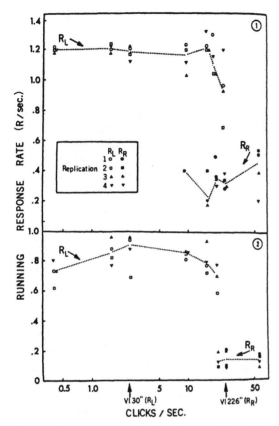

Fig. 7 Response rate "on" each lever for all four replications, for each rat. Each point represents the average response rate for all the trials in one session. Dotted lines connect mediance (from Migler & Millenson, 1969)

5 Critical slowing down

In discussing their results, Migler & Millenson pointed to some evidence indicating still other behavioural effects in the range of the critical probe values, viz. an increase in response latencies. This indeed was found in another stimulus generalization experiment (Cross & Lane, 1962). From the point of view of self-organization theory exactly this would be of some interest, since it is assumed, that the time needed for a system being in an instable phase to find a new stable state, increases, the closer the bifurcation is approached (critical slowing down).

For this reason, these findings shall be presented although Cross and Lane used human subjects, not animals. The authors trained their subjects to produce two levels of vocal pitch by humming (viz. 147 cps vs. 242 cps). After this was established, the subjects had to learn to assign these different vocal pitches to one of two levels of sound pressure (viz. 56 vs. 74 db) of a 5000 cps narrow band noise. Correct responses were reinforced by flashing a green light, which signaled the accumulation of points on a counter. The more points a person got, the higher was the final payment.

In the following generalization test the subjects were given 110 stimuli in a random order at 11 intensity levels arranged in 3 db steps from 50 to 80 db sound pressure level. This constituted the continuum of test stimuli (containing the discrimination stimuli, 56 db and 74 db). Figure 8 presents the results separately for 3 of the Ss. Similar to the aforementioned results of Migler & Millenson it can be seen that again the response rate in no way covaries gradually with gradual increase of the stimulus values from 50 db to 80 db. Rather, response probabilities of low pitch (R_1) were maximum over an extended range of test values around the original S^+ (56 db) and then sharply decline over a short range of transition between 65 db and 74 db. And also similar to Migler & Millenson's findings with rats is the fact that in this narrow range of transition the subjects produced not some kind of intermediate pitch but exhibited alternatingly R_1 and R_2 responses in the presence of one and the same test stimulus. This is clearly shown in the upper part of the figure, where these responses overlap.

Of especial interest, however, are the response latencies. Figure 8 shows, that they indeed increase inside the range of transition. They are negatively related to the relative response probability. If response probability is close to one or zero, than latencies were shortest. They were highest in the most instable range, which is presumably the middle between 65 db and 74 db. Since this seems to be a further quantitative measure of system behaviour in phase transitions besides behaviour fluctuations, it indeed could be worth while to replicate a similar study with animals. Anyway, the system dynamics which we are talking about is excellently shown by such a systematic increase in response latencies in this area, since critical slowing down and critical fluctuation are signs, that the system becomes destabilized and seeks new ordered states.

6 Hysteresis

Finally I will present data of a generalization experiment by G. Reynolds (1961), which in my opinion can be interpreted as showing still another synergetic effect in connection with phase transitions, namely, *hysteresis*. Hysteresis is frequently observed in phase transitions and may be interpreted as the temporary resistance of one stable state against the dynamics of formation of another stable state. Often mentioned as an example for hysteresis is horse gait. With increasing speed, horses fall into different movement programs, from walking to trotting to galloping. With decreasing speed the phase transitions are found at lower levels than with increasing speed. In perception

Self-Organizational Processes in Animal Cognition 197

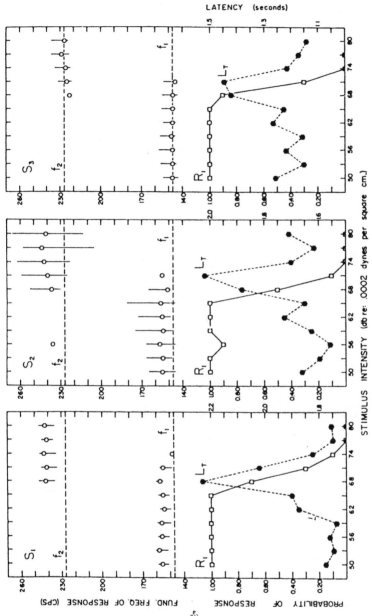

Fig. 8 Relations between response probability, latency, and topography in stimulus generalization. *Top:* Median Frequency in cps (circles) and the range (vertical lines) of high- and low-pitch responses as a function of stimulus intensity for each of three Ss. The dashed horizontal lines represent the vocal pitches previously differentiated. *Bottom:* Response probability (squares) equal the ratio of the number of low-pitch responses emitted to the number of stimulus presentations (10) at each intensity. The hexagons represent the average latency of high- and low-pitch responses to each stimulus intensity

hysteresis was first discovered by Max Wertheimer during the transition of one Prägnanz-stage to another. He called it *gestalt factor of objective set*.

Reynolds trained pigeons to peck a transparent response key located on one wall inside an usual skinner box. A black isosceles triangle was mounted on a white background behind this transparent key. Driven by a little motor, this triangle and the background slowly but *continuously* rotated clockwise (about twice the speed of the minute hand of a clock), the rotation being around an axis through the geometrical centre of the triangle. So the triangle successively passed through ten decants, each of 36°. This defined the ten different rotational positions, which constituted the stimulus continuum (see Fig. 9).

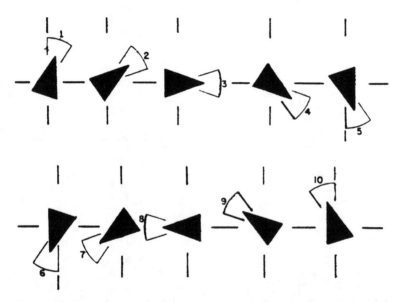

Fig. 9 The triangle in each of the ten successive decants of rotation and the reference numbers assigned to each decant (from Reynolds, 1961)

There were two experimental phases of interest in our context. In the first phase, pecking the key was reinforced (VI 90 sec.) only, when the triangle's apex was in the upper two decants nos. 1 and 10. Otherwise, responding was extinguished. In the second phase, conditions were reversed. Now responding was extinguished when the apex of the triangle was in decants nos. 1 and 10, in all other cases, reinforcement of responding occurred.

The results are depicted in Figure 10. There it can be seen, that again the continuous variation of the stimulus parameter leads to a discontinuous covariation of behaviour. Reinforcing pecking behaviour to decants 1 and 10 generally leads to a decreasing response rate, the further the triangle is rotated out of these two positions. Contrary to this, however, behaviour in

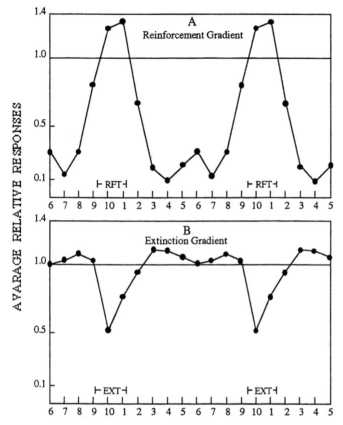

Fig. 10 The average relative generalization in each decant following reinforcement (A) or extinction (B) in only decants 10 and 1

decants 5 and 6 increases (see Fig. 10A). Correspondingly the reverse effect is observed after extinction to decants nos. 1 and 10. Now, the rate of responding increases with increasing rotation except that the rate is lower in decants 5 and 6 than in decants 4 and 7 (see Fig. 10B).

With regard to hysteresis, however, the data shown in Figure 11 are of especial interest. Figure 11 shows the cumulative recording of one pigeon's responding in each decant when responding was being reinforced in only decants 1 and 10. Keeping in mind that with this kind of recording increasing steepness of the curve means increasing response rate (i.e. a horizontal line evidences no responding), Fig. 11 shows the following: Some responding in decant 6, practically no responding in decants 7 and 8. Further rotation of the triangle through decant 9 leads to fluctuating stop-and-go behaviour. This kind of behaviour continuous in the beginning of decant 10 (which is reinforced). Starting from about the middle of decant 10, all through decant 1 and overshooting into most of the extinguished decant 2, there is a high

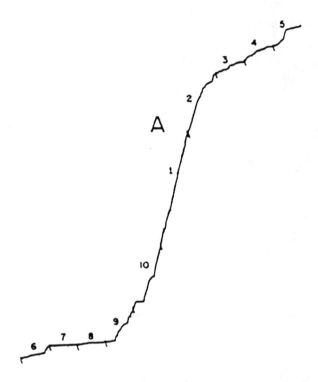

Fig. 11 Cumulative records showing the responding of one pigeon in each decant (numbers on the record), when responding was reinforced in only decant 10 and 1

frequent and steady go-behaviour, which at the end of decant 2 slowly starts to decrease again.

This slow and hesitating appearance of the conditioned behaviour with the triangle entering the reinforced decant no. 10 as well as the overshooting continuation of responding despite the triangles rotation out of decant 1 into the extinguished decant no. 2 seems to me to be very well be interpretable as an instance of hysteresis.

7 Conclusion and outlook

Testing the range of validity of a self-organization theory in animal cognition starts with looking for behavioural and phenomenological data, which already could be taken as evidence for the existence of phase transitions and related synergetic effects. As pointed out in the preceding sections, there seems to be such evidence. However, some questions are still open and some conclusions remain to be substantiated. In the experiment by Reynolds, for instance, one could have presented the animals with a triangle rotating in the opposite direction. Under the assumption of hysteresis effects in this case the pattern of responding should be the other way round. One further could have used different rotation velocities, maybe another kind of behaviour than a go/no-go behaviour. And in the Migler & Millenson paradigm one obviously should try to increase (or, in another condition, decrease) the stimulus parameter (click frequency) *continuously* in order to analyze the phenomena of fluctuation, critical slowing down and hysteresis more exactly. However, since all experiments reported on were conceptualized and run under quite different research questions, such a systematic test of the hypotheses which are of interest in this context understandably was not conducted and still has to be done.

Interesting as such experimentation would be, we prefer a different approach, however. If it is true that self-organization processes also occur in animal cognition, then there should also be multistability. In multistability the autonomy of order formation and non-equilibrium phase transition are revealed and consequently analyzable in their purest form. Furthermore, instable states of the system in phase transitions are run through and therefore are registrable many times. For this reason, our next aim is to try to show multistability with animals. Since we know that pigeons are able to see not only real movement, but also apparent movement, like the phi-phenomenon (Siegel, 1970, 1971; Emmerton, 1990), we plan to use Stroboscopic Alternative Movement (SAM) as the reference pattern.

References

Bischof, N. (1990): Phase transitions in psychoemotional development. In: H. Haken & M. Stadler (Eds.), Synergetics of Cognition, pp. 361 – 378. Berlin: Springer.
Cross, D.V. & Lane, H.L. (1962): On the discriminative control of concurrent responses: the relation among frequency, latency, and topography in auditory generalization. Journal of the Experimental Analysis of Behavior 5, 487 – 496.
Emmerton, J. (1990): Pigeons perception of complex movement patterns. In: M.L. Commons, R.J. Herrnstein, S.M. Kosslyn & D.B. Mumford (Eds.), Quantitative Analysis of Behavior. Behavioral Approaches to Pattern Recognition and Concept Formation, vol. 8, pp. 67 – 87. Hillsdale, NJ.: Lawrence Erlbaum.
Haken, H. (1977): Synergetics - An Introduction. Berlin: Springer.
Haken, H. (1984): Erfolgsgeheimnisse der Natur, Frankfurt/Main: Ullstein.
Haken H. & Stadler, M. (Eds.) (1990): Synergetics of Cognition. Berlin: Springer.
Honig, W.K. & Urcuioli, P.J. (1981): The legacy of Guttman and Kalish (1956): 25 years of research on stimulus generalization. Journal of the Experimental Analysis of Behavior 36, 405 – 445.
Kanizsa, G. & Luccio, R. (1990): The phenomenology of autonomous order formation in perception. In: H. Haken & M. Stadler (Eds.), Synergetics of Cognition, pp. 186 – 200. Berlin: Springer.
Kelso, J.A.S. (1990): Phase transitions: Foundations of behavior, In: H. Haken & M. Stadler (Eds.), Synergetics of Cognition, pp. 249 – 295. Berlin: Springer.
Köhler, W. (1958): Dynamische Zusammenhänge in der Psychologie. Bern: Huber.

Maturana, H.R. (1985): Erkennen: Die Organisation und die Verkörperung von Wirklichkeit. Braunschweig: Vieweg.
Metzger, W. (1954): Psychologie. Darmstadt: Steinkopff (2nd ed.).
Migler, B. & Millenson, J.R. (1969): Analysis of response rates during stimulus generalization. Journal of the Experimental Analysis of Behavior 12, 81 – 90.
Mostofsky, D.I. (Ed.) (1965): Stimulus Generalization. Stanford, CA: Stanford University Press.
Prigogine, I. (1985): Vom Sein zum Werden. München: Piper.
Rausch, E. (1952): Struktur und Metrik figural-optischer Wahrnehmung. Frankfurt/Main: Kramer.
Rausch, E. (1966): Das Eigenschaftsproblem in der Gestalttheorie der Wahrnehmung. In: W. Metzger & H. Erke (Eds.), Wahrnehmung und Bewußtsein. Handbuch der Psychologie, Bd. I/I, pp. 866 – 933. Göttingen: Hogrefe.
Reynolds, G.S. (1961): Contrast, generalization, and the process of discrimination. Journal of the Experimental Analysis of Behavior 4, 289 – 294.
Schmidt, S.J. (Ed.) (1987): Der Diskurs des Radikalen Konstruktivismus. Frankfurt/Main: Suhrkamp.
Siegel, R.K. (1970): Apparent movement detection in the pigeon. Journal of the Experimental Analysis of Behavior 14, 93 – 97.
Siegel, R.K. (1971): Apparent movement and real movement detection in the pigeon: Stimulus generalization. Journal of the Experimental Analysis of Behavior 16, 189 – 192.
Stadler, M. & Kruse, P. (1986): Gestalttheorie und Theorie der Selbstorganisation. Gestalt Theory 8, 75 – 98.
Stadler, M. & Kruse, P. (1991): Visuelles Gedächtnis für Formen und das Problem der Bedeutungszuweisung in kognitiven Systemen. In: S.J. Schmidt (Ed.), Gedächtnis, pp. 250 – 266. Frankfurt/Main: Suhrkamp.
Stadler, M. & Kruse, P. (1992): Konstruktivismus und Selbstorganisation: Methodologische Überlegungen zur Heuristik psychologischer Experimente. In: S.J. Schmidt (Ed.), Kognition und Gesellschaft, pp. 141 – 166. Frankfurt/Main: Suhrkamp.
Vetter, G.H. & Hearst, E. (1968): Generalization and discrimination of shape orientation in the pigeon. Journal of the Experimental Analysis of Behavior 11, 753 – 765.
Wertheimer, M. (1923): Untersuchungen zur Lehre von der Gestalt. Psychologische Forschung 4, 301 – 350.

Dynamic Models of Psychological Systems

H. Schwegler

Institute of Theoretical Physics and Center for Cognitive Sciences,
University of Bremen, D-28334 Bremen, Germany

Abstract: Two examples of a dynamic modelling of psychological systems in terms of differential equations are presented. The first one concerns time patterns of alcoholism on a medium time scale of the order of the magnitude of a month. The second example is a slight modification of Bischof's model of regulation of social distance on a time scale in which short-term irregular motions are averaged away. The different types of behaviour are discussed qualitatively.

1 Introduction

Because dynamic modelling of processes of perception is presented widely in other parts of this volume this is supplemented in this paper by a mathematical treatment of some phenomena in the field of behaviour. For a modelling of time dependent processes in psychological systems they have to be considered mathematically as "dynamical systems". One is looking for laws or rules that govern the time development and time behaviour of the system which can be formulated in terms of difference equations, differential equations, integrodifferential equations or even more sophisticated mathematical structures.

Even though fully exact quantitative descriptions of development and behaviour might not be achieved, at least a qualitative understanding of the different time patterns can thereby be supported. Some typical questions to be answered are: Is there an equilibrium state, and is it stable, such that it does not change spontaneously? Are there even several stable equilibria in competition, and what are their basins of attraction? Are there bifurcations of equilibria along changes of control parameters, such that equilibria disappear or new equilibria become possible under changing external conditions? What is, outside the equilibrium states, the dynamical development during time flow, and is a certain time pattern of development stable? In particular, are there limit cycles, that is stable, periodic (oscillatory) behaviours? Can chaotic behaviour emerge?

In this paper this type of investigation is illustrated by two examples. The first example concerns mathematical models of patterns of drinking which have been elaborated by an der Heiden, Schwegler and Tretter (1993). The second example is a discussion of the Zürich model of social motivation developed by Norbert Bischof (1989, 1993). In both cases the dynamics is formulated in terms of differential equations which for the sake of numerical integration can be converted into difference equations, of course. In terms of Haken's distinction (1994) my approach is a phenomenological one.

2 Models of dynamic patterns of drinking on a medium time scale

2.1 Primitive approach with one differential equation

The models of an der Heiden, Schwegler and Tretter (1993) consider phenomena of alcoholism which can be observed on a time scale of the order of the magnitude of a month. Thereby short time phenomena on a time scale of days, as the succession of abuse and hangover, are assumed to be averaged away. The essential variable describing the phenomena of interest is the mean alcohol consumption A in grams per week. In the language of synergetics (Haken, 1994) A is the order parameter.

Beyond the short and medium time scale phenomena there are changes on a long time scale of years, namely the development of the different types (or styles) of alcoholism, for example the change from conflict drinking to addiction. These are included in the models in the form of parameter changes which are not modelled dynamically (that means resulting from solutions of differential equations) but have to be put in by extra knowledge.

A simple ansatz of a differential equation for the mean alcohol consumption A reads

$$\frac{dA}{dt} = f - rA + g(A) \tag{1}$$

We have formulated it inspired by an episode of The Little Prince of Saint-Exupéry (1946). Following Saint-Exupéry's description of the drinker only very loosely we have elaborated some mechanisms mathematically oriented on psychiatric experiences (cf. Tretter, 1990).

The constant and positive driving term f describes influences which increase the alcohol consumption. They can be of psychic nature (such as the load of conflicts to be mastered) as well as of a social nature (such as the frequency of social occasions of drinking). The product $-rA$, with a reduction coefficient r, describes a reduction of alcohol consumption which is proportional to the consumption A itself. This can be caused simply by shame (here we differ from Saint-Exupéry's ideas that shame increases the alcohol consumption, giving rise to a vicious circle which is passed with exponentially increasing speed and leads to a collapse).

Instead of shame we consider another reason for self-enhancement of drinking. This is a deficit of competence for problem-solving (particularly with respect to the problems caused by alcohol itself), which is described by a sigmoid-shaped function $g(A)$ with a saturation value s, which is drawn in Fig.1. Because $g(A)$ increases the consumption, the saturation s does not limit the consumption A itself but only its increase dA/dt.

I have already pointed out that the simple models of an der Heiden, Schwegler and Tretter (1993) do not consider the short time scale variations of drinking behaviour. In a medium time scale theory they could be taken into account by adding a "noise term" of stochastic fluctuations to eq. (1).

The mathematical discussion of eq. (1) shows the following results. Depending on the ratio s/r of the saturation value s of competence deficit $g(A)$ and the reduction coefficient r we find two different types of solutions.

For sufficiently low ratios s/r there exists only one equilibrium state A_{eq} of stationary consumption (on an average) at a relatively low level. It is stable; that means the following: If, due to a short-term excess or abstinence, another mean consumption is realized the equilibrium value A_{eq} is re-established by the social and internal mechanisms incorporated in the differential equation. This is shown graphically in Fig. 2 a: the system relaxes to the equilibrium.

But if we increase the ratio s/r we reach a so-called bifurcation value beyond which the system behaves in a different way. Now there exist three equilibrium states A_1, A_{un} and A_2. Two of them, A_1 und A_2, are stable again, and their basins of attraction are separated by the unstable A_{un}. The stable equilibrium consumption A_1 is still at a relatively low level. However the system can stabilize alternatively at another equilibrium with a much higher mean alcohol consumption A_2. The graphs of the possible solutions of the differential equation are shown in Fig. 2 b as time-dependent functions $A(t)$.

A little more detailed discussion results in the "equilibrium surface" of Fig. 3 (upper part) and its projection, the so-called "bifurcation diagram" (lower part). In the upper part the possible equilibria A_{eq} are visualized graphically as a surface above the parameter plane (a two-dimensional "parameter space") spanned by the parameter ratios s/r and f/r. There are two different regions of this two-dimensional parameter plane: For parameter values in one region there exists only one equilibrium state (the case of Fig. 2a), the equilibrium surface has one sheet. For parameter values in the other region there exist three equilibrium states (the case of Fig. 2b), the equilibrium surface is folded and has three sheets. The points on the middle sheet represent the unstable values A_{un}, the points on the upper and lower sheets represent the two stable values A_2 and A_1 of the mean alcohol consumption. The part of the folded surface left to the s/r-axis is unrealistic because f and r and therefore f/r are always positive; this part is drawn only to give a comprehensive view of the surface.

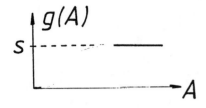

Fig. 1 Deficit of competence for problem solving (depending on the alcohol comsumption A)

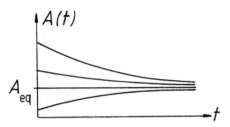

Fig. 2a Time dependent processes $A(t)$ in the parameter region with one equilibrium A_{eq} (stable)

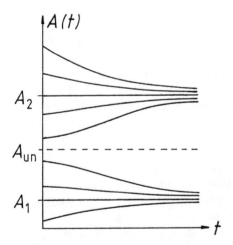

Fig. 2b Time dependent processes $A(t)$ in the parameter region with three equilibria. Only the equilibria A_1 and A_2 are stable

If we project the equilibrium surface onto the parameter plane the edges of the folded parts of the surface project to the v-like "cusp curve" in the lower part of the figure, which consists of the bifurcation points. For parameter values below the cusp (inside the v) there exist two stable equilibria, for all other parameter values (above the cusp, outside the v) only one. If we increase the value of the ratio s/r, keeping the other ratio f/r fixed, we recognize the situation already discussed in connection with Fig. 2: first there exists only

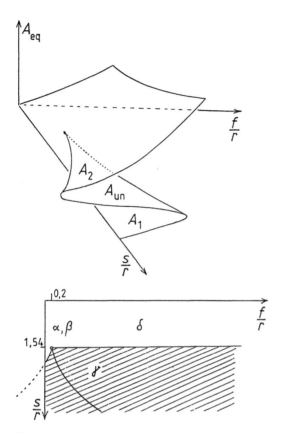

Fig. 3 Upper part: The equilibrium surface. Lower part: The bifurcation diagram with the cusp

one equilibrium state until we reach the bifurcation value, and then we have two stable equilibria.

The situation we have found is called the cusp catastrophe in catastrophe theory. This cusp has already been used by Moles (1986) in a discussion of alcoholism, but more phenomenologically without a dynamics in terms of a differential equation and with completely different interpretation of the parameters.

In the context of our interpretation, parts of the parameter plane can be identified with the different types of alcoholism in the classification of Jellinek (1960), see also Feuerlein (1989): In the left upper corner are the α- and β-drinkers which cannot be distinguished in the model. The α-, or conflict drinker is driven by the load of his conflicts whereas the β-, or occasional drinker is driven by the frequency of social occasions. Both driving forces are combined in the single driving coefficient f in our model. The alcohol consumption of α- and β-drinkers is between a low and a medium level.

If the occasions of drinking become abundant and are correlated with social urgency (as it is particularly the case in some professions like masons or brewery workers) and the alcohol consumption stabilizes at a much higher level, then we speak of the δ-, or habitual drinker. A similar behaviour can be caused by a high load of conflicts but if this is accompanied by an increased deficiency of competence for problem solving then the α- drinker glides into the γ-region. This could also happen after a long-lasting habitual drinking has lowered the competence of problem solving.

The γ-drinker is the addicted drinker in the narrow sense. There is only a very small interval of the parameter ratio s/r in which "hysteresis" occurs: here it is possible to fall from the upper to the lower sheet if, for a while, the social occasions and/or the load of conflicts, and therefore the coefficient f, could be diminished drastically. But below that small interval it is no longer possible to return from the upper to the lower sheet by that strategy. Then the only way of therapy must increase the competence of the patient for problem solving.

2.2 Improvement by a second differential equation

Whereas the primitive model gives a good qualitative description of four types of drinking in Jellinek's classification, it cannot explain the periodic drinking of the ϵ- (periodic) drinker. This, and also some slight variations of the other four types, emerge if we improve the model by the introduction of a second dynamic variable F, which we call frustration. We make the following ansatz of two coupled differential equations

$$\frac{dA}{dt} = h(F) + f - rA + g(A)$$
$$\frac{dF}{dt} = c - bA - aF$$
(2)

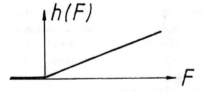

Fig. 4 The function $h(F)$

The piecewise linear function $h(F)$ shown in Fig. 4 describes the dependence of the alcohol consumption A on the frustration F. The time change of the frustration F is governed by the second of the eqs. (2). It shows a constant and positive driving term c, a reduction term $-aF$ depending on the frustration F itself, and another reduction $-bA$ which is brought about by alcohol consumption.

Figure 5 is a result of a numerical calculation, and it shows the trajectories in a "phase portrait". We have a modification of the results of section 2.1 in the region below the cusp where we previously had two stable equilibrium states. Now we have again two stable equilibria at alcohol consumption levels A_1 and A_2, but also at different values of the frustration. The higher consumption A_2 is combined with lower frustration which is decreased by the higher level of alcohol. Between the two stable equilibria we have an unstable one which is a saddle point.

Figure 6 shows another numerical result for larger values of the coefficients a and b. Here we find a limit cycle oscillation around the unstable equilibrium. This can be interpreted as the case of an ϵ-, or periodic drinker.

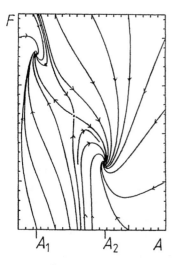
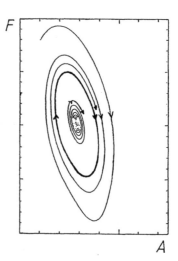

Fig. 5 Phase portrait with 3 equilibria (two stable ones and a saddle)

Fig. 6 Phase portrait with a limit cycle

3 The Zürich model of dynamical distance regulation

My second example is the Zürich model of regulation of social distance, which was developed by Norbert Bischof (1993). A person, or "subject", is considered who changes his distance from n surrounding persons or "objects" (see Fig. 7 a). A simple example is the spatial motion of a child in the presence of the mother, other relatives and perhaps some strange persons. More generally one can think about the motion of an animal in the presence of other animals of the same and of other species. The motion is governed by emotional mechanisms which are described in detail in Bischof's paper. Here I concentrate on the mathematical formulation of these mechanisms.

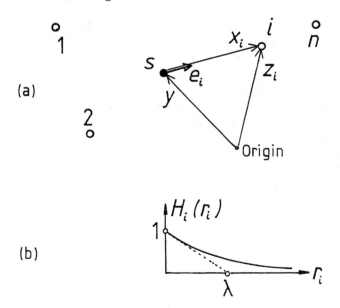

Fig. 7 a) A subject in a group of n objects. Visualization of the positions $\mathbf{y}$ and $\mathbf{z}_i$ and the relative positions $\mathbf{x}_i$. The $\mathbf{e}_i$ are the unit vectors in the direction from the subject to the objects. b) The proximity as a function of distance

It is assumed that the subject has two types of perceptions which are important for the motions we consider. First he perceives the positions $\mathbf{z}_i$ of the objects, and hereby their relative positions $\mathbf{x}_i = \mathbf{z}_i - \mathbf{y}$ relative to his own position $\mathbf{y}$ (cf. Fig. 7 a). The relative positions should be "known" to the subject in a decomposition $\mathbf{x}_i = r_i \mathbf{e}_i$. The factor r_i is the scalar distance and $\mathbf{e}_i$ is a unit vector in the direction from the subject to the object i.

From the scalar distance r_i the subject's brain "calculates" a psychological proximity $H_i(r_i)$. Slightly deviating from Bischof's assumption I assume this proximity to be an exponential function

$$H_i(r_i) = e^{\frac{-r_i}{\lambda}} \tag{3}$$

with a parameter λ (cf. Fig. 7 b). This deviation does not lead to essentially different results but it makes the discussion of the model much easier. The psychological proximity is between 0 and 1, $H_i \in [0,1]$, and because of movement of the subject H_i is time dependent.

Secondly the subject perceives the physiognomies of the surrounding objects, and his brain evaluates them by two characters, a relevance R_i and an intimacy F_i. It is assumed that both characters are scaled such that their values are between 0 and 1, that means $R_i \in [0,1]$ and $F_i \in [0,1]$. One can also define a strangeness $1 - F_i \in [0,1]$ complementary to the intimacy.

It is assumed by Bischof that the three types of characters, the constants R_i and F_i and the time dependent H_i, together with the directions $\mathbf{e}_i$, determine the motion of the subject completely.

3.1 Bischof's theory in terms of a differential equation

Bischof has formulated his theory in terms of a difference equation which corresponds to a cybernetic block diagram as shown in Fig. 8 (where I have performed some changes which are explained in the following). Instead of a difference equation I formulate a differential equation of motion because it can be discussed analytically in a much easier way. One can expect that the behaviour of the solutions does not differ significantly from that of the difference equations. For the sake of numerical integration the differential equation has to be converted into the difference equation again as usual.

Bischof's theory describes and explains the subject's motion on a medium time scale (analogous with the models of patterns of drinking) in which irregular excursions are averaged away. Thus, I have not formulated the dynamics analogously to Newton's mechanics with a second order equation (acceleration proportional to the force) as done by Bischof in the difference equation version. Instead, I prefer an ansatz with a first order differential equation (velocity proportional to the force), which formally corresponds to the mechanical motion under strong friction, but which in our case can be interpreted as the description of the motion on a medium time scale after an averaging over irregular spontaneous motions (called "ungedämpft" by Bischof).

Bischof made a proposal in his paper (1993) to model this "ungedämpft" motion by memory terms (referring to a concept of "emotional refuelling"). It seems to me doubtful whether the proposed memory terms give rise to the phenomenon which he wants to capture. In any case we can assume that these memory terms are not necessary for a description of the motion on a medium time scale. Therefore in Fig. 8 I have left out the memory effects, and in the lower left corner I have written the expression corresponding to a first order equation.

The first order differential equation for the dynamic variable $\mathbf{y}(t)$ reads

$$\frac{d\mathbf{y}}{dt} = \text{resultant "force"}$$
$$= \sum_i (A_S(\mathbf{y}) R_i F_i H_i(\mathbf{y}) \mathbf{e}_i + A_E(\mathbf{y}) R_i (1 - F_i) H_i(\mathbf{y}) \mathbf{e}_i) \quad (4)$$

Beside the characters R_i, F_i and H_i, and the unit direction vectors $\mathbf{e}_i$, it contains two global characters A_S and A_E without an object index i. They represent some type of more general sentiment or mood of the subject which is built by the brain by all of the object dependent characters R_i, F_i and H_i. Bischof defined them in the following way. First he defined a security S and an exitation E

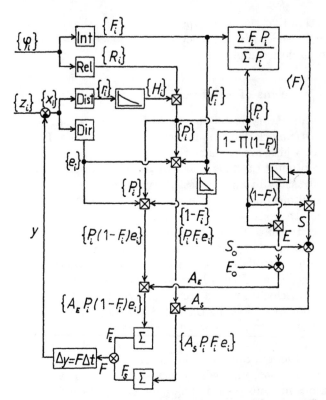

Fig. 8 Bischof's block diagram with my modifications. Symbols not explained in the text: φ_i physiognomies, $P_i = R_i H_i$, $<\ >$ averaged values, F_E, F_S, F forces

$$S(\mathbf{y}) = \{1 - \prod_i(1 - R_i H_i)\} \frac{\sum_i R_i H_i F_i}{\sum_i R_i H_i}$$
$$E(\mathbf{y}) = \{1 - \prod_i(1 - R_i H_i)\} \frac{\sum_i R_i H_i (1 - F_i)}{\sum_i R_i H_i} \qquad (5)$$

The somewhat complicated and nonlinear first factor in curly brackets was chosen in order to have S and E between 0 and 1 again, $S \in [0,1]$ and $E \in [0,1]$. The second factor in eq. (5) is an averaged intimacy, and an averaged strangeness, respectively. Then Bischof built two differences A_S and A_E with respect to a so-called dependence S_0 and an enterprise E_0

$$A_S(\mathbf{y}) = S_0 - S(\mathbf{y})$$
$$A_E(\mathbf{y}) = E_0 - E(\mathbf{y}) \qquad (6)$$

which are between -1 and 1, $A_S \in [-1,1]$ and $A_E \in [-1,1]$, and have the property

$$A_S(\mathbf{y}) = \begin{cases} > 0 & \text{for } S < S_0 \\ < 0 & \text{for } S > S_0 \end{cases}$$
$$A_E(\mathbf{y}) = \begin{cases} > 0 & \text{for } E < E_0 \\ < 0 & \text{for } E > E_0 \end{cases} \quad (7)$$

These global characters enter into the differential equation (4), positive values contribute to an approach or appetence, negative values to a removal or aversion.

The differential equation (4) can be studied in detail through numerical simulations. However, discussion of the equilibria and the directions of trajectories around them can be made analytically. We have two types of equilibria (given by putting the right hand side of eq. (4) to 0):
- intersection points of the contour lines $A_S = 0$ and $A_E = 0$.
- other equilibria brought about by a compensation of all terms.

Even to determine these we need some numerical calculations. A discussion of the positions of equilibria and of the basins of attraction would become much easier if the dynamics could be formulated in terms of a potential so that it would be a so-called gradient dynamics. Therefore one can ask if eq. (4) could be changed to a gradient dynamics, perhaps by only a slight modification.

3.2 Modification to a gradient system

This is possible, in fact, if we relinquish the soft nonlinearity of the first factor in eq. (5) and use the very slightly different linear version $\sum R_i H_i$ so that

$$\begin{aligned} S &= \sum_i R_i H_i F_i \quad , & A_S &= S_0 - S \\ E &= \sum_i R_i H_i (1 - F_i) \quad , & A_E &= E_0 - E \end{aligned} \quad (8)$$

Then we have a potential

$$U(\mathbf{y}) = \frac{\lambda}{2} \left(A_S^2(\mathbf{y}) + A_E^2(\mathbf{y}) \right) \quad (9)$$

which gives rise to the differential equation

$$\frac{d\mathbf{y}}{dt} = -\text{grad } U = A_S(\mathbf{y}) \sum_i R_i H_i F_i \mathbf{e}_i + A_E(\mathbf{y}) \sum_i R_i (1 - F_i) H_i \mathbf{e}_i \quad (10)$$

The function $\mathbf{y}(t)$ describes the motion of the subject which now proceeds along trajectories which are the projection of a symbolic point running down in the potential landscape (drawn above the position space, see Fig. 9) along the trajectory of steepest descent. The equilibria are the minima of the potential surface.

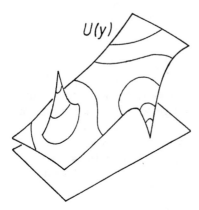

Fig. 9 The potential landscape (over the position space) with an attracting and a repelling object

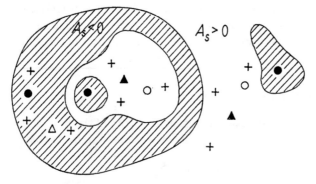

Fig. 10 Position space with attracting and repelling objects (open and filled circles), other attracting or repelling equilibria (open and filled triangles), and saddle points (crosses). The contour lines $A_S = 0$ consist of attracting equilibria, too

At the positions of the objects the potential surface has cone-like dents due to the proximity function $H_i(\mathbf{y})$. The potential $U(\mathbf{y})$ is not differentiable at these positions, which is somewhat similar to the electrostatic potential at the positions of charges. The objects can be maxima or minima if the cones are not too tilted. A maximum is an unstable equilibrium, it means that the subject runs away (aversion). A minimum is a stable equilibrium, where the subject can stay (appetence).

But there can exist equilibria at differentiable points outside objects as well which are defined by grad $U = 0$. These are stable if U is a minimum at the point, and unstable otherwise. In the general case all equilibria are isolated and there are no closed lines of equilibrium points as in the special case of the next section (cf. Figs. 10 – 12).

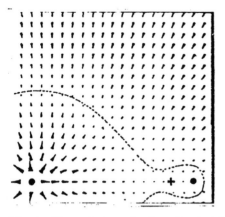

Fig. 11 Numerical results from Bischof (1989, 1993)

3.3 Further simplification

The discussion becomes even more simple, and the phase portrait more specific, if we have have a dynamics with either the A_S-term or the A_E-term only, for example:

$$\frac{dy}{dt} = A_S(y) \sum_i R_i H_i(y) F_i e_i \quad , \quad A_S = S_0 - \sum_i R_i H_i F_i \quad (11)$$

Then the contour lines $A_S = 0$ are closed lines of stable equilibrium points (Fig. 10).

At objects there are two possibilities: The potential surface can be so flat and tilted that it is not extreme, and therefore the objects do not act as (stable or unstable) equilibria. However, if the cones are not so tilted and are peaked to an extremum the object is either repelling (aversion) or attracting (appetence). In a connected part of the position space between two contour lines of equilibria there are one or more such objects, but always of only one and the same type (see Fig. 10, the attracting objects are shown as open circles, the repelling ones as filled circles).

Equilibria beside objects can be divided into the following three groups:
- The contour lines $A_S = 0$ consist of stable equilibrium points as already mentioned.
- In the regions $A_S > 0$ (where we have only attracting objects), all equilibria beside objects are either isolated maxima and therefore unstable (filled triangles in Fig.10), or unstable saddle points (crosses in Fig. 10).
- In the regions $A_S < 0$ (where we have only repelling objects) all equilibria beside objects are either isolated minima, and therefore attracting and stable (open triangles), or unstable saddle points.

Figure 12 shows some typical phase portraits. The lower three pictures visualize a bifurcation which takes place with changing S_0. The topmost of

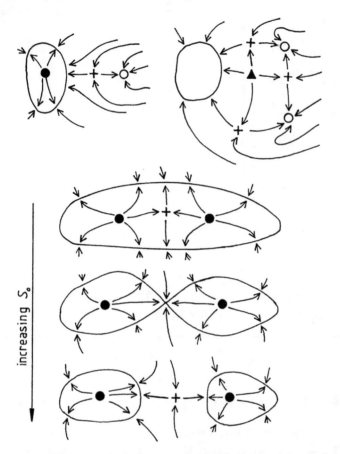

Fig. 12 Some typical phase portraits. The lower pictures visualize a bifurcation at the bifurcation value S_0^{bi}

these three is similar to a picture calculated by Bischof (Fig. 11) for a child in a surrounding with its mother and another person of much lower relevance. It shows a contour line of stable equilibria where the child can stay, the two objects are repelling and there is a point in between (called the "Buridan point" by Bischof) which must be an unstable saddle point.

References

An der Heiden, U., Schwegler, H. & Tretter F. (1993): Patterns of alcoholism in a mathematical model (in Press).
Bischof, N. (1989): Das Rätsel Ödipus. München: Piper (3rd ed.).
Bischof, N. (1993): Die Regulation der sozialen Distanz: Von der Feldtheorie zur Systemtheorie. Zeitschrift für Psychologie 1, 1993 (in Press).

Feuerlein, W. (1989): Alkoholismus - Mißbrauch und Abhängigkeit. Stuttgart: Thieme (4th ed.).
Haken, H. (1994): Some basic concepts of synergetics with respect to multistability in perception, phase transitions and formation of meaning. In this volume.
Jellinek, E.M. (1960): The Disease Concept of Alcoholism. New Haven: Yale University Press.
Moles, A. (1986): Vers un modèle systémique du passage du l'us à l'abus dans la consommation de l'alcool considéré comme un psychotrope. Humankybernetik 27, 33 – 43.
Saint-Exupéry, A. de (1946): Le Petit Prince. Paris: Gallimard.
Tretter, F. (1990): Sucht am Beispiel der Alkoholabhängigkeit. In: E. Pöppel & M. Bullinger (Eds): Medizinische Psychologie. Weinheim: Edition Medizin.

Part IV

Semantic Ambiguity

Ambiguity in Linguistic Meaning in Relation to Perceptual Multistability

W. Wildgen

Faculty of Languages, University of Bremen, D-28334 Bremen, Germany

Abstract: The central question assessed in this article is directly related to the topic of the conference: Is semantic ambiguity in some way related to perceptual multistability? In the first section a phenomenal classification of perceptual ambiguity is given (immediate bistabilities in perception, perception of textures, spatial rotation in mental imagination). In the second and third sections several types of semantic ambiguity and their relation to perceptual multistability are described. Section 4 discusses textual ambiguities and Sect. 5 shows a model of emotional ambiguity (bad1, bad2) based on catastrophe theory. In general the analogy between semantic and perceptual multistability is accepted and a transfer of methods and models seems possible. In the case of structural ambiguities an underlying chaotic process is postulated. As a general consequence a description of meaning in terms of scales and spaces is called for.

1 Three levels of perceptual multistability

In order to give an answer to the basic question of whether semantic ambiguity is in some way related to perceptual multistability, we must first describe the levels of perceptual multistability which could be relevant for such a comparison. We shall just enumerate some important forms of perceptual multistability.

1.1 Immediate bistabilities in perception

In 1832 the Swiss crystallographer L.A. Necker described the classical multistability of a three-dimensional regular geometrical object if it is represented in two dimensions (without depth deformation). Fig. 1 shows this classical case, called the Necker-cube.

If we take a shape in two dimensions with clear contours, we can observe a bistability in which the figure/ground distinction changes. Figure 2 provides two examples.

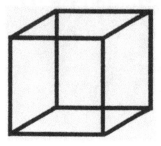

Fig. 1 The Necker cube

Fig. 2 (a) Rabbit ⟷ duck; (b) one cup ⟷ two faces

Other examples can be transformed from a non-ambiguous figure, through a set of intermediary pictures which are ambiguous, into an unambiguous figure. A classical example is shown in Fig. 3; it describes the transition:

old man ⟷ young woman

Perceptual multistability can also occur independently from semantic interpretation (from the recognition of an object). Rubin (1886–1951) analyzed the configuration made up of segments of a circle and rays (cf. Fig. 4).

If the rays are seen as figure, the circle segments are seen to belong to coherent circles in the background.

The first series of multistable figures is related to the mental reconstruction of a three-dimensional (meaningful) picture on the basis of the two-dimensional retinal distribution of luminosity.

Fig. 3 Old man ⟷ young women

Fig. 4 Rubin's figure (see text)

1.2 The perception of textures

The examples given in Figs. 1–4 are very convincing, as everybody can experience the effect: they are accessible to conscious experience. Our perception of textures, as Fig. 5 shows, is only accessible insofar as we experience a transition from disorder/homogeneity to order and inhomogeneity.

Julesz and his co-workers have shown that the recognition of textures is a phenomenon which appears at a presemantic and unconscious stage of perception. He proposes a statistical model in which statistics of first, second and higher order are relevant. If two textures are only distinguishable by statistics of third order, most animals cannot distinguish them. Many techniques of camouflage in the world of animals exploit this basic fact.

1.3 Spatial rotation in mental imagination

If textures involve very automatic, preconscious perceptual strategies, the rotation of pictures and objects in our imagination is a kind of conscious mental 'work' which is experienced as taking time and costing energy. Therefore, these processes point to a third level of multistability: *A set of different contours can be matched to one contour or not*. The fundamental bistability which is the result of mental rotation is, therefore:

– identity (through transformation)

Fig. 5 Distinguishing textures

– non-identity (through transformation)

In addition, we experience the effort of such a matching (some are easier, quicker, others are more complicated, slower). Fig. 6 shows some of the contours tested by Cooper and Shepard (1986, p. 124).

In order to answer our basic question, we must ask if one, several or all of these types of multistability have an analogue in semantic ambiguity in languages.

2 Ambiguity in linguistic meaning

At first impression semantic ambiguities seem totally independent from perceptual ones as the perceptual domain in language is that of phonetics and not that of syntax and semantics. The arguments for such an independence are the following:

a) Language, as a system of symbols, pertains to the level of consciousness (or even to a level beyond insofar as language is a system of conventions, which presuppose a social system from which these conventions emerge). The only level at which perceptual processes are related to language would be the level of conscious transformations analogous to the mental rotations described by Cooper and Shepard, cf. section 1.3.

b) The standard examples of multistable perception (in Fig. 1 to 4), i.e. level 1 in section 1, presuppose a three-dimensional object, whose perception is based on two-dimensional retinal input. If the test object is two-dimensional, the recognition of textures and the figure/ground distinction are the central

Fig. 6 Contours tested by Cooper and Shepard (1986)

problems. However, language is commonly considered to be a linear structure. Therefore, the problems of visual perception and the multistabilities in their solution are irrelevant for languages. Moreover, spatial objects are continuous (between border-lines), whereas language is considered to be discrete (in its units).

Both arguments are convincing if we accept their premises. We, therefore, must first critically assess these premises and ask:

- Does language and specifically the construction of meanings use unconscious, automatic and self-organized processes (as perception does)?
- Is linguistic meaning based on continuous scales and low-dimensional geometrical/topological representations?

As the reader can already guess, our answer to these questions is a positive one; the relation between perceptual multistability and semantic ambiguity does make sense. However, a positive answer to both questions is a challenge to structuralist semantics as it was established at the beginning of this century. We must, therefore, reassess the major facts of semantic ambiguity from a new perspective.

2.1 Ambiguities in the lexicon

Plato was already puzzled by the variability of meanings and tried to find the sources of "true" meaning by etymological reconstruction. After a long period of research in historical change (in the last century), we know that the origins of language are inaccessible to historical reconstruction; we must try a synchronic analysis of the phenomenon. We shall first look at the way lexicologists treat the problem.

Current lexicography documents the tremendous diversity of meanings. For instance Webster's Encyclopedic Dictionary distinguishes 24 specific meanings of the word 'eye' as a noun (and 21 uses in specific locutions) and 3 uses as a verb 'to eye'. Some of these are rather distant from the meaning of the perceptual organ called 'eye'.

Examples:
13.   The butt of a tomato
16.   The hole in a needle
23.   Winds and fair weather found at the centre of a severe tropical cyclone

Multiple meanings are also typical for other body parts such as *mouth*. The underlying processes which explain this diversity can be called metaphor (diffusion of meaning by similarity) and metonymy (transition from parts to wholes and vice-versa).

If no common historical origin exists (or if the commonality is hidden) the words are called homonyms (homophones, homographs); they have categorically different meanings (and are, therefore, called polysemous). The distinction between detectable and hidden identity operates on a scale of analyzability which is continuous. Thus the basic distinction between variants of meaning and different meanings is a continuous scale, which we can call: *scale of semantic categorization*.

Examples of polysemy (cf. Webster)

ear^1 1. the organ of hearing
         (subdivided into 22 specific meanings or uses)
ear^2 1. part of a cereal

It is very difficult to decide by the similarity of forms appearing under ear^1 if ear^2 has some similarity with ear^1. In the lexicon etymological knowledge is used for a clear distinction:

ear^1: lat: auris
ear^2: lat: acies

This etymological knowledge is mostly not accessible to the average language user and, therefore, irrelevant for his semantic 'perception' of the two words. Moreover, diachronic pathways can be complicated and multiply interrelated. (As language contact and lexical borrowing between Indo-European languages is a common phenomenon the same etymon can enter a specific language at different periods with different meanings.) The distinction between two words can, therefore, even appear if the etymon is the same as in[1]:

---

[1] OHG=old high German, OF=old French, Gmc=common Germanic

Table 1

bank[1]: 1. A long pile or heap; mass:
a bank of earth, a bank of clouds.
[ME: banke]; 16 variants and uses

bank[2]: 1. An institution for receiving, lending, exchanging and safeguard money.
[It. banca, OHG bank = bench]; 11 variants and uses.

bank[3]: 1. An arrangement of objects in a line or tiers.
[OF banc < Gmc; see bank[2]]; 10 variants and uses.

In French, Italian and Spanish two genders exist and differentiate two groups of meaning:

French: banc[1] (m) siège étroit
banc[2] (m) amas de sable
banque (f) entreprise commerciale.

Span.: banco[1] (m) asiento largo y estrecho,
banco[2] (m) establicimiento público de crédito
banca[1] (f) asiento de madera sin respaldo
banca[2] (f) comercio di dinero y crédito

Ital.: banco (m) seat
banca (f) institution

The Spanish example shows that both forms (banco, banca) have a similar differentiation; in Italian we find compounds with the meaning of 'banca' but the form of 'banco': bancogiro, banconota.

These few examples, which are symptomatic for the lexicon as a whole, show that lexical ambiguity is due to the following processes:

a) historical change in meaning and borrowing (from sources which have been fixed by historical changes),
b) meaning diffusion by metaphor and metonymy.

A further source of lexical ambiguity are the external (denotational) and the syntactic context. If we compare translations of the German verb 'aufziehen' (cf. Wunderlich, 1980, p. 30) into French and English we come up with a list of very divergent meanings in context.

Table 2

	transitive use			intransitive use	
French	(context)	English	French	(context)	English
lever	(curtain)	open	s'élever	(thunderstorm)	approach
hisser	(flag)	hoist			
monter	(picture)	mount	marcher	(the guard)	draw up
élever	(child)	raise	en cortège		
remonter	(toy)	wind up			
arranger	(meeting)	organize			
railler	(persons)	tease			

If lexical ambiguity in the shared lexical knowledge is explained by historical processes, individual meanings can be explained by processes of linguistic development. Labov and Labov describe the use of the words 'mama', 'dada' and 'cat' by their child between the 15th and the 17th month (cf. Labov, 1978, pp. 232–235).

mama: In the 16th month all members of the family could be called 'mama', the statistical trends towards the final designatum (the attractor of the process) were already clear: mother (67 uses), father (13), sisters and brothers (1, 7, 16, 13). In the 17th month the ambiguity was reduced to mother (420) and father (52) and disappeared thereafter.

dada: The word 'dada' was first used in the 17th month. It was ambiguous relative to father (89), brother Simon (10) and sister Sarah (1). Later the statistically dominant father became the unique person designated by 'dada'.

cat: This case is different from the two first ones as the final attractor of the meaning is not an individual (a member of the family) but a species. The child concentrates on specific features, the implicit definition of 'cat' is 'analytic'. Some criteria are preferred to others. For this child the roundness of the head seemed to be the dominant criterion.

As a consequence of this purely illustrative analysis of a semantic development, we can state two further trends.

c) The lexical designation of specific individuals starts with a statistical field with attractors and eliminates smaller attractors until one individual is left as designation.

d) The content of a classifying label is given first contours by preferences for certain features and ends up with a sharpening of these preferences. As a consequence the categorization remains vague and is oriented towards one (or a few) prototypes.

A central question for semantic analysis which follows from (d) is:

What are the underlying semantic qualities on which a preference scale is built? Are there 'inbuilt' preferences related to perceptual cues (form, colour, behaviour, etc.)?

2.2 Perceptual scales underlying lexical ambiguity

We shall report the results of a classical study done by Labov in order to prove our basic premise, that a perceptual continuum underlies semantic ambiguities. Additionally this example shows that the real or imagined context is a strong determinant in disambiguation.

Labov presented two-dimensional pictures of containers with a handle but with different depth and width. We shall only consider variations in diameter. The 24 test-persons had to label these pictures in two contexts:

n: neutral context: no further specification,
f: food-context: the test-persons had to imagine food in the containers.

Figure 7 shows the series of pictures, Fig. 8 the consistency profiles (% = of consistent responses).

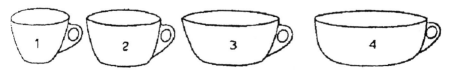

Fig. 7 Containers with handle (see text)

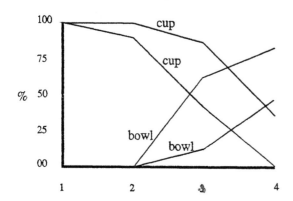

number of the container in the fig. above

Fig. 8 Labels assigned to the pictures in Fig. 7

In the food context, pictures 2 and 1 are already strongly ambiguous, picture 2 is by preference labelled as 'bowl', whereas in the neutral context the dominant labelling as 'bowl' occurs only in picture 4.

The general trend shown by these results motivates a further hypothesis on lexical ambiguities:

e) In language usage two factors govern the appearance and disappearance of semantic ambiguities:
   - underlying continuous scales (mostly based on perceptual, behavioural or emotional criteria),
   - imagined or real contexts of use.

2.3 Lexical ambiguity in dimensional adjectives

The domain of dimensional adjectives such as: long, deep, high, broad, etc. is specifically tuned to the perception of space and spatial features. We shall, therefore, discuss a specific lexical ambiguity which occurs when several such adjectives are used in order to characterize a three-dimensional object.

In a study on dimensional adjectives (DA) Lang (1987, p. 297) considered sentences of the following type:

$$x \text{ is } m_1 \text{ DA}, m_2 \text{ DA}' \text{ and } m_3 \text{ DA}''$$

where x is the subject of of the sentence (denoting a three-dimensional object), m_i are measurement expressions like 1 meter, 50 cm etc. and DA, DA', DA'' are different dimensional adjectives.

The ambiguity Lang discovered concerns the dimensional adjectives chosen to describe a board. The breadth of the board is either called "breit" (broad) or "tief" (deep). The second use is triggered by the context, in which the board is used as a window-bench and is seen (or imagined) as part of a three-dimensional window-case. If seen as a raw board in a stack, the DA "breit" (broad) is preferred. Talmy (1983) already gave a framework for the interpretation of this observation by Lang, as he distinguished several imaging systems. One of these is related to an imagined eye and refers to a body-centred perspective. This imaging system is triggered by the context "window-case", the dimension "breit" becomes "tief" (deep). As a further consequence the DA "breit" replaces "lang" (long). It seems that in the underlying hierarchy going from maximal to minimal we had first:

(neutral context)   lang   1 m
                          breit   30 cm

then the interpretation shifted to:

(window-case)   breit   1 m
                          tief   30 cm

This peculiar example of combined lexical ambiguity shows how mental imagery underlies the phenomenon of lexical ambiguity. In this particularly rich case two adjectives change their meaning (in context) in an interdependent way and we can assume that there is an underlying set of "imaging systems" (as Talmy proposed). The multistability concerns primarily the imaging systems and triggers a complicated redefinition of lexical items.

3 Morphological and syntactic ambiguities

In the Sect. 2.1 on lexical ambiguity we considered a unitary word (either a simplex, or a stem with bound morphemes). In this chapter we consider structures consisting of more than one free semantic unit. The source of ambiguities lies in the constructions and befalls the constructional meanings.

The analysis of morphological and syntactical ambiguities is, therefore, fundamentally different from the analysis of lexical ambiguities[2]. Morphological ambiguities are considered as transition phenomenon to syntactic ambiguities. Thus the rather free constructions in nominal composition (e.g. in German) are in many respects similar to syntactic constructions.

3.1 A basic list of constructional ambiguities

We start with a list of examples:

Table 3
a) nominal compound
   Mädchenhandelsschule (German). This compound has two possible readings which correspond to different bracketings of the three constituents
   1. girl (commerce school) = a commercial school for girls
   2. (girl commerce) school = a school for white slave commerce.
      This reading does not have any reference to existing institutions, but it is structurally possible.

b) fresh fruit market
   1. fresh market for fruit
   2. market selling fresh fruit

c) beautiful girl's dress
   1. a beautiful dress for girls
   2. dress belonging to a beautiful girl

d) some more convincing evidence
   1. some more evidence (which is convincing)

---

[2] It is clear that lexical ambiguities are inherited and partially eliminated by constructions

2. some evidence (which is more convincing)

e) flying planes (are/is dangerous)
  1. planes which fly (are ...)
  2. to fly planes (is ...)

f) the shooting of the hunters
  1. the hunters shoot
  2. the hunters are shot

g) amor Dei
  1. God loves
  2. God is loved

h) Dico Clodiam amare Catullum
  1. Clodia amat Catullum
  2. Catullus amat Clodiam

i) Jean fait manger les enfants
  1. les enfants mangent
  2. les enfants sont mangés

If we analyse these examples we conclude that two basic types of syntactic ambiguity exist.

(1) If we have two constituents of a syntactic construction, the relation between these constituents is either symmetric (this is an unstable borderline case) or asymmetric. In the normal case of asymmetry the left or the right constituent is the centre, the other the periphery. In most languages the place of the centre is typologically stable; therefore, this source of ambiguity is rather incidental. (But consider: is axe-hammer rather a hammer or an axe?) Is the proper name in 'uncle Watson', 'President Watson' the centre or the periphery of the noun phrase?

(2) If we have more than two constituents and if the principle of centre vs. periphery works we have already three possible constructions (C = centre, P = periphery)
  (a) P_1, P_2, C
  (b) P (= two constituents), C
  (c) P, C (= two constituents)

If more than three constituents are present the structural ambiguity increases very quickly. This source of ambiguity could make complex constructions impossible. Every natural syntax must, therefore, have a set of devices in order to limit this danger of disorganization. The main function of syntax can be seen in such a control of chaotic ambiguity. The major strategies are:

Ambiguity in Linguistic Meaning in Relation to Perceptual Multistability 233

a) The constituents are morphologically classified by bound morphemes, e.g. by case-forms. Syntax rules, which refer to case-classification, control the grouping of verbs with different types of objects, of subjects with their verb-phrase, of adverbs (if they are marked as such) with verbs, adjective with nouns,etc.

b) The order of appearance and the proximity of two constituents in an utterance can be a signal for a common construction (if no information of type (a) is given or if this information is insufficient).

c) Bound morphemes around the verbal stem can show the type of noun phrases which may specify the central meaning (this is a technique found in many American-Indian languages).

d) Discontinuities of stem and bound morphemes can frame a systematic construction (in German the auxiliary and the verb stem can mark the contours of a verb phrase).
   Example:
   Er *hat* den ganzen Kuchen *aufgegessen.*

e) Binding phenomena (congruence, pronominal coindexation) can relate constituents.

Syntactic ambiguity appears if these devices do not work. The situation is, therefore, basically different from perceptual multistability, where stability is the normal case; in syntax chaos would be the normal case and complicated filters reduce this chaos, but some islands of multistability remain (for contextual desambiguation).

In certain cases we have underlying (deep) semantic scales. We shall concentrate on these examples.

3.2 Semantic scales underlying syntactic ambiguity

The examples (e) to (i) all have a transitive verb as the centre of the construction:

(e) The gerundial construction takes 'planes' as patient, whereas the second reading replaces the gerund 'flying' by the (present participle) 'flying' in an attributive construction. The syntactic ambiguity operates on the border-line between *verbal* vs. *nominal* constructions.

(f) **shoot** (argument$_1$, argument$_2$) If we call the arguments:
   (1) agent (author, cause)
   (2) patient (affected, caused)
   we can say that in the first reading 'the hunters' are *agent*, in the second reading they are *patient*.

(g) **amare:**   'God' is agent or patient
(h) **amare:**   'Clodia' is agent or patient
              'Catullus' is agent or patient

(i) **manger:** 'les enfants' is agent or patient

We have to consider some basic linguistic and psycholinguistic facts:

Linguistic facts:

– The distinction between transitive and intransitive verbs is related to the syntactic phenomenon of passivization (passive-diathesis) and of pseudo-transitives.
– Different languages can be located on a scale ranging from:
  - ergative languages
  - mixed languages
  - accusative languages

Ergative systems mark the subject of the intransitive sentence in the same manner as the object of the transitive sentence. This category is called 'absolutive', whereas the subject of the transitive sentence is associated with a separate category called 'ergative', it represents the causer or actor.

Our western languages are mainly of the accusative type, where the subject of the transitive and of the intransitive sentence have the same case (called 'nominative'), whereas the object of the transitive sentences is separately marked by a case called 'accusative').

Psycholinguistic facts:

It is clear that the notion of cause (effect) and of agency (action) are important for the examples (f) to (i). A psychological analogue is the concept of 'phenomenal causality' analyzed by Michotte (1954). This psychological concept has as proto-typical image the transfer of energy between billiard-balls, or suspended balls. Fig. 9a shows this analogy, fig. 9b shows a typical configuration in Michotte's experiment.

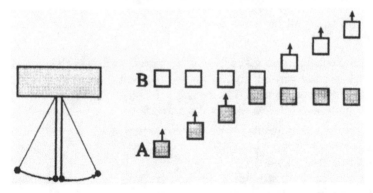

Fig. 9 Left: the mechanical prototype. Right: series of pictures shown by Michotte

The main features which give rise to 'phenomenal causality' are:

- the objects A and B must be seen as distinct figures not as parts of a whole,
- object A must be dominant in relation to object B; i.e. the behaviour of B must be somehow dependent on the behaviour of A but not vice versa,
- there must be variability, change in the situation perceived.

Ertel (1975, 100 ff) argues that these criteria can explain restrictions on passivization. If we take a process which by nature is rather symmetric, we can obtain different perspectives on it by different weights of phenomenal causality associated with the protagonists of the scene (P = prince, C = Cinderella).

Table 4

		P	C
a)	Cinderella was married by the prince	3	0
b)	The prince married Cinderella	2	1
c)	Cinderella married the prince	1	2
d)	The prince was married by Cinderella	0	3

The dominance shift happens between stages (b) and (c), whereas (a) reinforces the asymmetry of (b) and (d) reinforces the asymmetry in (c). Moreover, the sentences (a) and (d) allow for the elimination of the by-phrase, i.e. of 'the prince' in (a) and of 'Cinderella' in (d). The passive sentences can, therefore, be called pseudo-transitive. These facts allow for the acceptance of a (continuous) scale which connects both arguments in a transitive sentence.

With the semantic scale as background we can now explain the semantic ambiguities (f) to (i) as special types of multistability similar to perceptual multistability.

(f+g)   The ambiguities in (f) and (g) are due to the loss of information about what type of argument 'hunters' (f) and 'Deus' (g) have in genitive construction; whereas the ambiguities in (h) and (i) are due to the embedding of a clause into a syntactically dominant construction.

(h)   The construction in Latin called 'accusativus cum infinitivo' which is associated with verbs of saying ('verba dicendi') and others transforms the subject of the embedded clause into an accusative, whereby the arguments of the verb 'amare' become indistinguishable (word order is free in Latin).

(i)   The causative construction with 'faire' introduces a dominant cause (author, agent). The verb 'manger' is already pseudotransitive in the active clause, as the object can be left unspecified:
  – les enfants mangent (-)   (the children eat)
  – les enfants mangent un gâteau   (the children eat a cake)

The embedded clause 'manger les enfants' in reading i(2) behaves like a passive insofar as the agent can be left open. Thus two possibilities can be chosen:

(1) The agent in the embedded sentence, which is semantically under the control of the agent of the causative construction 'Jean' is left unspecified, in this case 'children' are the patients of the verb 'manger'; the children are eaten.

(2) The object of the embedded clause (the food) is left unspecified (as 'manger' is pseudotransitive); in this case the children are the subject of the embedded clause, they eat (something).

As a result of this short analysis we can say that syntactic ambiguities based on underling semantic scales function in the same way as perceptual multistabilities with an underlying linear parameter do.

4 Textual ambiguities

If syntactic ambiguities are already difficult to control, one would imagine that this tendency grows in the case of texts. Contrary to such an expectation ambiguity in texts was not even remarked as a relevant phenomenon by linguists. In the field of Gestalt psychology an experimental analysis of ambiguous texts was already proposed by Poppelreuter in 1912; Metzger (1982) completed this proposal. In Stadler and Wildgen (1987, pp. 106 – 117) these materials were reassessed. We shall only report some major results here.

In the first experiment (by Poppelreuter) two texts with different protagonists and antagonists are mixed. The two lines of the plot can be easily separated by the hearer, if he recognizes the two coherent thematic lines. The formal incoherences (e.g. pronouns do not fit the before mentioned nouns etc.) help to cut up the two stories. Most of the person tested could easily separate the texts.

In the second experiment (made by Metzger) the protagonists and antagonists are the same but the basic motivations and actions and the helpers which comment the events are different. Again local incoherences may help to separate the two plots. In retelling the story most of the hearers follow one of three strategies:

- they produce another rather incoherent story,
- they reorganize the story towards a coherent plot (which is rather independent of the plot in the original story),
- they eliminate all the elements of one of the underlying plots and thus disambiguate the mixed plot.

If textual ambiguity is of the type proposed by Poppelreuter and Metzger (many other types could be imagined) the ambiguity of textual meaning is defined by the existence of two (or more) consistent plots. The hearer has to find one or both of the plots. The situation is similar to a linear puzzle: one must assemble the pieces in order to regroup them into two different lines. The points of contact are defined by thematic cohesion and syntactic coherence.

Although the textual gestalt may be a very complicated object (e.g. the gestalt of a novel) it has in its kernel a linear plot or several interlaced plots around a thread.

A text is ambiguous if two different plots can be associated with the text. This corresponds to the bistability of a Necker-cube.

In a different sense a text may be ambiguous, if it either has a stable plot or is thematically chaotic. This corresponds to the problem of texture discrimination dealt by Julesz. Both types of textual ambiguity can be called global as they refer to the 'gestalt' of the whole text.

A text may have many local multistabilities, e.g. a neutral observer or a secondary person can turn out to be the central protagonist or antagonist, a line which seems to lead towards a resolution can turn out to contribute to the complication, a climax can turn out to be a secondary event; in fairy tales metamorphoses of central persons can happen, they take on different forms and characters (a man/a wolf) in different episodes of the tale. Nevertheless, and this is our central hypothesis, the basic dynamics remain similar throughout the whole domain of linguistic cognition and they are rooted in perceptual/motor and emotional multistability.

An example of semantic ambiguity with an underlying emotional scale is given by Poston (1987) and we shall give a short summary of the model he proposed.

5 The catastrophe theoretical model of a lexical ambiguity based on an emotional scale

The Black American use of 'bad' is ambiguous insofar as it can have a negative as well a positive meaning. If we start with a basic emotional parameter 'approval - disapproval', we can arrange the unambiguous adjectives: super, ok, so-so, lousy, awful, and the ambiguous adjective 'bad' as clouds around this axis.

Figure 10 (cf. Poston, 1987, p. 29) gives an initial representation of the overlapping of bad_1 (approval) and bad_2 (disapproval).

The bistability of the situation, the existence of a neutral zone, the sudden shifts (catastrophes) between bad_1 and bad_2, which depend on small changes in the context of its use, lead Poston to propose the catastrophe called 'cusp' as underlying schema. Fig. 11 shows a reorganization of the field of adjectives on the surface of critical points in the unfolding of the cusp.

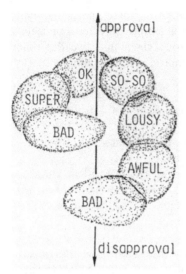

Fig. 10 Representation of the overlapping positive and negative meaning of the word 'bad' as used by Black Americans

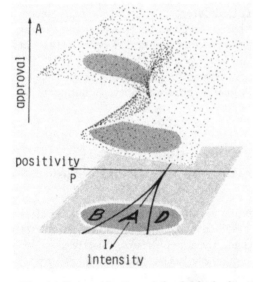

Fig. 11 Reorganization of the field of adjectives (see text)

In Fig. 11 the relevant external variables which govern the catastrophic and smooth changes in the field of adjectives are labelled P = Positivity, I = Intensity.
- If intensity (I) increases, the bistable situation appears, if it is low, no bistability occurs in: ok, so-so, lousy.

– If positivity changes from positive to negative, the interpretation of 'bad' changes suddenly, from bad_1 (positive, intensive) to bad_2 (negative, intensive).

6 Conclusion

In short:

(1) Semantic ambiguity very often has an underlying (linear) scale which is cognitively continuous. The qualities appearing on these scales are related either to perception or to emotion. This holds specifically for lexical ambiguities.
(2) Syntactical ambiguities are of another kind, as their background is the instability of syntactic productivity. Ambiguities appear as 'holes' in stabilizing filters. Some of the mechanisms for the delimitation of syntactic chaos are related to low-dimensional semantic spaces (e.g. to linear scales). Although the underlying qualities are more abstract, the organization of these ambiguities corresponds to that of perceptual multistability.

In further research the models for describing perceptual multistability (catastrophe theory, synergetics, chaos theory, etc.) should be applied to semantic ambiguity, if the assumption of a low-dimensional semantic space is plausible.

References

Basalou, L.W. (1987): The instability of graded structure: implications for the nature of concepts. In: U. Neissen (Ed.), Concepts and Conceptual Development: Ecological and Intellectual Factors in Categorization, pp. 101 – 140. Cambridge: Cambridge University Press.
Chung, S. (1977): On the gradual nature of syntactic change. In: C.N. Li (Ed.), Mechanisms of Syntactic Change, pp. 3 – 55. Austin: University of Texas Press.
Cooper, L.A. & Shepard, R.N. (1986): Rotation in der räumlichen Vorstellung. In: M. Ritter (Ed.), Wahrnehmung und visuelles System, pp. 122 – 130. Heidelberg: Spektrum.
Ertel, S. (1975): Gestaltpsychologische Denkmodelle für die Struktur der Sprache. In: S. Ertel, L. Kemmler & M. Stadler (Eds.), Gestalttheorie in der modernen Psychologie, pp. 94 – 105. Darmstadt: Steinkopff.
Gregory, R.L. (1987): The Oxford Companion to the Mind. Oxford: Oxford University Press.
Julesz, B. (1974): Cooperative phenomena in binocular depth perception. American Scientist 62, 32 – 43.
Labov, W. (1973): The boundaries of words and their meanings. In: J. Fishman (Ed.), New Ways of Analyzing Variation in English, pp. 340 – 373. Washington, DC: Georgetown University Press.

Labov, W. (1978): Denotational structure. In: Papers from the Parasession on the Lexicon. Chicago Linguistic Society 14, 220 – 260.

Lang, E. (1987): Semantik der Dimensionsauszeichnung räumlicher Objekte. In: M. Biernisch & E. Lang (Eds.), Grammatische und konzeptuelle Aspekte von Dimensionsadjektiven, pp. 278 – 458. Berlin: Akademie Verlag.

Marr, D. & Nishihaba, H.L. (1978): Representation and recognition of the spatial organization of three-dimensional shapes. Proceedings of the Royal Society of London, B 200, 269 – 294.

Metzger, W. (1982): Möglichkeiten der Verallgemeinerung des Prägnanzbegriffs. Gestalt Theory 4, 2 – 33.

Michotte, A. (1954): Die Theorie der phänomenalen Kausalität. In: A. Michotte, Gesammelte Werke, vol. 1, pp. 211 – 224. Bern: Huber.

Poppelreuter, W. (1912): Über die Ordnung des Vorstellungsablaufs. Archiv für die gesamte Psychologie 25, 208 – 349.

Poston, T. (1987): "Mister! Your Back Wheel's Going Round!". In: T.T. Ballmer & W. Wildgen (Eds.), Process Linguistics. Exploring the Processual Aspects of Language and Language Use, and the Methods of their Description, pp. 11 – 36. Tübingen: Niemeyer.

Shepard, R.N. (1984): Ecological Constraints on Internal Representation. Resonant Kinematics of Perceiving, Imagining, Thinking, and Dreaming. Psychological Review 91, 417 – 447.

Stadler, M. & Wildgen, W. (1987): Ordnungsbildung beim Verstehen und bei der Reproduktion von Texten. SPIEL (= Siegener Periodicum zur Internationalen Literaturwiss.) 6, 101 – 144.

Wunderlich, D. (1980): Arbeitsbuch Semantik. Königstein/Taunus: Athenäum.

The Emergence of Meaning by Linguistic Problem Solving

A. Vukovich

Institute of Psychology, University of Regensburg, D-93053 Regensburg, Germany

Abstract: The opinion that the acquisition of sign meaning in natural language and its actual realisation in interactions is, to an appreciable degree, the result of a problem-solving process can be traced back to the very beginning of linguistic theory in ancient times. The final accepted meaning of a string of symbols is elaborated by the Ss in such a way as to minimize incommensurabilities among primary meaning conceptions in a multistep approximation according to grammatical and empirical guidelines, with a special consideration of the so called communication scaffold of author, addressee, subject, situation and interaction-type. Rhetorical figures which signalize and carry out variations of the relationship between signs and designed concepts are presented for illustration. The generation of meaning in natural language is to be seen as a special case of the general readiness and interest in problem-solving of humans.

Before presenting his own theory of meaning acquisition by interactional games, Wittgenstein, in 1958, cites, at the very beginning of his "Philosophical Investigations", from a passage of the Confessions (400 A.D.) of Aurelius Augustinus. In these lines the Father of the Church speculates about the association of verbal signals and objects or events by parental reinforcement, according to the later dumb-bell model of symbols by Peirce, as an ordered pair of signals and meanings. In the same chapter (I;8) Augustinus had already conceived of the individual acquisition of meanings as the result of a problem solving process of the child which is supported by the ubiquitous rules of non-verbal communication: "For my elders did not teach me this ability, by giving me words in any certain order of teaching, (as they did letters afterwards), but by that mind which thou, my God, gavest me, I myself with gruntings, varieties of voices, and various motions of my body, strove to express the conceits of mine own heart, that my desire might be obeyed; but could not bring it out, either all I would have, or with all signs I would." There was no need for the holy man to go into details in this ancient text.

A later introspectionistic author of openhearted confessions, Jean Jacques Rousseau thought that parents and nurses should not make this problem-solving process during language acquisition too easy for the child (Émile 1762, I; 18): "I would have the first words he hears few in number, distinctly and often repeated, while the words themselves should be related to things which can first be shown to the child. That fatal facility in the use of words we do not understand begins earlier than we think." In this respect, the children in cities would have a more difficult time because of well meaning intentions gone awry. They are "brought up in a room and under the care of a nursery governess, do not need to speak above a whisper to make themselves heard. As soon as their lips move, people take pains to make out what they mean; they are taught words which they repeat inaccurately, and by paying great attention to them the people who are always with them rather guess what they meant to say than what they said ... Let the child's vocabulary, therefore, be limited; it is very undesirable that he should have more words than ideas, that he should be able to say more than he thinks. One of the reasons why peasants are generally shrewder than townsfolk is, I think, that their vocabulary is smaller. They have few ideas, but those few are thoroughly grasped."

Indeed, it might be beneficial to work under regular conditions with a small repertoire of basic concepts which can easily be redefined or transformed and to cope with the complexity of an object domain by combinations of elementary concepts, as one does in chemistry. It might be, for instance, more economical to use a limited scale of a few different grade positions – say ranging from ugly to beautiful – which can be broken down into many steps by quantifiers, like "little" and "very", or by relators, like "nearly" or "more than".

In a footnote of his influential publication, Rousseau points to another possible way to increase the meaning pool of a natural language. He maintains that single words change their meaning when they are embedded in different contexts: "I have noticed again and again that it is impossible in writing a lengthy work to use the same words always in the same sense. There is no language rich enough to supply terms and expressions sufficient for the modifications of our ideas ... In spite of this, I am convinced that even in our poor language we can make our meaning clear, not by always using words in the same sense, but by taking care that every time we use a word, the sense in which we use it is sufficiently indicated by the sense of the context, so that each sentence in which the word occurs acts as a sort of definition" (Rousseau, transl. by B. Foxley, p. 72).

This remark on the varying senses of different word uses illustrates a general thesis of Wolfgang Köhler: the properties of the whole determine the properties of its parts and often even determine which sub-sections of this whole constitute natural sections. The simulation of the linguistic coding and decoding of ambiguities and equivocations would have to take into account that peculiarity by a series of iterative loops. A first approximation is improved by the results which it furnishes; a third is based on the results of a second. Altogether, the psychology of language offers an easily-accessible field to demonstrate the necessity of the gestalt-theoretic orientation. The whole is something different than the class of its various elements; the meaning of a sentence is quite a different thing than the set of meanings of its isolated words.

Normally, the mental process which generates the meaning of an utterance starts without conscious agency. Meaning is a gift of nature which one hardly can refuse. Linguistic problem solving is our fate, mostly to welfare, sometimes to mischief. One is convinced, that whatever one says might be right or wrong, but it is rarely pure nonsense. Goethe mocks: "Upon the whole – to words stick fast! Then through a sure gate you'll at last enter the templed hall of certainty" (Faust I, 1990). The advice comes from the devil, of course, because it seems false and still is true. One cannot deal with texts without looking for sense in them and at least finding some plausible meaning. Fritz Giese, a German pioneer in Applied Psychology, proposed a diagnostic device to identify creative people: a series of 12 nonsense stories, each a kind of surrealistic lyric, which the subjects should remember after one minute reading time per story. If the persons were able to find a hidden sense in the adventurous combinations, they were supposed to be capable of remembering them better afterwards than persons who were dependent on their mere memory for senseless texts. The shortest of these short stories runs: "The city rioted, because the chimney sweep had brought the orange to the freight yard without buying a lemon for the return route", i.e. somebody does something senseless, but doesn't commit a further superfluous action and everybody is upset.

Of course, nowadays everyone is certain that there is no one-to-one relationship between words or sentences on the one hand, and the connotations understood by the receiver on the other. Hörmann (1976) discussed this topic at length under the headlines of "The Constancy of Meaning (Sinnkonstanz)" and "Levels and Vectors of Understanding". The comprehension of a person's utterances seems to be the result of a series of happy guesses led by some contextual hypothesis. To express this metaphorically: successful communication resembles the identification processes of a hunter, who infers, on the basis of few indicators gathered under unfavorable conditions of observation, the kind, size, and movement direction of his pursued prey.

To prevent misunderstanding and to provide texts which were suited for the interacting persons, the purpose of the communication, and the actual situation, ancient rhetoricians devised instruments for the clarification of the intentions of a speaker and to assure themselves of factual apperception as receivers. Undoubtedly, Augustinus was familiar with the pitfalls of understanding. He knew the disputes on the right denomination, the conventional meaning of phrases, and the combination of both aspects in the question of proper word selection. These elements of a lawyer's art had been traditional items of rhetorical training from Aristotle to Quintilianus. A remarkable insight by rhetorical scholars was the fact that clarity and distinctiveness – both components of the otherwise highly-valued perspicuities – are not always the primary goals of speech. One has to give the words and phrases just those meanings which help to achieve one's aims.

In some cases, the exact meaning of a single word embedded in a longer string of the text units can be derived from the meaning of the isolated words. This might be done by principles similar to the statistical analysis of variance, which assign each combination of two or more elements a special interaction term. This is, however, a luxurious, expensive, and rigid procedure. It seems to be more profitable to speculate about special tactics of coding, which determine the switch from one interpretation to another by moderator variables. For instance, a "hill" is an elevation of medium size, more or less round, of moderate steepness. When describing anatomical details, such an elevation is much smaller than when describing a landscape. The adequate frame of reference is activated by the content of the surrounding words. It displaces the trace lines of word meanings along the dimensions of the cognitive representation of their designations. There is, however, a bulk of more elaborate meaning transformations ready for use to explain various phenomena of meaning inconsistency. A short review will be given below.

In many cases the assessment of suitable meanings of actually-used words or of more comprehensive text units are not only detectable from outside by careful experiments and by thorough interpretation, but are also subject to deliberate actions of the speaker in everyday life situations. That is obvious, and sometimes it is conveyed expressively. There are recurring conspicuous patterns of information to be found in colloquial speech which regulate the relation between signals and their effects of communication. These patterns have the appearance and the functions of rhetorical figures. Rhetorical figures are patterns of information distinguished by their internal structure, by their connection points for external relations and by their pragmatic functions, i.e. mainly by their effects on the receiver. In the traditional catalogues of schemes, e.g. that of Lausberg (1960), many of them are not filed. They are, however, conceptualized in the same spirit. These augmentations of rhetoric might be described by the same verbal means and could be illustrated by parallel cases. At least some of the distinguishing processes meant by Rousseau might so be delineated in a more or less formal way:

– The technique of anchor-induced rescaling is illustrated by the following examples: "When one considers the mountains in the Himalaya the Alps are merely hills./ The fights in the convent during the French revolution were festive masses in relation to the discussions among our faculty members yesterday./ The trumpets of Jericho were soft whispers, in comparison to the outcry which the employees will make, if the employers break their mutual agreement." In these sentences the rescaling of concepts or designations is urged by newly introduced anchor stimuli or by changed boundaries of the object domain. If the set of the single words which includes "mountain" has to remain constant and the edges of variation are not to be reached by combinations of elementary signals like "huge mountain", then a widening of the factual range demands a new partition of the domain. When the total range of objects is diminished, however, a new partition is necessary, if the same number of distinctions among the object classes has to be made.

– Another technique for signalizing a special type of meaning coordination is the use of so-called meaning moderators: "This is – slightly exaggerated – a hit." Such phrases like "sugar-coated" or "in a straight forward manner" or "strictly speaking" function like the stops of an organ. Meaning moderators also are suitable for the protrusion into an area of otherwise unspeakable contents: "Conservatively speaking, that's a catastrophe./ In daring words: that was a very cautious attempt."

– The reference to the actual conditions of communication is a further way to qualify the meanings of words, phrases, and complete sentences in a different respect. This reference restricts the receiver's search for the right interpretation of verbal signals. It is quite common, and therefore easy to overlook. "On the left wing of our party they will call this treason against the working people./ Everything which you don't like you call awful. Aren't there any graduations to be made?" One presupposes that certain persons in a certain mood will use a certain sort of wording. One takes this into account when evaluating messages and correcting them accordingly. In practical cases, it is often difficult to separate personal views on a variety of subjects from a deviating use of the descriptional means, especially concerning differing concepts. For example, one has to take into account the temperament of the speaker and contextual changes in different generations. If diplomats speak of "an exchange of divergent opinions", they mean: "We quarreled. We could not find any accord."

Generally, there is an obvious endeavour to prevent the elements of verbal communication from any change in their communicative functions. Whenever possible, one claims an invariant meaning of words and sentences. This happens particularly in conversations among people who know each other well, who trust each other, who accept each other as competent in the field of their interaction, and who are accustomed to the actual situation of speaking. If the actual properties of this scaffold of communication are unknown, then the most probable values are to be inserted into them. If, under these conditions, the text evaluation shows significant deviations from the usual correlation between a certain type of signal and the object which it designates, then, instead of redefining the concepts or of replacing them by others, overtly or covertly, the values of the different components of the scaffold can be changed. One traces the deviations of word use back to some properties of the speaker. As causes of the observed divergence one considers an author's mode of speaking, his assessment of the listener, and the situation, some of his personal characteristics like capriciousness or irascibility, and so on. If these measures still do not make any sense, another candidate for finding a more logical meaning is the labeling of the actual designated act e.g. whether an utterance is to be held as irony, as an exaggeration, as a mistake, a white lie or some other type of saying serving a special purpose, e.g. to whittle an opponent down to size, to retreat from a hazardous confrontation, or to caricature for the sake of clarity. If it is impossible to reach the desired plausibility of propositions using such means, one has to exploit even other sources of variations. The set of techniques to solve problems of meaning generation is not limited. If no remedy could be found by all these techniques, the message was, perhaps, an acoustic misconception: "Really? Please say that again!"

The effects which the interplay of the different components of the communication scaffold has on the meaning of a message can be described by linear functions. The values of its arguments are combined in a compensatory structure: An outburst by the speaker might be a result of a bad temper or of overwhelming influence from the outside. One is routinely acquainted with compensatory concatenations; they are self-evident, as the following example of a dialogue on external events illustrates:

A: The farms in Europe are so tiny! On my farm in Australia, it took me half an hour to go by car from one end to the other.
B: Was the road that bad?
A: No, flat and smooth like the runway of an airfield.
B: Don't worry, I had such a slow car once, too.

The opponent reaches multistability using evasion on varying reference axes of a physical or mental coordinate system, respectively. The same takes place when trying to reveal the hidden meaning of a text. In both directions of an information exchange - when generating the signals to transfer a meaning, and when reconstructing the meaning out of the transferred signals - such a function can be used to arrive at a satisfying interpretation.

A considerable number of verbal patterns renders the mediation of meaning more difficult than necessary. The core of information could be expressed in a more comprehensible way. Those patterns complicate, however, the task of decoding in a pleasantly challenging manner. Challenging difficulties are the marks of problems.

A humble mean to achieve this goal is the description of bipolar features by unipolar-orientated denominations, and vice versa: attributes which are conventionally described by the distance to one pole of the variation are named by a word with the opposite meaning: "I'm twenty years young./ I'm as busy as an ice cream seller in January./ He is steady like snow in summer" (Vukovich, 1990). As in experiments on intersensory scaling, in everyday life one compares the intensities of qualitatively different variables. Abstractions of this kind are communicated as if such judgements would be the simplest task in the world: "In fact, he is not half as efficient as he seems to be at first glance" or "She is as mean as she is beautiful" instead of saying "She is very mean, and she is very beautiful, too."

Every surprising course of events marks a hurdle in the process of understanding. Simultaneously it attracts attention.

The horror of all disk-jockeys is melodies which continue after they were supposed to have already come to an end. Such patterns have counterparts in unwanted continuations of interactions: "I'll come to your party ... under no condition./ I can resist anything ... but temptation" (O. Wilde). If, by the continuation, even the grammatical category of a word is changed – from a verb to noun, for instance – then these propositions are nicknamed "garden path sentences". A related technique aims at leading the expectations of the listener in the wrong direction. It is the antithesis of an awaited comment which the speaker maintains in the end: "What does this fact teach us? I suppose it teaches us nothing. All that happened there was merely chance./ Isn't it dreadful? No. It might even be an advantage!"

The avoidance of nonsense and conflicts between rival components of meaning is, of course, the steering variable with the broadest, and additionally the most heterogeneous field of application. The partner has to speculate about conditions, associated circumstances, and the effects of events to get some reasonable hints at the intention of the author. The paradigm of nonsense is the internal contradiction.

Internal contradictions are strong incentives for attention. They demand solutions in themselves, and one is ready to search for such harmonizing interpretations spontaneously. In his Rhetoric, Aristotle already mentions the strange attractiveness of cognitive conflicts. Another motive might be to involve the addressee in an effort to find a harmonizing solution by himself, and to absorb the intellectual capacity of the listener so that he can not drift away in his fantasies. Examples are: "Less is more./ The first among equals./ Pampers babies are dry even when they are wet./ The worst experiences are the best./ No good deed will go unpunished." Taken literally these texts are chaotic or stupid. However, one does not take it so. One understands them easily ... after a little hesitation. Other sayings show more resistance against the spread of tidiness. The proverbial wisdom that wind and waves take the side of the better seaman belongs to this category, likewise: "The last shirt has no pocket./ Be what you are if you want to become someone else." A relevant category of traditional rhetoric, is the "oxymoron", that means the sharp blunt-edged: "beloved monster", a sometimes completing characterization of a sweetheart, or "military intelligence", seen from a pacifistic point of view, or a "dreadful idyl".

The attractiveness of internal contradictions have even led to the invention of false doubles to provoke interests: "Big business in small countries./ Large gains with low-calorie food."

Somewhere in the back of one's memory, one possesses many generalized, small everyday theories which can be used to analyse and synthesize the incomplete pieces of mutilated or distorted data. These are the basis for one's meaning experiences. They supply the stuff for an unbounded number of cognitive conflicts. Small efforts produce small effects, as a rule. If the contrary takes place, if small efforts produce great effects, it is therefore striking. Personal relations are governed by rules implied by Aristotle's concept of friendship, as described in the "Ethic for Nikomachos". This concept corresponds to the primary features of the later "Psychology of Interpersonal Relations" by Fritz Heider (1958): Friends have the same outlook; they prefer to live near to each other. If they do not do so, one is surprised. The facts call for an explanation. Perhaps, the description is to be corrected; perhaps, the friend was not a real friend, etc. According to a similar dynamic, victims are essentially punished twice, because – in the interest of community harmony – badly treated persons are seen as having deserved their fate.

An information scheme which shows the very conflict between striving after clearness and an increased chance for misunderstanding is the elucidating exaggeration (hyperbola): "The industry wants employees with the professional skills of fifty year-olds, who have the energy of thirty year-olds for the salary of teenagers." The author expects the receiver to know that he is exaggerating. Such an expression might be helpful, but it can always easily be misunderstood. Voluntarily or involuntarily, the receiver does not arrive at the decoding solution which the speaker had intended.

The fact that one is not bothered about intricate formulations might take root in the human proficiency in this respect. If one takes into account the difficulties of learning a foreign language, then it is obvious that the memory for sign-meaning relations is less developed than the problem-solving ability. One enjoys hide and seek of verbal meaning as one enjoys the resolution of entangled monograms. The reason might be the excitation of pure functional pleasure. This feature was pointed out already and illustrated by examples in other research areas by arousal theorists. One likes mild activation followed by the reward of success. One likes to solve riddles when it is possible to solve them in the majority of cases. Therefore one is also fond of some puzzles in interactions with unusual decoding demands. One is not only able to solve meaning problems; one is also ready to do so. Moreover, receivers who can read between the lines of obscure texts share a link with the speaker. In the long run, one perceives plain speech as a little boring. Some haziness in formulation, ambiguities, and contradictions are challenges for the addressee. This can be shown by looking at the structure of some jokes or funny wordings:

– One is amused, for instance, by proverbs and sayings, which are reanimated by slight changes. On Hamlet's footprints: "To pay or not to pay, that is the question./ Cleverness is all." Or the advertising idea of an airline: "AUA ... The Austrian way of flight".

– Another means to induce a little sophisticated decoding is a provoking mixture of cognitive categories in one single sentence as exemplified by the so-called complicated zeugma: "The lion is magnanimous and blond./ Personal respect and local distance separated them for years." A modern name for that time-honored scheme could be: "Unequal brothers from the same bassinet."

If understanding messages in natural language is to be seen as a problem solving process, then the system of explicatory concepts, from the gestaltists to its exemplification by the GPS of Newell, Shaw and Simon (1960), should be applicable. The process should start at an unsatisfying configuration and arrive at a more satisfying one. For this purpose steering variables, criteria of success, and diagnostic procedures to characterize each step of the meaning solution should be developed. Such a program brings up many questions: What are the communalities of meaning riddles? What kind of mental processes do they provoke, exactly? How are these riddles to be solved? At what point is one able to recognize that a certain piece of information is a solution for a special problem configuration? etc. The selection of proper transformations are to be based on try-outs or a systematic means–ends analysis, shortened and speeded up by expert system technology. There are trivial and rather complicated mappings:

The principles on which extrapolations in the course of a sense seeking process are based might be as simple as an increasing series, realized by grammatical comparatives: "Good, better, Coca Cola/ Cheap, cheaper, Sony". The object at the third position of this advertising argument is of supreme quality.

On other occasions the resolution of the meaning problem might be found by treating metaphors as plain speech, or by interpreting plain speech as it is meant metaphorically or allegorically: "For Germany there is no way out of Europe." Once upon a time such interpretations shed new light on several chapters of the Bible. The traditional figures of metonymy and synecdoche offer further schemes for escaping the impression of chaotic wording.

There are a group of steering variables or criteria which have to be fulfilled, if a certain transformation is to be accepted as a solution of the decoding process. A few are to be pointed out here: safeguarding the comprehensibility of utterances, coordinating the specificity of content, harmonizing the components of the communication scaffold, avoiding triviality, and avoiding offending other persons.

A rather global aspect is the transparency of formulations which can be broken down into a couple of subvariables like grammatical correctness, the simplicity of the vocabulary, clear relations between the subunits of extensive texts and so on.

In the preface of the optimistic "Tractatus Logico–Philosophicus", Wittgenstein (1921) noticed: "Its whole meaning could be summed up somewhat as follows: What can be said at all can be said clearly; and whereof one cannot speak thereof one must be silent." In the last chapter, the author closes his treatise cyclically with the second part of this famous dictum. In his system of "decimal figures as numbers of the separate propositions", the following sentence got the holy symbol 7: "Whereof one cannot speak thereof one must be silent." Perhaps it is principally or ideally true; practically, of course, it is wrong. There are so many concepts which are hard to express simply. Just the important insights or presentiments often begin as unclear concepts. It takes more than a few attempts to describe them precisely. It might be better, therefore, to dare the experiment of an approximate verbalisation and to catch the experiences in Gendlin's (1962) sense at the threshold of consciousness in the very form in which they appear. One of the reasons, however, why Wittgenstein's famous sentence is cited so many times might be its moral impact. It specifies a goal. Not only in philosophical reasoning does one feel obliged to choose words and to formulate sentences in such a way that the addressee is able to understand it. In colloquial speech, one follows this maxim, too. If the addressee is unable to understand his partner, he is – under normal conditions – free to ask: "What do you mean when you say ...?" or "More specifically, please!" or more polite "What should I imagine when you say ...?" The author who failed to express himself clearly then has to explain what he meant in a more detailed manner using different words. This principle functions well. It can even be used to interrupt the logorhetic stream of utterances of anxious or dominant people who do not end their communications of their own accord. A simple "What do you mean by ...?" helps one out of such a predicament. A recapitulation of the last utterance of the other - friendly, in one's own words, and reduced to the essential contents - as if to check whether one understood the speaker correctly, functions equivalently. Immediately afterwards, one has to change the subject of interaction. In everyday life, one is bound to clarity of message, in spite of the fact that it sometimes is difficult to achieve. However, people help one another to do so, even when it is against their own interests. Cross-examinations illustrate that case.

A special steering variable in the process of the meaning creation and meaning selection is the avoidance of triviality.

– If a politician talking on problems of international trade assures his listeners that "America is a continent", he does not intend to instruct his listeners on elementary geography. He can be sure that his addressees themselves know this plain fact. To rescue the message from triviality, one has to activate, however, all the meaning aspects which the word "continent" conveys. Then it makes sense. The politician wants, perhaps, to underline the great extensions of this territory, its relative isolation, the economic self-sufficiency of great land masses. This might be the core of his message, and so it is understood, except, perhaps by a clown, joking about the insights of a leading man.

– The avoidance of triviality is also the basic principle behind the construction of another rather conventional information scheme: If the name of the grammatical subject recurs in a predicative position within the same sentence, it does not indicate a tautology or a truism. Rather, it activates various meaning aspects of the critical expression. That is its purpose. "A man is a man./ Whisky is whisky./ A job is a job./ Vienna remains Vienna." Nobody will say: "So what?" or "What else?" One recognizes that by the first use of the word in question a certain case is identified only as something. When repeated, however, there is opportunity to activate the whole bundle of meaning aspects. A related type of paradoxical formulation runs "Dead always means very dead./ Married means totally married, sooner or later."

The last steering variable which will be mentioned here is the avoidance of offensive expressions; one wants to get around an altercation with the addressee and the responsibility for the interpretation should be left up to the receiver. Therefore, in critical situations one sometimes camouflages one's opinion and makes it more difficult for speaking partners to follow one's ideas. Being involved in controversial topics, one restricts, for instance, the message on an incomplete characterization and leaves the interpretation up to the addressee's judgement: "You may manage an institution well or the way our boss does it./ If you follow me you will succeed, otherwise it's all up to you." The most probable specification, naturally, is to submit the opposite meaning for the open place. However, that is up to the receiver.

An otherwise hurtful confrontation might be moderated by merely mentioning an opposing rule. The last step of inference is left up to the receiver.

A: It's a question of principles!
B: One can overdo discussing principles, too.
A: That's a Gothic work of carpentry!
B: Well, but you know rather simple-minded carpenters lived during that time, too.
A: You are not asked for an offering, but a donation.
B: Nobody can give more than he has.

A well-known strategy for seeking superficial consent follows the principle: "misunderstand me correctly". It goes to the utmost of conflict avoidance: the negotiators agree on the text of a contract, which is interpreted openly by at least one of the parties in a quite different way than by the others.

There are at least three reasons why it might be advantageous to subtilize and widen the repertory of meaning not by fixed interaction values of text-elements, but by problem-solving techniques:

– The application of these techniques affords a greater versatility together with less demand on our memory; the set of procedures can be enlarged without difficulty and without effecting the already available stock.

– The same techniques are useful for linguistic problem-solving as well as for problem solving in other subject areas. One domain can profit from the progress of the other.

– In both fields the course of action is steered – without leaning on any biological drive theory – to fulfil the requirements of the actual constellation (Wertheimer, 1945; Duncker, 1935) in the direction of greater harmony. Such a theory of the emergence of verbal meaning via problem solving would fit into a general theory of human behaviour.

Much work remains to be done before reaching a simulation model of the emergence of meaning in human conversation. Many people are eager to get a human surrogate for doing tedious work, but only few want a machine model to talk to; they prefer, in this case, the original. Even for translations from one language into another, the preliminary work of a simple computer device, which will be corrected by a native speaker afterwards, will suffice for many occasions. Still, for improving communication and for understanding its governing principles, rhetorical studies on linguistic problem solving might be worthwhile. Practically, such an approach resembles – as far as the enjoyment of the research worker is concerned – the collection of specimens of different types of plants and animals during previous centuries to get a reliable basis for scientific systematization. In the end, the explanations of the effects of rhetorical schemes should be formulated in terms of an information processing theory.

References

Augustinus, Aurelius (1631/1912): Confessiones, 400 A.D. Transl. by Watts, W. London: W. Heineman.
Duncker, K. (1935/1974): Zur Psychologie des produktiven Denkens. Berlin: Springer.
Gendlin, E.T. (1962): Experiencing and the Creation of Meaning. New York: Free Press of Glencoe.
Giese, F. (1925): Handbuch psychotechnischer Eignungsprüfungen, 236 f. Halle a.S.: Marhold.
Goethe, W. v.: Faust I.
Heider, F. (1958): The Psychology of interpersonal relations. New York: Wiley.
Hörmann, H. (1976): Meinen und Verstehen. Grundzüge einer psychologischen Semantik. Frankfurt/Main: Suhrkamp.
Köhler, W. (1920): Die physischen Gestalten in Ruhe und im stationären Zustand. Braunschweig: Vieweg.
Lausberg, H. (1960): Handbuch der literarischen Rhetorik. München: Hueber.
Newell, A., Shaw, J.C., & Simon, H.A. (1960): Report on a general problem-solving program for a computer. Information Processing: Proceedings of the International Conference on Information Processing.
Quintilian, M.F. (1975): Institutio oratoriae. Ausbildung des Redners. 2. Teil, Buch VII – XII. Darmstadt: Wissenschaftliche Buchgesellschaft.
Rousseau, J.J. (1762/1911): Émile. Transl. by B. Foxley. London: Dent.
Vukovich, A. (1990): Vorformen mathematischer Begriffe in der Alltagsrede. in: Vukovich, A. (Ed.): Natur - Selbst - Bildung, pp. 181 – 192. Regensburg: Bosse.
Wertheimer, M. (1945): Productive Thinking. New York: Harper & Brothers.
Witte, W. (1966): Das Problem der Bezugssysteme. In: Metzger, W. (Ed.), Handbuch der Psychologie, Bd. 1/1, Wahrnehmung und Bewußtsein, pp. 1003 – 1027. Göttingen: Hogrefe.
Wittgenstein, L. (1921): Tractatus logico philosophicus. In: Ostwald, W. (Hrsg.): Annalen der Naturphilosophie, Bd. 14, Leipzig: Unesma, 185 – 262.
Wittgenstein, L. (1958): Philosophische Untersuchungen. Oxford: Basil Blackville.

Part V

Models of Multistability

A Synergetic Model of Multistability in Perception

T. Ditzinger and H. Haken

Institute for Theoretical Physics and Synergetics, University of Stuttgart, Pfaffenwaldring 57/IV, D-70569 Stuttgart, Germany

Abstract: In this chapter we describe a mathematical model of a synergetic computer that represents a number of results found by psychophysical experiments in which multistable patterns are shown to subjects. Our model is derived from basic principles of synergetics where the concept of order parameters describing complex patterns was introduced. The differential equations describing pattern recognition depend on a set of parameters which may be interpreted as attention parameters. According to a psychological finding of Köhler (1940), we allow that these attention parameters saturate once a specific percept has been realized. By this the observed oscillations between percepts can be modelled. The lengths of the reversion times may be different for different percepts and depend on different parameters of the stimulus patterns. Our model is based on three parameters which can be put in direct relationship to the experimental data and a psychological meaning can be attributed to these parameters. In a generalization of our model we are also able to simulate some properties of the perception of the stroboscopic alternative movement (SAM), which was the topic of several experimental contributions to this conference.

1 Introduction

Original biological rhythms are endogenous in nature. Examples are provided by the EEG of a person at rest or sleeping, or by the periodic changes of sleep and wakefulness. In our contribution we wish to deal with a phenomenon which is partly endogenous and partly exogenous. We shall deal with the oscillations in the perception of multistable patterns by humans. A famous example is provided by the Necker cube (Necker, 1832) shown in Fig. 1, where a part of that figure is perceived either as front or back of a cube and the percept oscillates, i.e. the front is recognized for a while and then suddenly changes to be perceived as the back.

These changes occur involuntarily and after a rather well-defined period of time. These phenomena have intrigued both psychologists and laymen

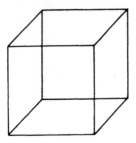

Fig. 1 The Necker cube

for a long time. In our present paper we wish to model a number of these phenomena by means of an approach based on synergetics (Haken, 1983, 1994). More specifically, we shall show how the three parameters on which our model is based can be determined by a comparison with psychophysical experiments.

2 Some properties of ambiguous pictures

Psychophysical experiments show that the perception of ambiguous pictures has a number of well-defined features which are described by Vickers (1972), Kawamoto and Anderson (1985), Kruse (1988), Ditzinger and Haken (1989). For the purpose of our present paper it will be sufficient to mention the following salient features:

1. The occurrence of oscillations
   When one alternative is perceived, it remains stable for a time T_1 and is then replaced by the other alternative which now stays stable for another time T_2. Then the first alternative appears again for time T_1, and so on. The results presented by different authors provide us with different results with respect to the mean reversion time $<T> = <T_1> + <T_2>$. Nevertheless the results are very well reproducible and reliable if they are obtained with the same subject, the same pattern, and the same influences of the surrounding, and if the method of measurement is the same.

2. The two alternatives in the interpretation of an ambivalent figure are, in general, not entirely equivalent.
   This becomes manifest by two different kinds of measurements. The stronger percept has a longer reversion time T_1 or T_2, and it is first perceived by a larger percentage of test persons. There is a functional relationship between these two results which can be derived by our model to be described below. In addition, one finds a strong increase in the total average reversion time of both alternatives if the inequality between the two percepts increases.

3. The reversion times exhibit some fluctuations which obey a specific distribution function which was measured by Borsellino et al. (1972). As is shown in Fig. 2, the curve of the distribution function increases strongly for small reversion times and decreases slowly for higher times, $T = T_1 + T_2$.

3 The model and its properties

Our model is based on the concept of the synergetic computer for pattern recognition (Haken, 1991). Its basic idea is the following: In a first step a number of prototype patterns are stored in the computer. When a test pattern, e.g. a part of one of the prototype patterns, is shown to the computer, the computer forms the so-called order parameters which, roughly speaking, are a measure of the overlap between the test pattern and the prototype patterns. These order parameters ξ_k, where k refers to the specific prototype pattern, are then subjected to a dynamics which pulls the order parameters to the minimum in a certain so-called potential landscape. Once the minimum is reached, the test pattern has been complemented to become one of the original prototype patterns. The dynamics is constructed in such a way that the test pattern approaches that valley in the potential landscape to which the test pattern had been closest in the beginning. If there are k prototype patterns, $k = 1,...M$ the equations of motion read

$$\dot{\xi}_k = -\frac{\partial V}{\partial \xi_k}, \quad k = 1,...M. \qquad (1)$$

The potential function V does not only depend on the order parameters ξ_k, but also on so-called attention parameters λ_k. When we insert the explicit function of the potential V into (1), the equations for the order parameters in the case of two percepts read

$$\dot{\xi}_1 = \xi_1 \left[\lambda_1 - C\xi_1^2 - (B+C)\xi_2^2 + 4B\alpha\xi_2^2 \left(\frac{1 - 2\xi_2^4}{(\xi_1^2 + \xi_2^2)^2} \right) \right] \qquad (2)$$

$$\dot{\xi}_2 = \xi_2 \left[\lambda_2 - C\xi_2^2 - (B+C)\xi_1^2 - 4B\alpha\xi_1^2 \left(\frac{1 - 2\xi_1^4}{(\xi_1^2 + \xi_2^2)^2} \right) \right] \qquad (3)$$

The individual terms have the following meaning:
λ_1 and λ_2 are the attention parameters
B and C are internal parameters which determine the speed at which a discrimination between the patterns occurs.
α is a parameter which takes care of the bias, i.e. if both patterns are perceived with equal strength, $\alpha = 0$. The interpretation of α by means of V has been described elsewhere (Ditzinger & Haken, 1989). Below, we shall elucidate its psychological meaning.

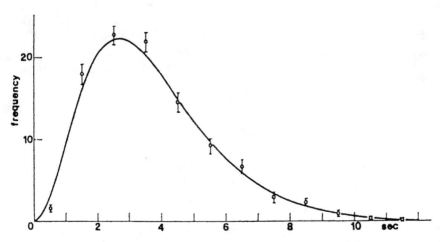

Fig. 2 Distribution function of reversion times. The frequency of the occurrence of a specific reversion time is plotted versus that time. (after Borsellino et al., 1982)

In the following we shall allow that the attention parameters may saturate and that they are, in addition, subject to fluctuating forces (Ditzinger & Haken, 1990). This is an assumption in accordance with previous assumptions in psychophysics, for instance with the suggestions made by Köhler (1940) (though we do not adopt his physical interpretation). In psychology there seems to be general agreement that the observed oscillations result from neuronal fatigue, inhibition or saturation. In accordance with this psychological model we establish an evolution dynamics for the attention parameters, which read

$$\dot{\lambda}_1 = \gamma \left(1 - \lambda_1 - \xi_1^2\right) + F_1(t) \tag{4}$$

and

$$\dot{\lambda}_2 = \gamma \left(1 - \lambda_2 - \xi_2^2\right) + F_2(t). \tag{5}$$

The right-hand-side depends on the strength of the percept, i.e. of the corresponding order parameters ξ_1, ξ_2. The attention to the pattern k will saturate if its corresponding perception strength ξ_k increases. The fluctuating forces F_1 and F_2 are assumed to have the following properties:

$$< F_k(t) > = 0, \qquad k = 1, 2 \tag{6}$$

and

$$< F_i(t) F_j(t') > = Q \delta(t - t') \delta_{ij}, \qquad i, j = 1, 2 \tag{7}$$

where $< ... >$ stands for the statistical average and Q is a parameter for the strength of the fluctuations. In the following we shall consider the parameters

α, γ and Q as parameters that can be chosen to match the experimental data. An analytical treatment of our equations shows that an oscillation between the two alternatives $k = 1$ and $k = 2$ can occur only under the conditions

$$B < 1, \tag{8}$$

$$\gamma < \frac{2B}{1 + B + 2C} \tag{9}$$

and

$$-\alpha_{crit} < \alpha < \alpha_{crit} \tag{10}$$

where the value of α_{crit} is given by

$$\alpha_{crit} = \frac{1 - B}{4B}. \tag{11}$$

If the value of α reaches the critical limits $-\alpha_{crit}$ or α_{crit}, the bias is so extreme that alternative 1 or alternative 2 is recognized with 100 percent, and an oscillation becomes impossible.

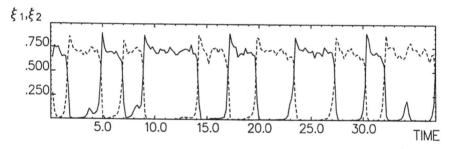

Fig. 3 Time variation of the order parameters ξ_1, ξ_2 with $B = 0.8$, $C = 1.0$, $\gamma = 0.1$, $Q = 0.001$, $\alpha = 0.0$. The solid line refers to one interpretation, the dashed line to another interpretation of the ambiguous picture. The oscillation of the perception is clearly seen. The results are based on our model described in the text

Figure 3 shows the time evolution of the order parameters ξ_1 and ξ_2 where the fluctuating forces have been taken into account. The characteristic changes between the two interpretations can be seen quite clearly. In addition, the fluctuations of ξ_1 and ξ_2 are also well visible. When we plot the frequency of occurrence of a specific duration of the percepts for $N = 10\,000$ oscillations, we obtain the curve of Fig. 4, which, at least qualitatively, corresponds very well to the curves measured by Borsellino et al. (1972). The dependence of the reversion time on the bias parameter α is shown in Fig. 5, where the same quantities as in Fig. 3 are used. In the upper part, the time evolution is shown for $\alpha = 0.02$, in the lower part for $\alpha = 0.04$. As one can see rather clearly, the individual reversion times $<T_1>$ and $<T_2>$ of the alternatives

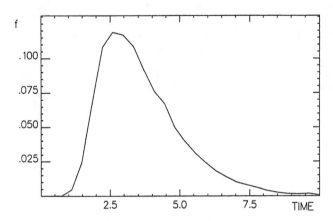

Fig. 4 Theoretical distribution function derived from our model with the same parameter values as in Fig. 3 for $N = 10\,000$ oscillations. The frequency of the occurrence of a specific reversion time is plotted against that time

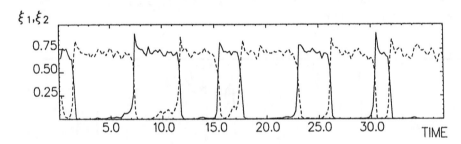

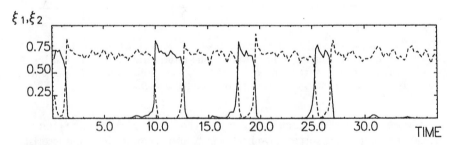

Fig. 5 The time dependence of the order parameters ξ_1, ξ_2 for two different bias parameters α (upper curve: $\alpha = 0.02$, lower curve: $\alpha = 0.04$)

1 and 2 are shifted in favour of the alternative 2 (dashed line), i.e. $<T_2>$ increases in agreement with the psychophysical results. One may also see that with increasing bias the total time $<T> = <T_1> + <T_2>$ increases. This property is also in full agreement with the properties found by psychophysical experiments. Let us now look at the histogram for $\alpha = 0.02$. Fig. 6 shows the distribution function for the reversion times T_1 and T_2 separately. Quite

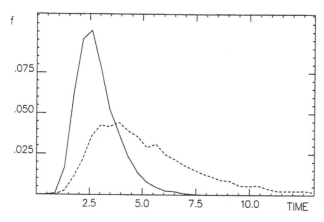

Fig. 6 The distribution functions are plotted individually for the reversion times T_1 (solid curve) and T_2 (dashed curve) with the same parameter values as in the upper part of Fig. 5 and $N = 10\,000$ oscillations

clearly the curves are discernable, having different positions of their maxima. A number of factors which influence the average reversion time $<T>$ are known from psychology. Some examples of them are listed in Table 1. In it we distinguish between the different effects caused by the properties of the stimulus pattern and by the internal or external conditions of the test person. It will be the aim of the following two sections to classify these different measurements by means of the model parameters α, γ and Q. We shall study the influence of the three parameters on the reversion times and present a psychological interpretation.

4 The Borsellino Experiments

First we wish to describe an experiment by Borsellino et al. (1982), who studied the influence of different sizes of the pattern, e.g. of the Necker cube, on the reversion times of the percept for different subjects. In Figure 7 the measured average reversion times are plotted versus the visual angle for different observers. After a long plateau for small angles with comparatively constant values of the average reversion times, a strong increase of the measured time for the so-called slow observers, i.e. for observers with large reversion time, can be registered. For the other (fast) observers the reversion time is little affected by the visual angle.

A key to model this kind of behavior can be found by interpreting another result of Borsellino et al. In their investigation the frequency distribution of the reversion times was measured as a function of different visual angles for the same person. For small visual angles the distribution function is rather narrow, whereas for increasing visual angles a pronounced increase of the

Table 1 The dependence of reversion time on various conditions according to different authors

Conditions	Influence on time	Authors
of the stimulus pattern		
Bias	Increase	Künnapas (1957), Oyama(1960)
Visual angle	Increase	Borsellino et al. (1982), Dugger and Courson (1968)
Complexity	Increase	Gordon (1903), Donahue and Griffitts (1931), Porter (1938), Ammons and Ammon (1963), Wieland and Mefferd (1966,1967)
Incompleteness	Increase	Babich and Standing (1981), Cornwell (1976)
Flicker	Decrease	Heath et al. (1963)
of the test person		
Knowledge, Intension	Control	Pelton and Solley (1968), Girgus et al. (1977)
Instruction	Control	Bruner et al. (1950)
Creative Capacity	Decrease	Bergum and Bergum (1979), Klintman (1984)
Hypnotic Susceptibility	Decrease	Wallace et al. (1976)
Age	Increase	Botwinick et al. (1959), Heath and Orbach (1963)
Exercise	Decrease	Adams (1954)
Heat, Noise	Decrease	Heath et al. (1963)
Reward, Punishment	Control	Johanson (1922)
Diverting Concentration	Increase	Reisberg and O'Shaugnessy (1984)
Length of observation time	Decrease	Brown (1955)
Change in retinal position	Control	Babich et and Standing (1981)
Muscle Tension	Decrease	Malmo (1959), Thetford and Klemme(1967)

width and of the average value can be observed. This leads us to correlate the visual angle Θ with the parameter Q of our model. This result may be interpreted by the idea that with increasing visual angle there will be a greater probability for the influence of the surrounding on the attention parameters.

We still have to interpret the differences of the Borsellino results in Fig. 7 with respect to the individual subjects. To this end we identify the parameter γ with the recognition speed of the test person. This interpretation of the parameter γ can be realized in Fig. 8, where the distribution functions of the simulation for 3 different values of γ are shown. These results are in good agreement to measurements of the distribution functions for different test persons by Borsellino et al. (1972). With decreasing values of γ the mean values and the widths of the distribution functions increase, i.e. the smaller the value of γ, the smaller is the individual reversal rate. Therefore the identification of the parameter γ with the individual recognition speed of the test person becomes clear.

In this way the measurements by Borsellino et al. (1982) shown in Fig. 7 can be fully simulated by our model. In Fig. 9 the average reversion time $<T>$ is plotted against the strength Q for different values of γ. As may be seen, after a rather long plateau with comparatively constant values of the reversion times, a phase of rapid increase is visible especially for small values of γ, which means slow observers. High values of γ are little affected by increasing values of Q, as in the Borsellino experiment. In our simulation we

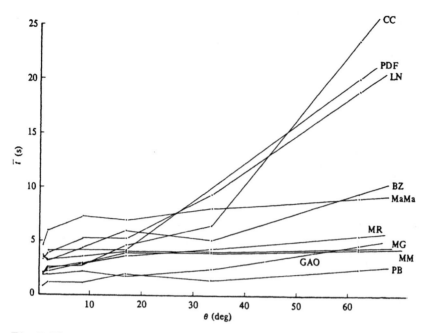

Fig. 7 The mean reversion times $<T>$ are plotted versus the visual angle for several subjects. (according to Borsellino et al. 1982)

used $N = 10\,000$ oscillations in order to get a good average of the reversion times. In comparison, the experimental run lasted with a small number of estimated from 100 to 400 oscillations. Because of this there are no more intersections of the curves for the different subjects as in the measurement, especially for the 'middle' and 'low' regions of the values of Q.

5 The Künnapas and Oyama Experiments

In the next step of our analysis we wish to identify the bias parameter α on a psychological level. We use an experimental study by Künnapas (1957) who measured the average reversion times when ambiguous patterns are looked at where the bias may be changed. Therefore he used a series of Maltese crosses as in Fig. 10. These patterns can be interpreted either as a black cross on a white background or conversely as a white cross on a black background. Künnapas increased the angle of the black segment step by step (from the left to the right in Fig. 10). In this way the weight between the two interpretations is dramatically changed. The smaller area will be preferred, i.e. it will be interpreted as the pattern to be seen in the foreground.

Figure 11 shows the results of the measurements by Künnapas for different types of subjects where the average reversion time of the alternative black

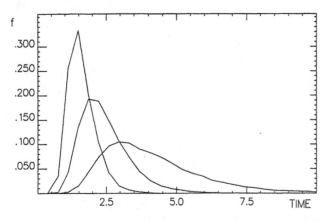

Fig. 8 Theoretical distribution functions for 3 different values of γ: $\gamma_1 = 0.2$ (left curve), $\gamma_2 = 0.14$ (middle), $\gamma_3 = 0.09$ (right curve). The other parameter values are the same as in Fig. 4. The width and the mean values of the distribution functions increase with decreasing γ

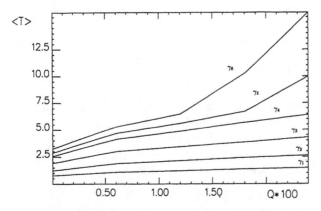

Fig. 9 The average reversion time as calculated as a function of Q with $B = 0.8$, $C = 1.0$, $\alpha = 0.0$, $N = 10000$ and different sizes of γ: $\gamma_1 = 0.3$, $\gamma_2 = 0.2$, $\gamma_3 = 0.14$, $\gamma_4 = 0.11$, $\gamma_5 = 0.1$, $\gamma_6 = 0.09$

cross on white background is plotted versus the angle δ of the black segment from 5 to 85 degrees. The strong increase of the reversion time starting at about 85 degrees is particularly evident. In order to model this property, we identify the angle of the black segments with the bias angle α. The result of this simulation is shown in Fig. 12, where the average reversion time $< T_2 >$ is plotted versus α for different values of γ. In addition to the qualitatively excellent agreement of these results with the values of Künnapas, there is a strong indication for a linear relationship between the angle δ of the black segments and α of the following kind:

Fig. 10 The Maltese crosses (Vickers, 1972) see text

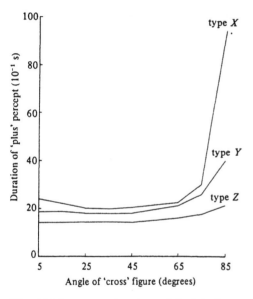

Fig. 11 Mean reversion time of the 'black cross' percept versus angle of cross figures in degrees for different types of observers. (after Künnapas, 1957 and Vickers, 1972)

$$\alpha = \alpha_{crit}\left(\frac{\delta - 45^o}{45^o}\right) \qquad (12)$$

or, equivalently,

$$\delta = 45^o\left(\frac{\alpha}{\alpha_{crit}} + 1\right) \qquad (13)$$

because the increase of $<T_2>$ occurs at a position corresponding to 85 degrees when the bias is gauged in the way of Fig. 12.
A good confirmation of the identification of the bias parameter α with the opening angle of a Maltese cross sector is given by an experiment of Oyama

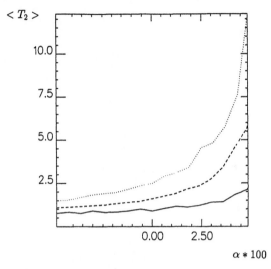

Fig. 12 Theoretical average reversion times versus bias parameter α by means of our model with the following parameters: $B = 0.8$, $C = 1.0$, $Q = 0.0025$, $\gamma = 0.14$ (dotted), $\gamma = 0.2$ (dashed), $\gamma = 0.3$ (solid)

(1960). He used a slightly different stimulus pattern compared to Künnapas with only six sectors, as can be seen in Fig. 13. The two alternating figures are shown at the bottom of Fig. 13, named the α-figure and the β-figure. Oyama (1960) investigated the relative dominance of the α-figure, which is given by

$$R_\alpha = 100 \frac{<T_\alpha>}{<T>}. \tag{14}$$

The results of two different measurements are shown in Fig. 14, where the relative dominance R_α is depicted versus the opening angle δ of the α-sectors. The dominance of the α-figure is highest when the angle δ is smallest and decreases linearly as the angle increases in the following way:

$$R_\alpha = 50(2 - \frac{\delta}{60°}). \tag{15}$$

In Fig. 15 the results of our simulation are shown. The relative dominance $R_1 = 100 \frac{<T_1>}{<T>}$ is depicted versus increasing bias parameter α. As in the Oyama experiment a linear relation holds, which also can be shown analytically:

$$R_1 = 50(1 - \frac{\alpha}{\alpha_{crit}}). \tag{16}$$

In comparison to the experimental relation (15) this is a strong evidence for a linear relation between the bias parameter α and the sector opening angle δ of the Maltese cross with six sectors:

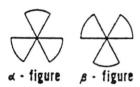

Fig. 13 Maltese cross with six sectors with the two alternatives α-figure and β-figure used by Oyama, (1960)

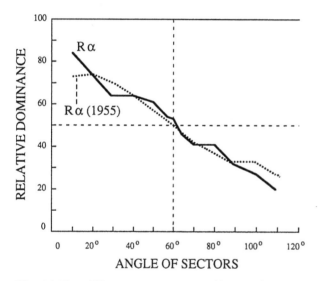

Fig. 14 Two different measurements (Oyama & Torii, 1955; Oyama, 1960) of the relative dominance R_α of the α-figure as a function of increasing opening angle of the α-sector. (after Oyama, 1960)

$$\alpha = \alpha_{crit}\left(\frac{\delta - 60°}{60°}\right). \tag{17}$$

By means of the relative dominance and the relation (16) we have the possibility of a direct measurement of the bias parameter α:

$$\alpha = \alpha_{crit}(1 - \frac{R_1}{50}). \tag{18}$$

For the three parameters α, γ and Q of our model, we have thus found the following correspondence:

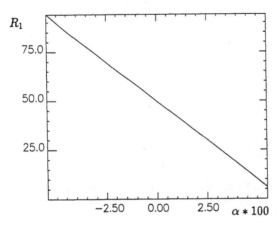

Fig. 15 Theoretical relative dominance R_1 versus bias parameter α with $\gamma = 0.1$, $B = 0.8$, $C = 1.$, $Q = 0.0005$ and $N = 300$ oscillations

- γ is a parameter specific for the individual persons and represents speed of oscillations
- α is a parameter which depends on the object and weighs the alternatives. In the measurements of Künnapas the bias of the alternatives is changed by changing the opening angles of the sectors of a Maltese cross. Another factor of influence which can also be simulated by our model is the orientation of a Maltese cross.
- Q is a parameter depending on the object irrespective of the bias of the alternatives. In the comparison to the measurements of Borsellino we assigned Q the size of a Necker cube.

This allows us to interpret Table 1 in the following way: The psychological factors which directly influence the test persons can be correlated to the parameter γ. The parameters α and Q refer to the rest of the psychological factors presented in Table 1. In addition to the two considered cases with different values of Q (Borsellino experiments) or different values of α (Künnapas and Oyama experiments), numerous possibilities of combinations of these two effects can be thought of, depending on the observed object. For instance, it may be possible to measure the effect of a mixture of Maltese crosses with different opening angles and different visual angles for different persons and to predict the mean reversion times. In addition, such a prediction is possible for the distribution function of the reversion times.

Up to now our model has been concerned only with static multistable patterns. In the following section we introduce a generalization of our model in order to simulate phenomena of the perception of apparent motion patterns.

6 Stroboscopic Alternative Movement (SAM)

By means of our model we simulate two properties of SAM measured by Kruse, Stadler and Wehner (1986). The stimulus pattern are four spots located in an upright rectangle. Two diagonally positioned spots are always flashed simultaneously and then replaced by the remaining two and so on, as was shown by Kruse, Strüber and Stadler (1994) or Kanizsa and Luccio (1994) in this symposium. After sight fixation to a point in the middle, usually two different movement impressions can be seen: either a horizontal or a vertical movement of two spots. Oscillations between these two different alternatives occur as a function of the frequency ω of the flashing of the spots. Kruse et al. (1986) measured the influence of different stimulus frequencies ω on the rate of apparent change (RAC), which is the reciprocal value of the averaged reversion time $<T>$. In Fig. 16 the z-transformed measured rates of apparent change are plotted versus the stimulus frequency ω. A linear relation between the RAC and the stimulus frequency can be seen.

We introduce the phenomenon of SAM into our model in the following way. According to the measurements there is an influence with the frequency ω to the attention of the observer. Therefore we add a periodic force with amplitude A and frequency ω to the attention parameter equations (4), (5):

$$\dot{\lambda}_1 = \gamma\left(1 - \lambda_1 - \xi_1^2\right) + F_1(t) + A\cos(\omega t) \tag{19}$$

and

$$\dot{\lambda}_2 = \gamma\left(1 - \lambda_2 - \xi_2^2\right) + F_2(t) + A\cos(\omega t). \tag{20}$$

Now we are able to compare the properties of our model with the measurement. In Fig. 17 the theoretical RAC is depicted versus ω in arbitrary units for the amplitude $A = 0.05$. The other parameter values are the same as in the preceding simulations. A similar linear relation between the RAC and the parameter ω as in the measurement is apparent.

Kruse et al. (1986) also studied an effect of adaptation in SAM. Therefore their experiment was divided into two parts. In the first part an adaptation frequency $\omega_0 = 2.2$ Hz of the SAM is presented to the test persons. In the second part of the measurement a stimulus pattern with the test frequency ω is presented. The values of the test frequencies are randomly chosen between 1.2 and 3.2 Hz. The results for 4 subjects with 15 experimental runs each are shown in Fig. 18. Again the z-transformed RAC is plotted versus the test frequency ω. As can be seen, there is a clear peak of the RAC in the case of equality between adaptation frequency and test frequency ($\omega_0 = \omega = 2.2$ Hz). With increasing distance of the test frequency to the adaptation frequency the RAC approaches the linear trend shown in Fig. 16.

In order to simulate this experiment we add a further periodic force, corresponding to the adaptation frequency ω_0 to the attention parameter equations (19), (20) with the amplitude $A_0(t)$:

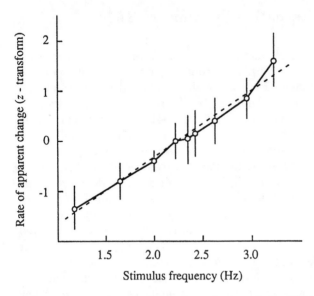

Fig. 16 Measured rate of apparent change of the SAM versus the stimulus frequency. (after Kruse et al., 1986)

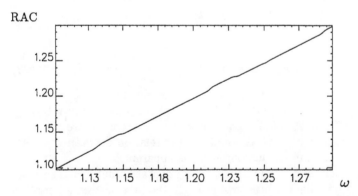

Fig. 17 Theoretical rate of apparent change versus ω in arbitrary units with $A = 0.05$, $\gamma = 0.1$, $\alpha = 0.$, $B = 0.8$, $C = 1.$, $Q = 0.001$ and $N = 1000$ oscillations

$$\dot{\lambda}_k = \gamma\left(1 - \lambda_k - \xi_k^2\right) + F_k(t) + A\cos(\omega t) + A_0(t)\cos(\omega_0 t), \quad k = 1, 2. \quad (21)$$

We assume the amplitude A_0 to be time dependent and decreasing to 0 in the course of time. To simulate the measurement of Kruse et al. (1986) we choose a certain value A_0 of the adaptation frequency somewhat smaller than the value of the amplitude A of the test frequency. The result of the simulation with $A = 0.05$, $A_0 = 0.04$, and an adaptation frequency $\omega_0 = 1.2$ is shown in Fig. 19. The theoretical RAC is plotted versus the parameter ω in arbitrary units. Again a good agreement to the measurement is visible. In the case of

A Synergetic Model of Multistability in Perception   271

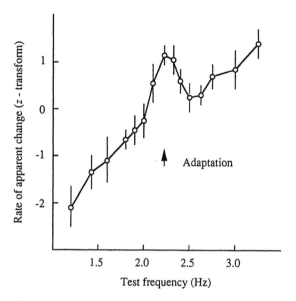

Fig. 18 Measured RAC in the test period after adaptation with 2.2 Hz plotted versus test frequency ω. (after Kruse et al., 1986)

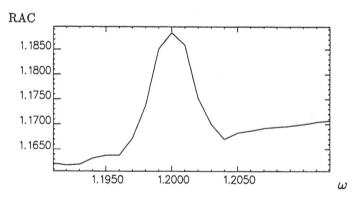

Fig. 19 Theoretical RAC after adaptation with $\omega_0 = 1.2$, $A_0 = 0.04$, $A = 0.05$. The other parameter values are the same as in Fig. 17

$\omega = \omega_0$ again a peak of the RAC can be seen and with increasing distance of ω to ω_0 the linear relation of the RAC to ω holds as in Fig. 18.

There is a multitude of further psychophysical measurements which have been simulated by our model, such as hysteresis in perception, habituation and long scale time dependences, the probability of first occurrence of a certain alternative, latency times, and the general case of multistable patterns with more than two alternatives.

7 Conclusion

We presented a model based on synergetics which is capable of describing a number of known psychophysical phenomena which occur during the perception of ambiguous patterns. By means of three model parameters α, γ and Q we were able to present the classification of the observed psychological factors influencing the average reversion times. In this way we were able to present a psychological interpretation of these parameters and to model psychophysical measurements. We generalized our model for the perception of static ambiguous patterns to the perception of apparent movement patterns. By means of this generalization it became possible to simulate properties of the perception of the stroboscopic alternative movement.

References

Adams, P. (1954): The effect of post experience on the perspective reversal of a tridimensional figure. American Journal of Psychology 67, 708 – 710.

Ammons, C.H. & Ammons, R.B. (1963): Perspective reversal as affected by physical characteristics of Necker cube drawings. Proceedings of the Montana Academy of Sciences 23, 287 – 302.

Babich, S. & Standing, L. (1981): Satiation effects with reversible figures. Perceptual and Motor Skills 52, 203 – 210.

Bergum, B.O. & Bergum, J.E. (1979): Creativity, perceptual stability, and self-perception. Bulletin of Psychonomic Society 14, 61 – 63.

Borsellino, A., De Marco, A., Allazetta, A., Rinesi, S. & Bartolini, B. (1972): Reversal time distribution in the perception of visual ambiguous stimuli. Kybernetik 10, 139 – 144.

Borsellino, A., Carlini, F., Riani, M., Tuccio, M.T., De Marco, A., Penengo, P. & Trabucco, A. (1982): Effects of visual angle on perspective reversal for ambiguous patterns. Perception 11, 263 – 273.

Botwinick, J., Robbin, J.S. & Brinley, J.F. (1959): Reorganization of perceptions with age. Journal of Gerontology 14, 85 – 88.

Brown, K.T. (1955): Rate of apparent change in a dynamic ambiguous figure as a function of observation time. American Journal of Psychology 68, 358 – 371.

Bruner, J., Postman, L. & Mosteller, F. (1950): A note on the measurement of reversals of perspective. Psychometrika 15(1), 63 – 72.

Cornwell, H. (1976): Necker cube reversal: sensory or psychological satiation? Perceptual and Motor Skills 43, 3 – 10.

Ditzinger, T. & Haken, H. (1989): Oscillations in the perception of ambiguous patterns. Biological Cybernetics 61, 279 – 287.

Ditzinger, T. & Haken, H. (1990): The impact of fluctuations on the recognition of ambiguous patterns. Biological Cybernetics 63, 453 – 456.

Donahue, W.T. & Griffitts, C.H. (1931): The influence of complexity on the fluctuations of the illusions of reversible perspective. American Journal of Psychology 43, 613 – 617.

Dugger, J. & Courson, R. (1968): Effect of angle of retinal vision on the rate of fluctuation of the Necker cube. Perceptual and Motor Skills 26, 1239 – 1242.

Girgus, J., Rock, I. & Egatz, R. (1977): The effect of knowledge of reversibility on the reversibility of ambiguous figures. Perception and Psychophysics 22(6), 550 – 556.
Gordon, K. (1903): Meaning in memory and attention. Psychological Review 10, 267 – 283.
Haken, H. (1983): Synergetics - An Introduction. Berlin: Springer.
Haken, H. (1991): Synergetic Computers and Cognition. Berlin: Springer.
Haken, H. (1994): Some basic concepts of synergetics with respect to multistability in perception, phase transitions and formation of meaning. In this volume.
Heath, H.A., Ehrlich, D. & Orbach, J. (1963): Reversibility of the Necker cube: II. Effects of various activating conditions. Perceptual and Motor Skills 17, 539 – 546.
Heath, H.A. & Orbach, J. (1963): Reversibility of the Necker cube: IV. Responses of elderly people. Perceptual and Motor Skills 17, 625 – 626.
Johanson, A.M. (1922): The influence of incentive and punishment upon reaction time. Archives of Psychology 54, 1 – 53.
Kanizsa, G. & Luccio, R. (1994): Multistability as a research tool in experimental phenomenology. In this volume.
Kawamoto, A. & Anderson, J. (1985): A neural network model of multistable perception. Acta Psychologica 59, 35 – 65.
Klintman, H. (1984): Original thinking and ambiguous figure reversal rates. Bulletin of the Psychonomic Society 22(2), 129 – 131.
Köhler, W. (1940): Dynamics in Psychology. New York: Liveright.
Kruse, P. (1988): Stabilität, Instabilität, Multistabilität, Selbstorganisation und Selbstreferentialität in kognitiven Systemen. Delfin 11, 35 – 58.
Kruse, P., Stadler, M. & Wehner, T. (1986): Direction and frequency specific processing in the perception of long-range apparent movement. Vision Research 26, 327 – 335.
Kruse, P., Strüber, D. & Stadler, M. (1994): The significance of perceptual multistability for research on cognitive self-organisation. In this volume.
Künnapas, T. (1957): Experiments on figural dominance. Journal of Experimental Psychology 53, 31 – 39.
Malmo, R.B. (1959): Activation: A neuropsychological dimension. Psychological Review 66, 367 – 386.
Necker, L. (1832): Observations on some remarkable phenomenon which occurs on viewing a figure of a crystal or geometrical solid. The London and Edinburgh Philosophical Magazine and Journal of Science 3, 329 – 337.
Oyama, T. & Torii, S. (1955): Experimental studies of figure-ground reversal: I. The effects of area, voluntary control and prolonged observation in the continuous presentation. Japanese Journal of Psychology, 26 (Japanese text, 178 – 188, English summary, 217 – 218).
Oyama, T. (1960): Figure-ground dominance as a function of sector angle, brightness, hue, and orientation 60(5), 299 – 305.
Pelton, L. & Solley, C. (1968): Acceleration of reversals of a Necker cube. American Journal of Psychology 81, 585 – 589.
Porter, E. (1938): Factors in the fluctuations of fifteen ambiguous phenomena. Psychological Record 2, 231 – 253.
Reisberg, D. & O'Shaughnessy, M. (1984): Diverting subjects' concentration slows figural reversal. Perception 13, 461 – 468.
Thetford, P. & Klemme, M. (1967): Arousal and Necker cube reversals. Psychological Record 17, 201 – 208.
Vickers, D. (1972): A cyclic decision model of perceptual alternation. Perception 1, 31 – 48.

Wallace, B., Knight, T. & Garrett, J. (1976): Hypnotic susceptibility and frequency reports to illusory stimuli. Journal of Abnormal Psychology 85(6), 558 – 563.

Wieland, B.A. & Mefferd, R.B. (1966): Effects of orientation, inclination, and length of diagonal on reversal rate of Necker cube. Perceptual and Motor Skills 23, 823 – 826.

Wieland, B.A. & Mefferd, R.B. (1967): Individual differences in Necker cube reversal rates and perspective dominance. Perceptual and Motor Skills 24, 923 – 930.

Concepts for a Dynamic Theory of Perceptual Organization: An Example from Apparent Movement

G. Schöner[1] and H. Hock[2]

[1] Institute for Neuroinformatics, Ruhr-University of Bochum, D-44801 Bochum, Germany
[2] Department of Psychology, Florida Atlantic University, Boca Raton, FL, USA

Abstract: To address the problem of cooperativity in coherent motion perception we investigate the hypothesis that perceptual organization is governed by dynamic laws that reside at the level of macroscopic perceptual variables. Theoretical concepts are provided to deal both with intrinsic tendencies of perceptual organization and the specificational power of the stimulus. The theory aims at (a) identifying lawful aspects of perception in relation to how percepts persist and how perceptual change comes about, and (b) providing operational language elements with which perceptual theories can be constructed in such a way as to enable direct experimental test of propositions about perceptual organization. A central idea is that temporal stability is an essential and non-redundant property of organized percepts. The validity of the conceptual framework can be evaluated by testing specific predictions for multistable percepts that include the occurrence of hysteresis and its dependence on rate of stimulus change, the loss of stability near points of perceptual change and characteristic switching time distributions for spontaneous reversals. We provide an exemplary model of the perceptual organization of apparent motion both to demonstrate how propositions about perceptual organization can be formed and evaluated and to critically test the theoretical framework through comparison of the theoretical predictions with recent experiments on the dynamic properties of multistable percepts. More generally, we discuss the conceptual consequences of the dynamic theory for the issues of categorization, invariance, and top-down processes in perception.

1 Introduction and overview

Recent psychophysical experimentation has provided evidence for cooperativity in motion perception (Williams, Phillips & Sekuler, 1986, Ramachandran & Anstis, 1987, Nawrot & Sekuler, 1990). These results pose afresh the problem of perceptual organization (Dodwell, 1978; Pomerantz & Kubovy, 1986; Rock, 1986), that is, the problem of how and which relationships are formed among components of a pattern that is sufficiently stable over time to be observable in perceptual experience. The idea is that the perception of coherent motion cannot in certain cases be understood simply in terms of detected elementary stimulus information, but that an understanding of intrinsic organizational mechanisms and their consequences for the stability of the resultant pattern is required to explain the emergence of a coherent percept. Ultimately, these intrinsic mechanisms may represent the prejudices our perceptual system formed in the course of evolution and reflective of the properites of our visual environment. Current theoretical concepts which deal with this problem are limited to rather general references to neurophysiological mechanisms of local inhibitory and excitatory interactions and to correlative computational mechanisms. On the other hand, cooperativity and stability are central concepts in the fields of nonlinear dynamics and the theory of pattern formation (Haken, 1983; Guckenheimer & Holmes, 1983; Collet & Eckmann, 1990). Language elements from the dynamic theory of coordination patterns (Schöner & Kelso, 1988a; Schöner & Kelso, 1988b) are potentially fertile for the theoretical development of new ideas regarding cooperativity in the emergence of coherent percepts for two reasons: (1) In the dynamical language, intrinsic tendencies of cooperation and the organizational constraints specified by the stimulus can be expressed as different contributions to one dynamics; (2) Experimental tools have been developed for measuring the dynamic properties of patterns. These tools render the concepts of a dynamic approach operational and enable both testing the validity of the approach as well as testing concrete modelling assumptions. Thus, if experimental evidence has been obtained that perceptual pattern formation and change does exhibit dynamical behavior, then propositions regarding the nature of intrinsic mechanisms may usefully be formulated in this theoretical language. These propositions could form the basis for a consistent dynamical description from which the observable dynamical behavior of the system can be derived. This should, in turn, lead to direct experimental tests of the propositions. Ultimately the language elements of dynamic theory could be used to arrive at "classical" perceptual theories that explain "what is perceived when" (cf., e.g., Koffka, 1935).

In this chapter we propose a number of concepts that may provide the first steps towards a dynamic theory of perceptual organization. The central concept is temporal stability: Fundamentally, *stability (or lack of it) can be viewed as indicative of how close a pattern is to becoming unrealizable in experience*[1]. Stability can be measured in a variety of ways, for instance, through response variability, discrimination, or through the probability of a perceptual switch in situations where such switches occur spontaneously, e.g., in multistable percepts. Multistable percepts provide an important testbed

---

[1] Note the important distinction between temporal stability and invariance (which in the perception literature is sometimes also called stability): Temporal stability refers to the persistence of a percept in time under constant conditions. Invariance designates the persistence of a percept under a change of a parameter of the stimulus.

for the conceptual framework because mapping ambiguous perceptual situations onto multistable dynamics has a number of measurable consequences including hysteresis and its dependence on the rate of stimulus change, loss of stability near points of perceptual change, and characteristic distributions of spontaneous reversal times. Because any model based on the language elements of the dynamical theory leads to these predictions, their experimental test enables evaluation of the conceptual framework of the theory.

The paper is structured as follows: In the next section we define the theoretical language in general terms. In Section 3 an example involving the perceptual organization of apparent motion is modelled with the help of the dynamic concepts. In Section 4 dynamic effects as predicted from the theory are discussed and compared to direct experimental tests by Hock, Kelso and Schöner (1993). Section 5 concentrates on one particular aspect of multistable perception within the same model system: the statistics of spontaneous reversals which we derive from the stochastic dynamical system and compare to experimental data. In Section 6 we discuss the implications of the theory for the issues of wide versus narrow sense ambiguity, top-down influences on perceptual organization, selective adaptation, and categorization in perception.

2 Some concepts for a dynamic theory of perceptual organization

Treating perceptual organization as a dynamic phenomenon means putting the stress on cooperativity, that is, on the dependence of perceptual organization on the "state" of the perceptual system. This view places processes of perceptual change centrally and aims to account for the properties of percepts[2] with respect to their persistence in time and their spontaneous or induced change in time. To make the idea concrete and fruitful, three tasks must be solved: (a) The percept must be characterized in a manner and on a level relevant to perceptual organization. A key step is therefore to introduce variables, $\mathbf{x}$, (below referred to variably as *collective* or *perceptual variables*) that characterize the relationships that can be formed. (b) Aspects of the stimulus that are specific to the organization of the percept must be parametrized and expressed as part of a dynamics of the perceptual variables. (c) Intrinsic organizational tendencies must be expressed as another part of the perceptual dynamics and aspects of the stimulus that are not specific to the organization of the percept must be parametrized and included in this part of the dynamics[3].

---

[2] We use the words percept, organized percept, and perceptual pattern quite interchangeably in the sense of sets of relationships among relevant components.
[3] The general framework of this approach resembles recent work on the coordination in biological movement which has shown that the coordination activity of the

The choice of perceptual variables may be based on empirical insights or may be guided by results from computational theory. In the worked-through example (Sects. 3 and 4) the perceptual organization of apparent motion (AM) in displays with multiple flashing point lights is characterized by two variables, r and ϕ, for each pair of lights. The first describes the existence (presence or absence) and the second the nature (e.g., apparent motion or synchronism) of a relationship.

Once a level of description and a set of perceptual variables have been found, perceptual dynamics are defined as equations of motion of these variables[4]. Note that these dynamics are not necessarily related to explicit time dependence of the stimulus. In fact, the solutions of the dynamics most interesting to us are attractor solutions of the dynamics[5] onto which we map the observed stable percepts. However, a central tenet of the theory is that a percept is more than a particular pattern described by particular values of the collective variables. The percept is also characterized by its temporal stability, by the nature of the process of eliciting this percept upon providing the stimulus, and possibly by the nature of the process of switching into this percept under constant stimulus conditions from other possible percepts, all of which are properties of the pattern dynamics. In other words, we cannot deal with the eventual percept in isolation, but must consider as well the local environment of a percept, the ways to arrive at a particular perceptual organization, and the manners in which that particular organization can give way to other percepts. The stimulus is assumed to act on the dynamics by setting parameters and defining forces, so that given a perceptual situation and given a stimulus, the dynamics are completely and uniquely determined (unlike the percept!).

We distinguish two types of stimulus-dependence of the perceptual dynamics, based on the degree to which stimulus information is specifying or not to the perceptual organization in question. Since Gibson's crusade for the specificational prowess of the stimulus (e.g., Gibson, 1966) a wealth of experimental and theoretical studies has proven his point that the array of physical stimulation impinging on the retina or the other sensory surfaces severely constrains the set of physical situations that may have given rise to the stimulation array and that these constraints can be detected by the per-

---

central nervous system takes the form of low-dimensional equations of motion of coordination patterns, which are characterized by collective variables (Schöner & Kelso, 1988a). We emphasize, however, that the implied analogy of movement coordination and perceptual organization is a purely formal one. As the theoretical concepts discussed here are elaborated the limits of this analogy become apparent.

[4] Mathematically, these dynamics are modelled as differential equations (including stochastic forces, see below).

[5] Attractor solutions are invariant solutions (unchanging in time) to which the system relaxes from an entire set of initial conditions. These solutions account for the persistence of a percept in time under the influence of sensory input that may include stochastic components and the presence of internal fluctuations. The simplest attractors are asymptotically stable fixed points.

ceptual system (a survey is given, for instance, in Ullman, 1979; Marr, 1982). We deal with those aspects of the stimulus that are specific with respect to the perceptual organization problem in question (that is, as characterized by the chosen perceptual variables) in two steps: (a) Express these aspects of the stimulus as a specified pattern of perceptual organization and characterize it in the same coordinates as used to describe (both existence and nature of) the relationships in the organized percept. (b) The specifying information thus characterized is made part of the dynamics, defining forces that attract the perceptual pattern toward the specified patterns. A formal dynamic equation indicating this would be:

$$\dot{\mathbf{x}}_t = c^{\text{env}} \; \mathbf{f}^{\text{env}}(\mathbf{x}_t) \tag{1}$$

where $\mathbf{f}^{\text{env}}$ is a function defined such as to attract $\mathbf{x}$ to the patterns required by the stimulus, and the parameter c^{env} is to indicate the strength of this attraction (relative to other influences to be discussed now).

A second contribution to the perceptual dynamics expresses intrinsic organizational tendencies, which are assumed to exist independently of the presence of any particular stimulus. We call this second type of dynamics *intrinsic dynamics*, denoted formally by $\mathbf{f}^{\text{intr}}(\mathbf{x})$. The stimulus comes in because the strength of such intrinsic tendencies (relative to each other and relative to the specifying information) may depend on stimulus properties that do not by themselves (that is, without taking reference to an intrinsic tendency) constrain a perceptual pattern. The parameters of the intrinsic dynamics are therefore also stimulus-dependent. Consider, for example, the proximity rule for perceptual grouping: although distance does not by itself specify the formation of a group, the strength of the tendency to group elements depends on their distance (cf., e.g., Ullman, 1979; Burt & Sperling, 1981; Dawson, 1991). Generally, in our language, the task of perceptual theory may be viewed as that of identifying intrinsic dynamics, the solutions of which describe a large class of possible percepts.

Putting it all together, the perceptual pattern evolves under both types of dynamics, expressed formally as the sum of the two contributions[6]:

$$\dot{\mathbf{x}}_t = \mathbf{f}^{\text{intr}} + c^{\text{env}} \; \mathbf{f}^{\text{env}}(\mathbf{x}_t) \tag{2}$$

Two limit cases indicate how these two contributions interact: (a) If the stimulus specifies a pattern that is in agreement with the intrinsic tendencies, then the two contributions overlap and thus cooperate. The resulting pattern is highly stable. (b) If, on the other hand, the pattern as specified by the stimulus is in conflict with one or several intrinsic tendencies, then attracting and repelling forces overlap leading to competition between the contributions to the perceptual dynamics. The resulting attractor solutions correspond to less

---

[6] Note that here additivity does not reduce generality because the individual terms are defined operationally and are not restricted to be independent as mathematical functions.

stable patterns and multistability may arise leading to spontaneous pattern change.

To deal with spontaneous pattern change in perceptual organization, fluctuations must be included in a dynamic description (cf. Schöner & Kelso, 1988b, for a discussion). Fluctuations can be taken into account by introducing stochastic forces that act on the dynamics and represent variables unaccounted for on the chosen level of description. They can be modelled mathematically by additive gaussian white noise if we assume that the noise sources acting on the pattern dynamics are (1) many, independent degrees of freedom (2) which are correlated over short times compared to the typical times of the pattern dynamics and (3) that do not depend in a singular fashion on the collective variables (see, for example, Horsthemke & Lefever, 1984; for a discussion of stochastic modelling). In the formal equation of the perceptual dynamics fluctuations appear as stochastic forces:

$$\dot{\mathbf{x}}_t = \mathbf{f}^{intr} + c^{env} \, \mathbf{f}^{env}(\mathbf{x}_t) + \mathbf{q} \, \xi_t \tag{3}$$

where ξ_t is a vector of independent, gaussian white noise processes with zero mean and unit variance that act with strength $\mathbf{q}$.

In the theoretical language used here, ambiguous stimuli may give rise to multistable dynamics, that is, multiple coexisting attractors[7]. Multistable situations can be used as a tool for elucidating intrinsic dynamics because stability can be measured very efficiently in these situations: (a) The switching rate out of a percept or back into a percept as well as the average residence time in a percept are measures of the stability of the different attractors relative to each other (cf. Schöner & Kelso, 1988b). In recent experiments the probability of switching out of a percept was used in this sense (Hock et al., 1993). (b) The time course of perceptual change induced intentionally or by unspecific cues may likewise reveal relative stability. (c) Susceptibility may be estimated if perceptual switches can be induced by providing context that favours one of the percepts (cf., e.g., Ramachandran & Anstis, 1987): The relative ease of such induced switches (e.g., in terms of switching rates, or in terms of the amount of context required) is indicative of susceptibility. (d) A further measure of stability can be based on the following observation: The more stable a pattern, the less effective is a perturbation of a fixed size. This could be implemented by measuring the probability with which switches are induced by a fixed perturbation procedure (e.g., showing intermittently an unambiguous version of a display) for different stimulus parameters.

Multistable percepts also provide a testing ground to evaluate the validity and delineate the range of applicability of the basic concepts of the dynamic

---

[7] Note that although reversible perception has occasionally been referred to as multistable perception (e.g., Attneave, 1971) it has remained open whether the concept of stability was purely a metaphor or related in a testable way to the concept of temporal stability in dynamical systems.

approach because a number of precise predictions can be made by the theory that do not depend on the details of the mathematical modelling: (1) Whenever a percept changes abruptly as stimulus parameters are changed smoothly, hysteresis is predicted. Hysteresis can be measured when a stimulus series varying these parameters is presented in an ordered fashion. The point of perceptual change in the series depends then on the direction of presentation with a positive overlap of the regimes where the two percepts are possible. Such hysteresis has been observed in experiments on perceptual organization a number of times (for example: Metzger, 1941, Chapter 3; Williams, Phillips & Sekuler, 1986; Hock, Kelso & Schöner, 1993). (2) That hysteresis is caused by an underlying multistable dynamics can be shown by testing for the dependence of the size of the hysteresis loop on the rate of change of parameter. In a dynamic theory the size of the hysteresis increases as the rate of parameter change increases (see Sect. 4 for detailed discussion). (3) At the boundary of the hysteresis, stability is lost. This loss of stability can be observed, for instance, through an increase in the probability of switching out of the pattern that is about to disappear. This can be measured by presenting the stimulus series up to a point and estimating the switching probability while the stimulus is kept constant at that point. In different runs the distance of this point from the boundary of the hysteresis is varied. The prediction is that the switching probability increases as this point approaches the boundary of the hysteresis. (4) The statistics of spontaneous switches among different percepts for a constant stimulus can be predicted. In the dynamic theory spontaneous switches are due to stochastic excursions of a dynamical system across boundaries of basins of attraction. As a consequence, there is no particular typical time scale for residence time in one percept, leading to a roughly Poissonian probability distribution of residence time. This contrasts with other explanations of spontaneous reversals such as satiation or selective adaptation, which lead to a finite preferred residence time. The probability that a least one switch has occurred increases linearly with time initially and then saturates in dynamic theory, while it is of sigmoidal form in the alternative theories (cf. Sect. 5).

3 Dynamic model of perceptual organization of apparent motion: An example

An example from the perceptual organization of apparent motion (AM) is discussed both to illustrate how the theoretical language can be used to build concrete models of perceptual organization and to set the stage for a comparison of the predictions of the theory with recent experiments (Hock et al., 1993). The example is based on Gestalt work dating back, at least, to von Schiller (1933) (see Koffka, 1935, Chapter VII, and Metzger (1975), Chapter XVII, for early review; Hoeth (1966) for extensive review and experimentation; Kruse, Stadler and Wehner (1986), Ramachandran and Anstis (1987); Hock, Kelso and Schöner (1993) for recent experimentation). The basic perceptual situation is illustrated in Figure 1. A number of different lights are arranged in a fixed spatial geometry. The lights are flashed with on/off times in a range adequate for the possibility of AM among a number of lights. The conditions in different experiments concern different relative timings among the various lights. Only one complete on/off cycle of all lights may be presented, or a periodic series of on/off cycles.

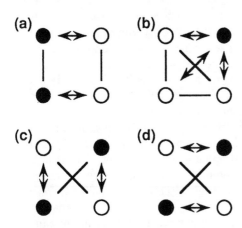

Fig. 1 Different relative timing conditions in displays of four lights at the corners of a square. Lights that are on simultaneously have the same filling (either black or white), while lights which are on alternatingly have different fillings. Perceived apparent motion (AM) is indicated by arrows, perceived simultaneity is indicated by a solid line. No line is drawn in the absence of a particular perceived relationship

In Fig. 1 basic forms of perceptual organization of AM are illustrated for three different relative timing conditions in the same rectangular arrangement (motion quartet). (a) When the left and right pairs alternate, a horizontally moving "stick" (vertical pair of lights that are perceived as moving synchronously) is seen. Note that diagonal AM, which is possible if only two diagonal lights are alternated, is not perceived in the motion quartet. (b) When one point alternates with the three others one possible percept involves "splitting" motion: one point splits into three points. When the points of light forming the two diagonals are alternated ((c) and (d)) the possible percepts include: (c) vertical movement of two lights to the left and right which are perceived to move synchronously and (d) horizontal movement of two lights at the top and bottom which again are perceived as moving synchronously. This ambiguous display, sometimes called *stroboscopic alternative motion*, had originally been studied by von Schiller (1933) and many authors since. In summary, perceptual organization of AM takes place in the sense that various relationships among different flashing lights may be formed, but it is also possible that no particular relationship emerges. If a relationship is perceived it takes one of two forms: Either AM is seen between two lights, or the two lights are perceived to move synchronously or near synchronously, so that a spatial connection between the lights is suggested, for example, as part of a structure that moves as a whole.

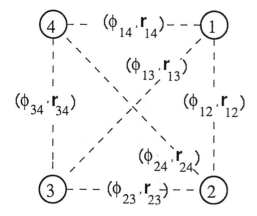

Fig. 2 To index the perceptual variables we number the four points from 1 to 4 in clockwise order. Each variable carries a double index for the two points the relationship of which it expresses. The resulting phase space has 12 dimensions

The variables. To capture the existence and nature of perceived relationships we assume that only pairwise relationships matter. We describe the two aspects – existence and nature of a relationship – by separate variables for each pair (i,j) of light points. The existence is coded into an amplitude type variable, $x_{ij} \in [-\infty, \infty]$ with $r_{ij} = |x_{ij}| \approx 0$ indicating no relationship exists and $r_{ij} = |x_{ij}| \approx 1$ indicating the existence of a relationship.[8] The nature of the relationship is expressed in terms of the *perceived* relative timing between two lights. Such relative timing can be mathematically described by relative phases, $\phi_{ij} \in [-\pi, \pi]$ (mod 2π). The following limit cases (applying for the case that a relationship is perceived: $r_{ij} \approx 1$) illustrate the meaning of this perceived relative timing variable: A relative phase value close to zero expresses the relationship of synchrony: the two points are part of a structure. Relative phase values close to π (anti-phase) express (symmetric) alternation: AM takes place between the two points. Intermediate phase values indicate non-symmetric AM, e.g., at $\phi_{ij} = \pi/2$, sequential motion is seen from i to j. Because six pairs, (i,j), can be formed out of four points we have a phase space of twelve dimensions spanned by six amplitudes, x_{ij} and six relative phases ϕ_{ij}. Figure 2 illustrates how these variables are indexed. Note, that we have tacitly assumed a particular symmetry of relationships within pairs: $\phi_{ij} = -\phi_{ji}$, $r_{ij} = r_{ji}$.

---

[8] The reason we introduce a real valued variable of which only the absolute value carries meaning is of a technical nature: On the one hand it is convenient to assess a strength of a relationship in terms of a positive quantity because there is no cancellation of strengths. On the other hand it is less convenient to formulate dynamics in terms of such positive quantities because these involve boundary conditions. This will become clearer below.

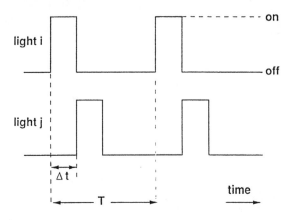

Fig. 3 The relative timing measure between two lights for periodic stimuli is based on the cycle time, T, and the latency, Δt, between the lights and is cast into a relative phase, $\psi = 2\pi \Delta t/T$, by normalization

Contribution to the pattern dynamics expressing the specifying aspect of the stimulus. The relative timing, Δt_{ij}, between two lights (i,j), is specific to the perceived relative timing, ϕ_{ij} and can be cast into the same type of variable by normalizing to cycle time, T (see Figure 3): $\psi_{ij} = 2\pi \Delta t_{ij}/T$ is the *specified relative phase* and hence specifying information. The amplitude dimension of specifying information is based on the following assumption: If required and perceived relative timing match closely ($\phi_{ij} \approx \psi_{ij}$) then the stimulus specifies the existence of a relationship ($r_{ij} \approx 1$), else the stimulus specifies that no relationship exists ($r_{ij} \approx 0$). Note that given three relative phases, e.g., $\psi_{12}, \psi_{13}, \psi_{14}$, all six relative phases of the stimulus are specified, e.g., $\psi_{23} = \psi_{13} - \psi_{12}$, $\psi_{24} = \psi_{14} - \psi_{12}$, and $\psi_{34} = \psi_{13} - \psi_{14}$, due to the physical definition of relative timing. The stimulus therefore specifies a subspace in the ϕ-directions. By implication we allow that unphysical relative timings may be perceived, e.g, $\phi_{23} \neq \phi_{13} - \phi_{12}$. This yields the extra degrees of freedom that are necessary to distinguish, for instance, the ambiguous percepts (c) and (d) in Figure 1.

Casting specificational information into dynamics, we define forces that erect attractors at the required relative phases and the specified value for the amplitude variable. These contributions to the vector field are illustrated in Fig.4 (a,b), where the rates of change, $\dot\phi_{ij}$ and $\dot r_{ij}$, are plotted as a function of the respective variables. At fixed points the rate of change vanishes so that in these plots we may identify fixed points as zeros. Intuitively, we may read off the stability of such fixed points from the slope of the rate of change at the zero. A positive slope indicates an unstable fixed point: A small deviation to the right of the fixed point, for instance, leads to a positive growth rate of the deviation taking the system further away from the fixed point. Negative slope identifies a stable fixed point. In fact, the inverse slope at the fixed point is

Dynamic Theory of Perceptual Organization 285

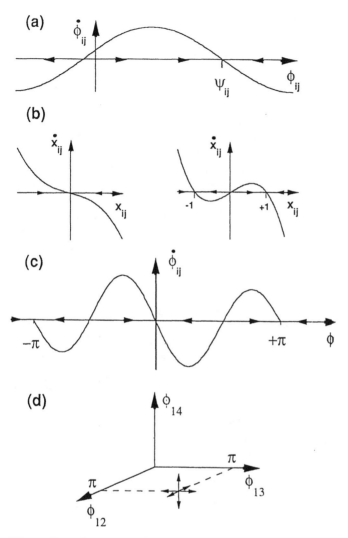

Fig. 4 Contributions to the vector-field of the perceptual dynamics illustrated by plotting individual force components: (a) Specifying information defines an attractor at $\phi_{ij} = \psi_{ij}$. (b) Specifying information to the amplitude variables is shown in two cases. Left: An attractor at $x_{ij} = 0$ is defined if $\phi_{ij} \approx \psi_{ij}$. Right: Attractors at $x_{ij} = \pm 1$ are defined if $\phi_{ij} \not\approx \psi_{ij}$. The two cases are parametrically related through the match function acting as a bifurcation parameter in a pitchfork bifurcation. (c) Two contributions to the intrinsic dynamics defining attractors at $\phi_{ij} = 0$ and $\phi_{ij} = \pm\pi$. Panel (d) shows a direct projection of the vector-field rather than a force plot. The third intrinsic tendency is modelled through repellors, e.g., as illustrated at $(\phi_{12} = \pi, \phi_{13} = \pi, \phi_{14} = 0)$

a direct measure of stability, the relaxation rate[9]. The functional forms for the dynamics are listed in the Appendix.

Intrinsic dynamics. Intrinsic tendencies of perceptual organization account for how the organized percept is formed given the constraints imposed by the specificational aspects of the stimulus. For instance, the stimulus underlying percept (a) in Figure 1, has both diagonal and horizontal pairs alternating, so that anti-phase relative timing is specified: $\psi_{13} = \pi = \psi_{14}$. AM is only seen horizontally, however, and no relationship is perceived among diagonal lights.

We postulate three intrinsic tendencies: (a) *tendency to see synchrony*, (b) *tendency to see symmetric AM*, and (c) *tendency not to see splitting motion of elements when they are synchronous with other elements ("parts of a structure")*. The first two tendencies are obvious: When two elements are related by synchronism they are perceived as moving as part of a structure (not necessarily rigid). In percept (a) of Fig.1, for instance, a relationship of synchronism may be perceived between the top and bottom lights, which are then moving jointly left right. This percept persists if slight asynchronisms are introduced into the display (percept of a "wobbly stick" moving left-right). The relationship of symmetric AM is perceived as AM between the two lights such that the to and the fro component are equal in velocity. This is regularly the case in periodic displays, even when the relative timing of two elements is not exactly anti-phase. These two tendencies can be expressed as forces acting on each relative timing variable separately. In Fig. 4 (c) such forces are illustrated. In isolation from all other influences, the forces erect attractors at $\phi_{ij} = 0$ and $\phi_{ij} = \pi$.

The third intrinsic tendency introduces interaction among several perceived relative timings. Consider again Fig. 1 (a): The second tendency allows for perceiving symmetric AM movement both horizontally and diagonally, leading to each light performing splitting motion. The first tendency leads to perceiving top and bottom lights simultaneously to form a structure. The third tendency account for the suppression of splitting motion in this case. Note that splitting motion is not generally suppressed, but only for elements that are also in relationships of synchronism. Thus, for instance, the splitting of a single element as in Fig. 1 (b) is still possible[10]. The third

---

[9] This simplified discussion is valid in one dimension only. In more dimensions, for example, the twelve dimensions of the present dynamics, a fuller treatment of stability theory including diagonalization is necessary.

[10]This assumption differs from those of other models of perceptual organization of AM (Ullman, 1979; Yuille & Grzywacz, 1988; Dawson, 1991), all of which include some form of splitting suppression. Also, our treatment of synchronism is different from all of these approaches, while the distance dependence of the second intrinsic tendency is shared by all approaches including ours. Dawson (1991) postulates another intrinsic tendency to organize AM so that different velocity vectors align maximally. Because our formulation is sufficient for the restricted cases considered, we need not take this tendency into account. We

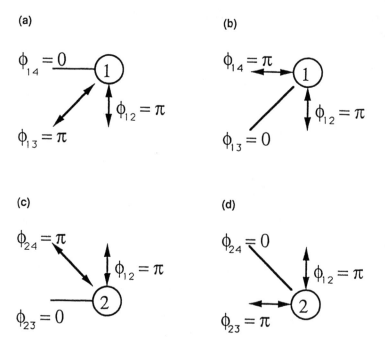

Fig. 5 Four configurations (a to d) that lead to a repulsive force in the equation of motion of ϕ_{12}

intrinsic tendency can be expressed mathematically as a destabilizing force erecting a repellor at those combinations of relative phases that correspond to splitting motion of elements of structures. It is sufficient to consider three relative phases at a time, with one expressing synchronism and two expressing motion, all with respect to one vertex element. Consider, for example, the interactions of ϕ_{12}. Fig. 5 lists all repelling configurations relevant to ϕ_{12}. In Fig. 4 (d) we illustrate how a repellor[11] is erected by sketching the vector-field in a relevant three-dimensional subspace of the twelve-dimensional dynamic phase space. The mathematical form of the intrinsic dynamics is discussed in the Appendix.

To account for variability in perceptual performance, spontaneous switches among percepts, and escape from unstable percepts, random perturbations of the perceptual dynamics must be explicitly dealt with (cf. Section 2). Mathematically, we turn the perceptual dynamics into a stochastic dynamic

---

clarify, however, that it has not been our goal here to provide the most general set of such intrinsic tendencies (limiting our contribution to "classical" theories of perception). Instead, the model is meant as an exemplary implementation of the dynamic concepts.

[11]Strictly speaking, a 9-dimensional hyperplane of repellors is erected, the other variables all being marginally stable. In the presence of the other forces, this hyperplane collapses into a single repellor.

system by introducing stochastic forces (see Appendix for details). We note that while spontaneous perceptual changes can be understood as stochastic switching among multiple attractors in such models (cf. Schöner & Kelso, 1988b), care must be taken to verify experimentally whether switches are in fact governed by stochastic rules or alternatively, may arise on a deterministic basis, for instance, from selective adaptation. This distinction can be made based on concepts like return maps of switching times (see Section 5).

The strengths of the intrinsic tendencies depend on unspecific aspects of the stimulus, in particular, its spatial and absolute temporal structure. Dependence on distance between elements and on absolute timing has been well-studied for AM (Korte, 1915; Neuhaus, 1930; for recent review see Anstis, 1986). Burt and Sperling (1981) based an estimate of "stimulus strength" on spatial and temporal distances at which perceptual change comes about in an ambiguous (multistep) AM display. While the concrete numbers given by Burt and Sperling cannot be directly used for the two-step displays discussed here, their methodology could be used to determine operationally the parameter values of the intrinsic dynamics. For the purpose of the exemplary simulations in this article we have used rough estimates of adequate model parameters as a function of distance based on the experiments by Hock and colleagues (1993). The general idea is, of course, that over larger spatial and absolute temporal distances the intrinsic tendency to see symmetric AM becomes weaker. There is less information on the first intrinsic tendency, which we have treated analogously. We did not assume any distance dependence for the third intrinsic tendency.

Putting it all together, we check the consistency of our modelling assumption by demonstrating that the percepts of Fig. 1 are in fact stable solutions. Analytic work, however, becomes quickly intractable, so that we take recourse to numerical demonstrations. In Fig. 6 the results of a simulation illustrate that all of the four basic percepts depicted in Fig. 1 are stable solutions of the complete dynamics[12]. The graph shows the temporal evolution of the six relative timing variables, $\phi_{12}, \ldots, \phi_{34}$ (left column) and the six amplitudes, $r_{12}, \ldots, r_{34}$, expressing the existence (if near 1) or absence (if near 0) of a relationship. The parameters of the intrinsic dynamics depending on geometry were kept constant, while the temporal order of the stimulus light points was changed as indicated by the diagrams at the top of the graph. These pictograms also illustrate the resultant solutions during the various intervals (same conventions used as in Fig. 1). Because understanding this graph facilitates comprehension of further results in Section 4 we discuss it in some detail. The angular character of the relative phase values is taken into account by identifying π and $-\pi$. When a relative phase fluctuates around π, the resultant trajectory is split in our graphical representation: Small random

---

[12]The parameters A_{ij} and B_{ij} must be chosen such that $B_{ij} + 2A_{ij} < 0$ and $B_{ij} - 2A_{ij} < 0$; The diagonal coupling must be smaller than the horizontal and vertical couplings.

Dynamic Theory of Perceptual Organization   289

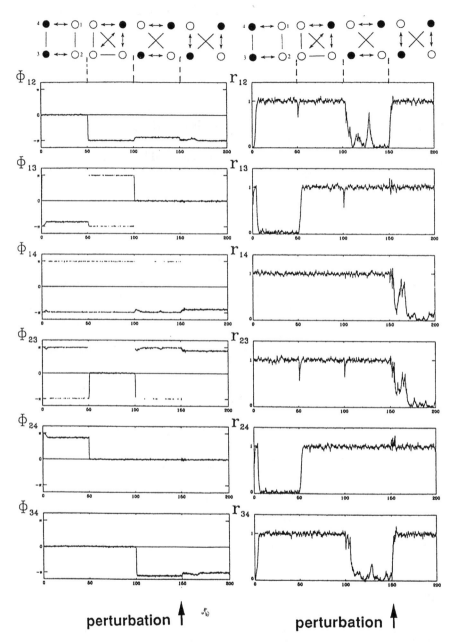

Fig. 6 A numerical simulation of the dynamic model (see Appendix) in which the three basic stimuli of Figure 1 are presented sequentially as indicated on top. For each stimulus the corresponding percept is quickly stabilized. The ambiguous percept (for third stimulus) is perturbed at $t = 150$ by increasing the relative strength, c_{ij}^{env} of the specifying information for the vertical relationships ($(1,2)$ and $(3,4)$) for 5 time units. This leads to a switch to the other possible percept, which persists after the perturbation is switched off thus demonstrating the bistability of the dynamics in that regime. The parameters of the intrinsic dynamics were kept fixed during the entire simulation at $A_{12} = A_{14} = A_{23} = A_{34} = 1$, $A_{24} = A_{13} = 0.1$, $B_{ij} = -2A_{ij}$, $c_{ij} = 3$, representing a square arrangement of the lights

excursions below π are plotted in the upper segment, small random excursions above π are mapped by a modulo-2π operation onto values slightly above $-\pi$ and are plotted in the lower segment. Note that after a change of stimulus (at the times designated by vertical dashed lines) all variables rapidly relax to the new stable state. The stability of the percepts is evidenced by the persistence of solutions in the presence of small fluctuations. In the last two intervals, the stimulus giving rise to the bistable percepts (Fig. 1, (c) and (d)) is applied. We demonstrate the bistability of the perceptual dynamics in this case by applying an external perturbation (external force driving the system away from its stable state) where indicated. This induces a switch to the other possible percept (cf. also Section 4).

(a)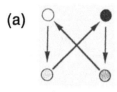

(b)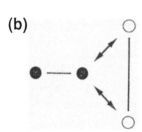

Fig. 7 Two displays that test the intrinsic dynamics postulated in the model. In (a), left and right lights are alternating, but top and bottom are phase delayed by $\pi/2$. The dynamics produces the sequential apparent motion pattern as indicated by the one-headed arrows. This stimulus was used by Finlay & Caelli (1979) who reported the percept shown here. The stimulus in (b) severely test the interaction rule: can the middle point perform splitting motion to the two vertically spaced points if it is synchronous with the leftmost point? In informal observations we found that splitting is never observed when the leftmost point is visible, but is easily obtained if that point is dropped from the display

Note that these consistency tests involve all three intrinsic tendencies, including the interaction rule, which is, for instance, responsible for the suppression of the diagonal relationships in percept (a) of Figure 1. A number of other tests are possible to assess the performance of the model in the sense of a classical theory, that is, in terms of the ability to correctly predict organized percepts for stimulus timings and geometries not explicitly used in defining the model. We list a few such tests: (a) By introducing in display (a) of Fig. 1 a slight delay between the upper and lower lights on both left and right ($\psi_{12} = -\psi_{34} = \Delta\psi$ small) we may test for the persistence of the perception of synchronism. In the model a relationship continues to exist with perceived relative timing detuned from synchronism only slightly. In informal observations we found that a spatial relationship was still seen ("wobbly stick" moving horizontally) in this kind of display. (b) A similar detuning of the alternation pattern of left and right lights in the same display, for example, as $\psi_{14} = \psi_{23} = \pi + \Delta\psi$ ($\Delta\psi$ small) allows to test the tendency to see symmetric apparent motion. The model predicts persistence of AM, although slightly detuned from symmetry. Experimentally, this is a common observation. In fact, the appearance of symmetric to and from movement may persist over a range of detunings. (c) A more striking effect is produced if the relative timing is increasingly detuned until a phase delay of $\pi/2$ exists between upper and lower elements. In this case the model predicts consecutive apparent motion (corresponding to relative phases of $\pi/2$) in a sequential pattern illustrated in Fig. 7 (a) in the form of a figure 8. This percept was reported by Finlay & Caelli (1979) and also observed by us informally. (d) A direct test of the interaction rule is possible by working with displays that clearly offer the possibility for splitting movement from parts of a structure. An example is shown in Figure 7 (b). In informal observations we never found splitting when the synchronism between the two horizontal points and between the two vertical points was maintained. In fact, one may observe the breakdown of splitting if the leftmost point is first hidden and then uncovered.

4 Dynamic effects in the perceptual organization of apparent motion

To study the signatures of underlying perceptual dynamics we consider the ambiguous stimulus Figure 1 (c) and (d).

Bistability. In the theory the two percepts correspond to two attractors that are stable at the same time (bistability). Therefore, both solutions may be observed depending on initial conditions. This was demonstrated in the simulation of Figure 6 where one particular pattern emerged initially. We then induced the other pattern by increasing the relative strength of the specifying information for the intended relationships during a short time interval, after which the parameters returned to their previous values. This led a perceptual switch to the other possible percept. Switches among the two percepts may also occur spontaneously due to perturbations of the patterns by random influences. The phenomenon of spontaneous reversal of ambiguous percepts has been studied experimentally since the early days of Gestalt theory and we shall address the statistics of such switches in detail in Section 5. Experimentally, Hock, Kelso and Schöner (1993) established bistability in a series of experiments where the spatial layout of four point lights was varied randomly from trial to trial. The manipulated parameter was the aspect ratio, the ratio of height and width of the rectangle. For aspect ratios near one, i.e., a square configuration, both percepts were perceived in different trials.

Hysteresis. By varying the aspect ratio the ambiguous percept can be biased towards either of the possible percepts (cf. Metzger, 1941, Chapter 3). For a flat and wide rectangle, vertical motion is perceived, while a tall and thin rectangle leads to the perception of horizontal AM only. Hysteresis can therefore be established by preparing a series of stimuli in which the aspect ratio of the rectangle is varied from one extreme to the other. At either end of the stimulus series the system is monostable. Therefore, as the series is presented sequentially in time the initial percept persists in the bistable regime. Switching to the other percept occurs only as the initial pattern loses stability at the far end of the regime of stability. If the stimulus series is presented in the reverse order the opposite happens. As a result, the points in the series where a perceptual switch occurs depend on the direction in which the series is presented. In particular, a positive overlap of the series exists in which the percept depends on the direction of presentation. The amount of overlap is referred to as the size of the hysteresis.

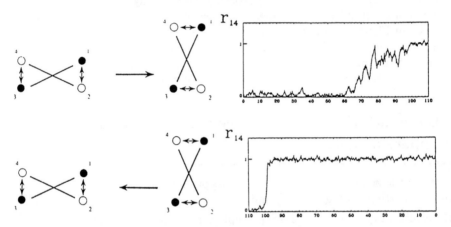

Fig. 8 Hysteresis in the model: For the ambigous stimulus a stimulus series is sequentially presented in which the aspect ratio of the motion quartet changes from flat to tall and reversely as indicated by the pictograms to the left. This is achieved by changing model parameters A_{ij} and B_{ij} every 10 time units such that the model aspect ratio B_{12}/B_{14} varies from 1.22 to 0.82 while the ratios A_{ij}/B_{ij} and B_{13}/B_{12} are kept constant. To the right the temporal evolution of r_{14} is shown to illustrate the change of perceptual organization. Top: Stimulus change from flat to tall, time increasing to the right. Bottom: Stimulus change from tall to flat, time increasing to the left. The switching point among the two percepts depends on the direction of change with an overlapping region in which the percept persists. The amount of overlap is the size of the hysteresis

In Figure 8 the hysteresis effect is illustrated in the theoretical model. The stimulus series are defined by varying the coupling constants to represent the variation in aspect ratio. Hock, Kelso and Schöner (1993) observed hysteresis experimentally using the type of stimuli described. When performing such an experiment an additional problem must be overcome: Because stimuli are not presented in random order, response bias may lead to the persistence of a reported percept along the series. To eliminate response bias, Hock et al. (1993) developed a paradigm in which only one response is made during an entire run-through. These run-throughs are varied in depth, however, so that a determination of the switching point is possible.

Dependence of hysteresis on rate of stimulus change. The dynamic origin of a hysteresis phenomenon can be proven through a clear-cut test. In a dynamic theory the size of the hysteresis depends on the rate at which the stimulus series is presented: The faster we run through the series, the larger the size of the hysteresis. Intuitively, this may be understood as follows: If we go through the stimulus series slowly the system may spontaneously switch out of its initial state due to fluctuations as soon as we enter the bistable regime. If the series is presented at a faster rate the system is less likely to perform such spontaneous switches in the short time interval during which the system is bistable. Ultimately, for sufficiently fast change of stimulus we shall see the full range of the bistability as hysteresis without any spontaneous switches, so that the size of the hysteresis is maximal.

In Figure 9 we demonstrate this phenomenon in simulations of the dynamic model. We use the same stimulus series as in Figure 8 but at a lower rate of change. Note that switches now occur earlier in the bistable regime leading to a smaller size of the hysteresis.

A more thorough understanding of this aspect of dynamic theories relies on the notion of time scales relations (cf., for example, Schöner, Haken & Kelso, 1986; Schöner & Kelso, 1988b; for treatment in another context compare Lugiato et al., 1989). The basic problem is that in a stochastic dynamical system that has multiple attractors, spontaneous transitions among the different basins of attraction occur if we observe the system on a sufficiently long time scale, τ_{obs}. In fact, for $\tau_{obs} \gg \tau_{equ}$ the system is governed by a stationary probability distribution of the dynamic variables. In a multistable system this is a multi-modal distribution with peaks at the different underlying attractors.[13] In another limit, the system has relaxed to a particular attractor, but does not switch randomly among multiple attractors (which is what we have tacitly assumed up to now). This limit applies if we observe the system on a time scale τ_{obs} with $\tau_{rel} \ll \tau_{obs} \ll \tau_{equ}$, where τ_{rel} is the (local) relaxation time to the attractor in question.[14] The size of the hysteresis depends on a further time scale, τ_{par}, the typical time in which we change the control parameter of the system (here: the stimulus).[15] If we change parameters sufficiently fast: $\tau_{rel} \ll \tau_{par} \ll \tau_{equ}$, the system stays in its local attractor and we can observe the full hysteresis. If we change the parameters more slowly: $\tau_{rel} \sim \tau_{equ}$, spontaneous switches occur as we go through

---

[13]Formally, τ_{equ} is defined as the time it takes for a typical initial distribution to relax to the stationary distribution.

[14]The formal definition of τ_{rel} is the time it takes for the system to relax to the attractor from a nearby initial condition. We always have $\tau_{rel} \ll \tau_{equ}$.

[15]The formal definition of τ_{par} is somewhat difficult technically because it requires expressing the parameter change in units of τ_{rel}. Here we are interested in relative statements comparing different rates of parameter change, so that we may gloss over these technical details.

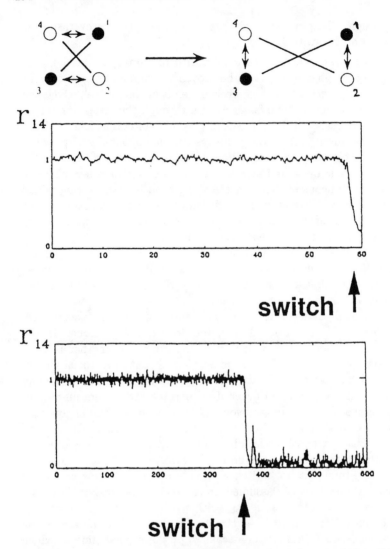

Fig. 9 Rate-dependence of hysteresis in the model: On top the left half (from square to flat rectangle) of the hysteresis simulation of Fig. 8 is shown again for comparison, time now increasing to the right. In the bottom panel the same parameter series is run through, but now the stimulus is changed only every 100 time units. Clearly, the switch to the other percept occurs much earlier due to fluctuation-induced spontaneous switches

the bistable regime and the size of the hysteresis is reduced. Theoretically, in the limit $\tau_{par} \gg \tau_{equ}$, hysteresis may disappear because the system is with overwhelming probability in the most stable state at all times. Note that in experiments the time scale of parameter change, τ_{par}, may be changed, say increased, either by presenting the same stimulus series at a higher rate, or, while keeping the rate constant, by making larger steps on the parameter scale at each instance of change. This fact can be used to exclude a constant reaction time as the cause of an effect of τ_{par}.

In their experiments, Hock and colleagues (1993) chose the first procedure. The predicted increase of the size of the hysteresis as the stimulus series was run through faster was indeed observed.

Loss of temporal stability at the boundaries of the hysteresis. In the theory, at each boundary of the bistable regime one of the percepts loses stability. This implies that the corresponding attractor loses its capacity to attract, so that relaxation to the attractor becomes slow as we approach the boundary of the bistable regime. Near the instability, small fluctuations can drive the system away from this solution and the system may relax to the other stable solution. Instabilities of this type are also referred to as bifurcations in the theory of dynamical systems or phase transitions in physical theories.

In the hysteresis paradigm, loss of stability can be observed. Due to the persistence of the initial percept it is possible to reach the less stable states near the boundary of the hysteresis. The prediction of dynamic theory is therefore that the stability measures (cf. Section 2) indicate loss of stability as the (far) boundary of the hysteresis is approached. The probability of spontaneous switching is such a (relative) stability measure. The simulations in Fig. 11 and 12 illustrate this effect: Spontaneous switching out of the less stable state near the boundary of the hysteresis occurs on average earlier than switching in the symmetric case. Hock, Kelso and Schöner (1993) tested this prediction experimentally by presenting stimulus sequences that led subjects through the hysteresis, but stopped changing the aspect ratio at different points in the hysteresis regime. The percentage of spontaneous switches to the other percept during the final "idle" condition was used to estimate the spontaneous switching probability. The result was a clear enhancement of this switching probability at constant stimulus parameter in those conditions approaching the boundary of the hysteresis. Therefore, Hock, Kelso and Schöner (1993) have established that temporal stability is lost during abrupt perceptual change induced by continuous change of stimulus parameters.

5 On spontaneous perceptual switching

The study of spontaneous switches among possible percepts of ambiguous stimuli is an important tool to investigate the underlying dynamics of perceptual organization. Experimentally, such switches have been the focus of numerous studies since the early days of Gestalt psychology (for example, Brown, 1955; Fisher, 1967, Lindauer & Baust, 1974; DeMarco et al, 1977).

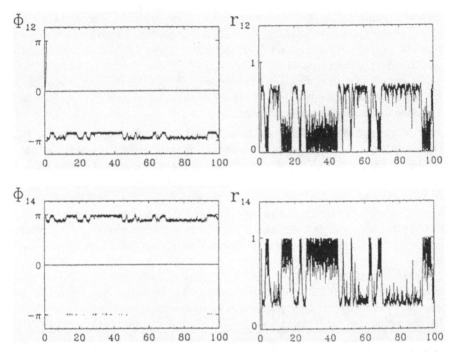

Fig. 10 Spontaneous switches occur due to fluctuations in a long simulation of the bistable dynamics modelling stroboscopic alternative motion. (For these simulations the dynamical model was simplified by eliminating amplitude variables: $r_{ij} = \alpha(\phi_{ij})$ and by using symmetry arguments)

In the present theory, ambiguous stimuli lead to multistable perceptual dynamics. Spontaneous perceptual switches occur due to fluctuations of the perceptual variables that occasionally take the system from one basin of attraction to the other. Incidents of such switches are shown in the simulation of Figure 10. It is important to realize that these switches occur within the perceptual dynamics and do not involve any additional types of processes such as adaptation or learning. This contrasts with explanations invoking special processes such as satiation (Köhler & Wallach, 1944), saturation of attention (Ditzinger & Haken, 1989; 1990), or synaptic modification in a neural network model (Kawamoto & Anderson, 1985). It is clearly an important question whether perceptual switching observed experimentally can consistently be described by stochastic switching in a multistable perceptual dynamics, as in the present theory, or whether there are underlying deterministic processes that govern switching. In this section we point out a number of measures and methods of analysis that may help to clarify this question.

We begin by discussing spontaneous switching in the dynamic theory. To be concrete, consider a bistable situation where switching can be defined as changing from one basin of attraction to the other. Suppose such switches occur at times $T_1, T_2, \ldots$, from which we may calculate the *residence times*, $\Delta T_n = T_n - T_{n-1}$. Switching is governed by the stochastic dynamics and we may use measures such as mean first exit time or mean first entry time to calculate the statistics of switching (see Schöner & Kelso, 1988b, for an introduction and Gardiner, 1983, for thorough review). Briefly, the idea is as follows: If the system is in one basin of attraction at some fixed time t_0, consider the time, t, when the system first switches out of this basin of attraction. This time for the first switch is a random variable and its distribution, $v(t)$, can be calculated theoretically from the stochastic dynamic equations. An analogous switching time distribution can be calculated for switching in the opposite direction. In a direct simulation of the stochastic dynamics we may view the series of residence times, Δt_n, as a sampling of these distributions (which, in this context, might more adequately be called residence time distributions). Specifically, the even numbered events comprise a sampling of switching times in one direction, the odd-numbered events a sampling of the switching times in the opposite direction. In some experiments, the times at which spontaneous reversals occur are measured directly (e.g., DeMarco et al., 1977), and their distributions are analyzed. These distributions can be directly compared to the theoretical ones.

Another way to look at the switching statistics is based on the following observation: The switching time distribution, $v(t)$, defines the probability density for the system to switch in the time interval $[t, t + dt]$. Therefore, its integral,

$$P_{\text{switch}}(T) = \int_{t_0}^{T} v(t) dt \tag{4}$$

is the probability that at least one switch has occured in the entire time interval $[t_0, T]$. This switching probability can also be directly estimated in experiments (Hock, Kelso & Schöner, 1993; see below) and thus likewise allows for comparison of theory and experiment.

The correlations within the switching processes can be visualized by generating the *return map of residence times*, that is, studying the functional relationship of ΔT_{n+1} versus ΔT_n. A purely random switching process does not lead to a functional relationship: for a given value of ΔT_n, there are many different entries for ΔT_{n+1} and with increasing sample size the area above the abscissa of a plot of the return map is filled more or less evenly. This can be contrasted with theories that involve underlying deterministic processes (see below). Such return maps may be obtained from theoretical simulations, but in the same manner also from experiments, and hence offer a further link of theory and experiment.

Analytical calculations of the switching statistics are actually quite difficult beyond the simplest one-dimensional systems (cf. Gardiner, 1983). We take again recourse to numerical simulations, which we now base on a simplified dynamical model to ease the computational load (see caption of Fig.10).

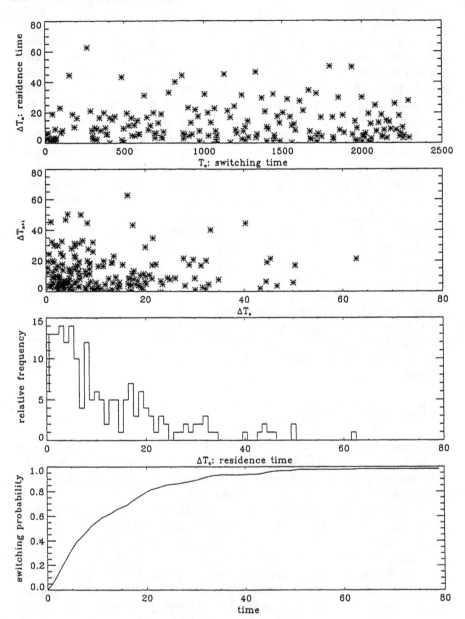

Fig. 11 Switching time statistics from model simulations of the motion quartet in the square geometry

The data in Figure 11 are compiled from long simulations in a symmetric situation (horizontal and vertical interactions identical). In this case switching time statistics in the two directions are equivalent (unbiased ambiguity) and we may join odd- and even-numbered switches in the statistics. The residence times, ΔT_n plotted as a function of the time of switching, T_n (Figure 12 top panel) reveal the stochastic nature of the switching process. The return map of residence times, $\Delta T_{n+1} = \text{function}(\Delta T_n)$, displayed in the second panel likewise illustrates this stochasticity: The different time intervals are randomly distributed in the panel T, $(\Delta T_n, \Delta T_{n+1})$, and no functional relationship emerges. A histogram of residence times estimating the residence time distribution, $v(t)$, is shown in the third panel. Note the roughly Poissonian shape of this distribution. The corresponding integral estimating the switching probability, $P_{\text{switch}}(t)$, is plotted in the bottom panel. Switching probability starts a roughly linear increase from zero and then levels off while approaching the upper bound of 1. The results in the third from top panel may be compared to De Marco et al.'s (1977) data, while the switching probabilities match well those measured by Hock, Kelso & Schöner (1993).

We introduce bias in the perceptual dynamics by changing the parameters of the intrinsic dynamics to reflect a rectangular arrangement of lights with an aspect ratio different from one. In Figure 12 the results of a simulation are shown in which parameters reflect a tall rectangle, favouring horizontal movement. Odd- and even-numbered switching times, i.e., switching into and out of the, say, horizontal motion percept, are now dealt with separately and are marked by different symbols. Note the strong difference in mean and distributions of residence time. The return map shows, however, that even so no causal relation between switches into and out of a particular pattern exists. In the switching probability we find that the slope of the initial linear increase is an indicator of the attractivity of the corresponding attractor. The differences in statistics between the two directions of switching can be used, therefore, to assess the relative stability of the two possible perceptual states (cf. Schöner & Kelso, 1988b). Note the characteristic scaling of the width of the switching time distribution with its modus: The larger mean switching time leads to a longer tail of the distribution compared across the two directions of switching. The corresponding characteristic for switching probability is the scaling of the initial slope with the length of the transient to probability one. This feature was observed by Hock, Kelso & Schöner, 1993, when comparing spontaneous switching at different aspect ratios.

We contrast this analysis of spontaneous switching in a stochastic dynamics of perceptual variables with proposals that a deterministic process underlies switching. In a recent model by Ditzinger and Haken (1989), for instance, a variable called attention parameter, obeys oscillatory dynamics for ambiguous stimuli. Such oscillations lead to a simple functional relationship in the residence time return map: For symmetric ambiguity (both percepts having the same attractiveness in the model) a single fixed point emerges,

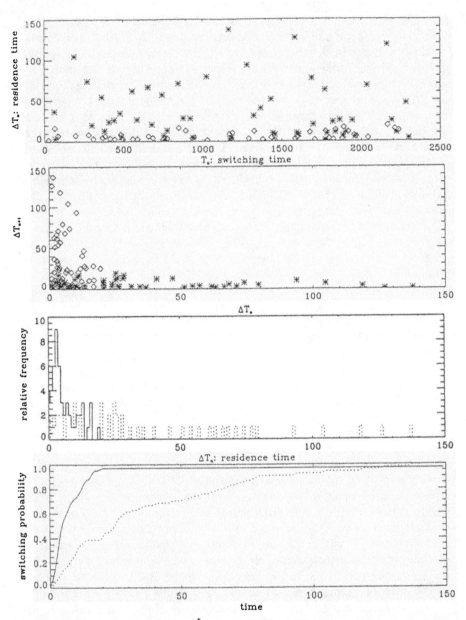

Fig. 12 Switching time statistics from model simulations of the motion quartet in a tall rectangle geometry favouring horizontal movement. The switching times to the horizontal percept are marked by diamonds and the switching times to the vertical percept by asterisks. Switching time distributions are solid for residence in vertical movement and dotted for residence in horizontal movement

that is, switching always occurs after a constant time interval. With bias toward one of the percepts, the return map alternates between two points, that is, each pattern has a definite residence time, but that time differs for the two patterns. In the presence of noise these functional relationships are smeared out, but a definite time scale for the occurrence of switches remains (Ditzinger & Haken, 1990).

In experiments by Hock and Voss (1990) spontaneous switching was studied in the following paradigm: In a given trial subjects reported only at the end of the trial if they had experienced at least one reversal. The length of these trials was varied, however, so that responses could be analyzed for switching probability as a function of duration of exposure (calculated simply as percentage of trials of a given length, in which a switch occurred). This switching probability can be directly compared to the theoretical $P_{\text{switch}}(t)$. It was found that switching probabilities increased initially linearly in time and levelled off for larger times, remarkably similar to the theoretical probabilities in Figures 11 and 12. At different aspect ratios the slope of the initial linear increase of switching probability varied, reflecting different amounts of bias towards one of the percepts (as determined from percentage of trials in which a particular percept is reported). This experiment therefore supports the present theory accounting for spontaneous switching purely in terms of the multistable perceptual dynamics. Note, that switching time distributions with a definite time scale (non-Poissonian), as obtained, for example, in Ditzinger and Haken's (1990) model lead to roughly sigmoid switching probability time courses that do not match well with these experimental results. A more detailed analysis of spontaneous switching, in particular, invoking return maps, may serve to resolve this issue definitely.

Classical studies on reversible figures and ambiguous perception recorded switching times directly by having subjects make a response whenever a switch occurred. Although a number of methodological questions are associated with this procedure it is interesting to note that the resulting distributions of residence time (to be compared to $v(t)$ here) are Poissonian in shape with a steep initial increase and a long tail, quite similar to the theoretical examples shown above (DeMarco et al., 1977).

It is conceivable, of course, that two types of switching mechanisms are present, so that switching can occur stochastically within an underlying multistable perceptual dynamics, but also can be induced by deterministic processes like satiation or learning. There are clear hints that satiation-like phenomena occur on a larger time scale of dozens of seconds and minutes, while perceptual stability assessed in terms of relaxation time must be of the order of seconds or smaller. If such a separation of time scales holds, then the concepts of dynamic theory can be used to address long-term effects, for instance, in terms of a slow dynamics of parameters of the perception dynamics (see, Schöner, 1989, for an example from movement coordination dynamics). This leads to two types of switching processes: One due to fluctuations in a multistable system occurring on a very wide range of time scales, and one due to the underlying slow parameter change and occurring on a definite time scale. In principle, these two processes can be differentiated by a time series analysis of the switching or residence times on different time scales. The contribution of the perceptual dynamics can be observed at fixed level of adaptation, for example, by observing over short times early in adaptation (as in Hock, Kelso & Schöner, 1993) or by observing in the fully adapted regime (as in Kruse, Stadler & Strüber, 1991). The contribution of the adaptation process can be observed when the stimulus material is temporally structured on the longer time scale (as in Experiment 5 of Hock, Kelso & Schöner, 1993; cf. Sect. 6 Discussion) or by looking at the evolution of the response over the longer time scale at constant stimulation (this requires repeated response during exposure, cf. Kruse, Stadler & Wehner, 1986; Kruse, Stadler & Strüber, 1991).

6 Discussion

The central hypothesis argued here posits that perceptual change (re-organization) is governed by dynamic laws. Based on this hypothesis we have developed a number of concepts for a dynamical theory of perceptual organization: (1) Organized percepts are described in terms of collective or perceptual variables that express existence and nature of relationships among components of a percept. (2) Aspects of the stimulus that are specific to the organization of the percept (specifying information) are parametrized by the same type of collective variables and define contributions to a perceptual dynamics, that is, forces in the equations of motion of the perceptual variables that attract toward the specified percept. (3) Intrinsic dynamics capture the active contribution of the perceptual system to perceptual organization. These concepts represent a theoretical, but operational language[16]. Temporal stability of percepts, measurable in terms of variability of perceptual response, of probability of perceptual change for ambiguous stimuli, of discrimination, and in other ways, is the most important element of the language. Essentially, stability can be viewed as indicative of how close a percept is to becoming unobservable. Experimental evidence for the validity of the stability concept can be obtained by testing a number of concrete predictions that arise for multistable percepts: Hysteresis and its dependence of rate of stimulus change, the linkage between distance from the boundary of hysteresis and the probability of perceptual change, as well as the switching statistics of spontaneous perceptual changes. In the context of a simple, but well-studied model system we

---

[16]The mathematics are related to those used by Poston and Stewart, 1978, in an earlier attempt to link nonlinear dynamics and perception. However, the formalization of the problem of perceptual organization, the manner in which mathematical concepts are related to measurement, as well as the nature of the questions asked are quite different.

have demonstrated these predictions theoretically, and through comparison to recent experiments (Hock et al., 1993) concluded, that the language of the dynamical theory of perceptual organization is an adequate one.

These theoretical concepts constrain "classical" theories of perception but do not by themselves explain "what is perceived when". The example furnished demonstrates, however, that it is possible and useful to formulate theories that address this question using the concepts of the dynamic approach. Essentially, the task of constructing such theories consists of finding adequate levels of description and forming propositions about intrinsic dynamics. The experimental paradigms and stability measures can be used as tools to evaluate such propositions. In this respect, multistable percepts can be used as experimentally accessible windows into the more fundamental ambiguity of perception that results from the fact that the proximal stimulus information does not uniquely specify its physical cause for a wide class of behavioral situations. From the viewpoint of dynamic theory, two types of ambiguous stimuli must be distinguished: On the one hand, stimuli may be ambiguous in as far as they do not completely specify the physical cause giving rise to a particular physical stimulation (wide sense ambiguity). It has been argued that fundamentally all stimuli are ambiguous in this sense, which makes the information processing task of the nervous system non-trivial (Ullman, 1979; Marr, 1982). On the other hand, some stimuli can be phenomenally perceived in multiple, clearly distinguishable ways, each of which appears as a unique and reproducible percept (narrow sense ambiguity). Most of the Gestaltist's demonstrations of ambiguity are based on ambiguous stimuli in this latter sense. In the language used here, ambiguous stimuli in this sense generate multistable perceptual dynamics, while ambiguous stimuli in the wide sense do not necessarily lead to multistability.

Attempts to formulate general "classical" theories of perception and explain "what is perceived when" must face challenges from the observation that "top-down" influences on perceptual organization such as described by semantics, perceptual learning, and selective adaptation may change the effective intrinsic dynamics in different behavioral situations. The stress on functional descriptions and the implied closeness of the description to the behavior of perceiving puts the dynamic approach into a strong position to face these challenges. For illustration consider the phenomenon of selective adaptation:

It has been shown in numerous systems that adaptation is specific to the organization of a percept, and is clearly separated by time scale from the process of perceptual organization per se. While adaptation typically occurs on the time scale of dozens of seconds, organized percepts stabilize on the time scale of seconds or parts of seconds (cf. Hock, Kelso & Schöner, 1993). Therefore, the two processes can be decoupled in a dynamic description. Here, we sketch how adaptation can be dealt with in a dynamic theory[17]. Briefly, the idea is that adaptation acts on the perceptual dynamics in the form of changing the relative strength of different contributions. Because adaptation is slow, this change can be viewed as adiabatic[18] change of parameter so that, essentially, at any point during the process of adaptation the system has a well-defined perceptual dynamics that can be assessed either by measuring locally in time or by controlling the temporal structure of the stimulus material on the fast and the slow time scales. The change occurs only with respect to the realized percept, in other words, the process of adaptation depends on the perceptual variables or state. We may assume that under adaptation the pattern that is being perceived slowly loses stability. This hypothesis can be tested directly by assessing the stability of an organized percept during adaptation. In multistable percepts this process ultimately leads to perceptual change, as is observed experimentally. If adaptation to all possible percepts in a multistable situation is allowed, the prediction is that spontaneous reversal rates increase due to the decreased relative stabilities. This has often been reported (e.g., Brown, 1955; Kolers, 1964; Kruse, Stadler & Strüber, 1991).

In Hock, Kelso and Schöner (1993) adaptation was observed in the following manner: Stimulus series leading through a perceptual hysteresis were presented in two orders, ascending and descending. If these two conditions were presented in randomized fashion, hysteresis was observed as expected. However, when the conditions were blocked, e.g., first, many trials in ascending order, then, many trials in descending order, and all data were pooled, the hysteresis effect was diminished and, in some subjects, abolished. Detailed analysis revealed that on the ascending trials there was much more exposure to one pattern than to the other, with the opposite bias existing on the descending trials (the effect being aggravated by the particular experimental design of varying depth of presentation). The observation is clearly consistent with our hypothesis: In the randomized conditions, where hysteresis is observed, the decreases of stability of both percepts may lead to enhanced spontaneous reversal rates, but the effect on the boundaries of the hysteresis is canceled.[19] In the blocked condition, however, the block of ascending conditions decreases the stability of one, but not of the other pattern and hence shifts the right boundary of the hysteresis to the left (taking ascending as going from left to right). In the block of descending conditions the oppo-

---

[17]The discussion is based on ideas first developed for problems of coordination dynamics, where learning and recall of coordination patterns are examples of "higher" processes that act on the movement behavior described by dynamics (Schöner, 1989; Schöner, Zanone & Kelso, 1992).

[18]Adiabatic change of parameter is change that occurs on a much slower time scale than the time scale of the dynamics onto which the parameters act.

[19]There may be a small decrease of the size of the hysteresis due to a change in time scale relationships, cf. Sect. 4, but that effect is much weaker than effects resulting from a shift of the perceptual boundaries.

site happens, leading to a shift of the left boundary of the hysteresis to the right. Overall, the hysteresis is strongly reduced in size. Theoretically, it may become negative, which can be excluded for the randomized design.

A final note concerns what the dynamic concepts have to say about the issue of categorization. A fundamental problem of perceptual organization is to understand how continuously varying sensory information may give rise to distinct classes of percepts. For instance, AM between two flashing lights can be perceived over a whole range of timings and spacings and is clearly distinguished from flicker motion and succession (Finlay & Caelli, 1979). However, the AM perceived within that range has a perceptual quality that is common to all percepts across different parameter settings. A dramatic effect of this type is the categorical perception of speech (see Repp, 1984, for review). In that case, when sufficiently uniform stimuli are used, subjects discriminate only poorly between different acoustic stimuli that approximate one particular utterance but vary continuously in an acoustic parameter such as voice onset time. The discrimination peaks at the category boundary, however, where subjects start to report hearing another utterance in identification tasks. The issue of categorization has given rise to the notion of invariance and considerable effort has gone into identifying invariants that could account for the persistence of categories as parameters are varied (see, for instance, the contributions to Perkell & Klatt, 1986). From the point of view of the dynamic theory, invariance is related to symmetries (for a discussion in the context of coordination patterns, compare Schöner, Jiang & Kelso, 1990). The constitutory property of perceptual categories is, however, the existence of boundaries between categories, at which perceptual change occurs. In dynamic theory, these boundaries are associated with instabilities: The idea is that the stable solutions of the perceptual dynamics, that is, the organized percepts, may depend continuously on the parameters of the dynamics, and thus, on the stimulus. When different contributions come into conflict instabilities can occur that lead to abrupt changes of the organized percept as some aspect of the stimulus is changed continuously. Hence, the perceptual dynamics, while based on graded measures and therefore on a sub-symbolic level itself, gives rise to classes of percepts demarcated by points of abrupt perceptual change. The perceptual dynamics therefore account for the generation of symbolic information. In a way, the generation of categories from continuously scalable stimulation is based on temporal stability and loss thereof rather than on invariance: Within categories the percept as described by the collective variables may vary in a graded fashion as stimulus parameters are varied. Categories are well-defined by the condition that the temporal stability of percepts within a category is preserved as stimulus parameters are changed. One may say, therefore, that categorization can be accounted for in the dynamic theory even when continuous changes of the percept occur within categories. This may well be the case in most examples of category formation if sufficiently rich stimuli are administered. The dynamic view of categorization can be tested by detecting the predicted loss of stability at category boundaries with the methods discussed in this article.

7 Appendix: The dynamic model for perceptual organization in apparent motion quartets

Specifying information. The following functional form assures the attraction of ϕ_{ij} to ψ_{ij} and fulfills the periodicity [invariance under $\phi_{ij} \to \phi_{ij} + 2\pi$] and symmetry [invariance under $\phi_{ij} \to -\phi_{ij}$] requirements:

$$\dot{\phi}_{ij} = -c_{ij}^{env} \sin(\phi_{ij} - \psi_{ij}) \tag{5}$$

with $c_{ij}^{env} > 0$. This dynamics is illustrated in Fig. 4 (a).

The dynamics for the amplitude variables, x_{ij}, must realize symmetric point attractors at $x_{ij} = \pm 1$ (so that $r_{ij} = 1$) if the corresponding relative phase variable is sufficiently close to its specified value, ψ_{ij}. Else, a single point attractor at $x_{ij} = 0$ must exist. A possible functional form for such dynamics is:

$$\dot{x}_{ij} = \alpha(\phi_{ij})x_{ij} - |\alpha(\phi_{ij})|x_{ij}^3 \tag{6}$$

where $\alpha(\phi_{ij})$ is positive if ϕ_{ij} is close to a meaningful value and negative else. This is the normal form of the corresponding pitchfork bifurcation (see, e.g., Guckenheimer & Holmes, 1983). For $\alpha > 0$ the stationary solutions $x_{ij} = \pm 1$ are the only attractors, while for $\alpha < 0$ the state $x_{ij} = 0$ is the only attractor. The switching between these limit cases can be controlled by the match function, $\alpha(\phi)$. Its functional form is chosen here to provide convenient control over the range of valid matching:

$$\alpha(\phi_{ij}) = \alpha_0 \tan[h\{\cos[2(\phi_{ij} - \psi_{ij})] - m_0\}] \tag{7}$$

in the interval $\phi_{ij} \in [\psi_{ij} - \pi/2, \psi_{ij} + \pi/2]$ with $\alpha(\phi_{ij}) = -1$ elsewhere. Here, $\alpha_0 > 0$ is a constant, h is the steepness and m_0 the threshold of a sigmoid function. This dynamics is illustrated in Fig. 4 (b).

Intrinsic dynamics. The first two tendencies – tendency to see synchronism and tendency to see symmetric AM – are expressed mathematically as a vector-field with attractors at $\phi_{ij} = 0$ and $\phi_{ij} = \pi$. Again using periodicity and symmetry, the lowest order (in a Fourier series) guaranteeing these attractors is:

$$f_{ij}(\phi_{ij}) = A_{ij} \sin(\phi_{ij}) + B_{ij} \sin(2\phi_{ij}) \tag{8}$$

where $B_{ij} < 0$. The ratio B_{ij}/A_{ij} expresses the relative strength of the two forces. These forces are illustrated in Figure 4.

The third intrinsic tendency – tendency to see no splitting motion of elements that are synchronous with other elements – is accounted for in terms of repelling forces. To define a functional form for such forces consider the equation of motion of, say, ϕ_{12}. This relative phase interacts with ϕ_{13} and ϕ_{14} on one side and ϕ_{23} and ϕ_{24} on the other. In each case a repelling force is defined if $\phi_{12} \approx \pi$ and the other two phases are at 0 and π respectively. Figure 5 illustrates all configurations that must lead to repulsive forces. A simple functional form complying with these and the periodicity and symmetry requirements is:

$$\begin{aligned} c_{12} \cos^2(\frac{\phi_{12}-\pi}{2}) [& r_{13} \cos^2(\frac{\phi_{13}-\pi}{2}) r_{14} \cos^2(\frac{\phi_{14}}{2}) \\ & + r_{13} \cos^2(\frac{\phi_{13}}{2}) r_{14} \cos^2(\frac{\phi_{14}-\pi}{2}) \\ & + r_{23} \cos^2(\frac{\phi_{23}-\pi}{2}) r_{24} \cos^2(\frac{\phi_{24}}{2}) \\ & + r_{23} \cos^2(\frac{\phi_{23}}{2}) r_{24} \cos^2(\frac{\phi_{24}-\pi}{2})]. \end{aligned} \quad (9)$$

Here we have multiplied the various contributions with the corresponding amplitude variables because only existing relationships are to participate in the interaction. A few mathematical transformations lead to the simplified expression:

$$\begin{aligned} \frac{c_{12}}{4}(1 - \cos(\phi_{12})) [& r_{13}r_{14}\{1 - \cos(\phi_{13})\cos(\phi_{14})\} \\ & + r_{23}r_{24}\{1 - \cos(\phi_{23})\cos(\phi_{24})\}]. \end{aligned} \quad (10)$$

The complete dynamics are obtained by adding the contribution from specifying information and the various intrinsic contributions.

Stochastic forces. We employ the standard assumptions: (a) many independent noise sources; (b) noise sources correlated over short times compared to the time scales of the proper dynamics; (c) noise sources regular, that is, not zero at a particular value of the dynamic variables. Hence, we may model noise contributions by additive stochastic forces, each of which is an independent gaussian white noise process (cf., e.g., Horsthemke & Lefever, 1984). The variances of each stochastic force, $Q_{\phi,ij}$ and $Q_{x,ij}$, are additional model parameters.

Acknowledgements

Work reported in this chapter was begun while the first author was at the Center for Complex Systems, FAU, Boca Raton, Florida with support from NIMH (MH 42900-01). In Bochum, Germany, the work was supported by the Ministerium für Wissenschaft und Forschung des Landes Nordrhein-Westfalen and the Deutscher Akademischer Austauschdienst. We would like to thank Drs. S.L. Fonseca e Castro, H. Haken, J.A.S. Kelso, P. Kruse, H.A. Mallot, H. Shimizu, M. Stadler, and Y. Yamaguchi for helpful discussions.

References

Anstis, S. (1986): Motion perception in the frontal plane. In: K. R. Boff, L. Kaufman & J.P. Thomas (Eds.), Handbook of Perception and Human Performance, pp. 16 – 1 to 16 – 27. New York: Wiley.

Attneave, F. (1971): Multistability in perception. Scientific American 225, 62 – 71.

Brown, K.T. (1955): Rate of apparent change in a dynamic ambiguous figure as a function of observation time. American Journal of Psychology 68, 358 – 371.

Burt, P. & Sperling, G. (1981): Time, distance, and feature trade-offs in visual apparent motion. Psychological Review 88, 171 – 195.

Collet, P. & Eckmann, J.P. (1990): Instabilities and Fronts in Extended Systems. Princeton, NJ: Princeton University Press.

Dawson, M.R.W. (1991): The how and why of what went where in apparent motion: modeling solutions to the motion correspondence problem. Psychological Review 98, 569 – 603.

DeMarco, A., Pennengo, P., Trabucco, A., Borsellino, A., Carlini, F., Riani, M. & Tuccio, M.T. (1977): Stochastic models and fluctuations in reversal time of ambiguous figures. Perception 6, 645 – 656.

Ditzinger, T. & Haken, H. (1989): Oscillations in the perception of ambiguous patterns. Biological Cybernetics 61, 279 – 287.

Ditzinger, T. & Haken, H. (1990): The impact of fluctuations on the recognition of ambiguous patterns. Biological Cybernetics 63, 453 – 456.

Dodwell, P.L. (1978): Human pattern and object perception. In: R. Held, H.W. Leibowitz & H.-L. Teuber (Eds.), Handbook of Sensory Physiology, vol. 8, Perception, pp. 523 – 548. Berlin: Springer.

Finlay, D. & Caelli, T. (1979): Frequency, phase, and colour coding in apparent motion. Perception 8, 595 – 602.

Fisher, G.H. (1967): Measuring ambiguity. American Journal of Psychology 80, 541 – 557.

Gardiner, C.W. (1983): Handbook of Stochastic Methods for Physics, Chemistry and the Natural Sciences. Berlin: Springer.

Gibson, J.J. (1966): The Senses Considered as a Perceptual Systems. Boston: Houghton Mifflin.

Guckenheimer, J. & Holmes, P. (1983): Nonlinear Oscillations, Dynamical Systems, and Bifurcations of Vector fields. New York: Springer.

Haken, H. (1983): Synergetics – An Introduction. Berlin: Springer (3rd ed.).

Hock, H.S., Kelso, J.A.S. & Schöner, G. (1993): Bistability and hysteresis in the organization of apparent motion patterns. Journal of Experimental Psychology: Human Perception and Performance 19, 63 – 80.

Hock, H.S. & Voss, A. (1990): Spontaneous pattern changes for bistable stimuli: Evidence against neural satiation. Paper presented at the 31st Meeting of the Psychonomic Society, New Orleans, LA.

Hoeth, F. (1966): Gesetzlichkeit bei stroboskopischen Alternativbewegungen. Frankfurt/Main: Kramer.

Horsthemke, W. & Lefever, R. (1984): Noise-induced Transitions. Berlin: Springer.

Kawamoto, A.H. & Anderson, J.A. (1985): A neural network model of multistable perception. Acta Psychologia 59, 35 – 65.

Koffka, K. (1935): Principles of Gestalt Psychology. New York: Harcourt, Brace & World Inc..

Köhler, W. & Wallach, H. (1944): Figural after-effects: an investigation of visual processes. Proceedings of the American Philosophical Society 88, 269 – 357.

Kolers, P.A. (1964): Apparent movement of a Necker cube. American Journal of Psychology 77, 220 – 230.

Korte, A. (1915): Kinematoskopische Untersuchungen. Zeitschrift für Psychologie 72, 194 – 296.

Kruse, P., Stadler, M. & Strüber, D. (1991): Psychological modification and synergetic modelling of perceptual oscillations. In: H. Haken & H.P. Koepchen (Eds.), Rhythms in Physiological Systems, pp. 299 – 311. Berlin: Springer.

Kruse, P., Stadler, M. & Wehner, T. (1986): Direction and frequency specific processing in the perception of long-range apparent movement. Vision Research 26, 327 – 335.

Kubovy, M. & Pomerantz, J.R. (Eds.) (1981): Perceptual Organization. Hillsdale, NJ: Erlbaum.

Lindauer, M.S. & Baust, R.F. (1974): Comparison between 25 reversible and ambiguous figures on measures of latency, duration and fluctuation. Behavior Research Methods and Instrumentation 6, 1 – 9.

Lugiato, L.A., Broggi, G., Merri, M. & Pernigo, M.A. (1989): Control of noise by noise and applications to optical systems. In: F. Moss & P.V.E. McClintock (Eds.), Noise in Nonlinear Dynamical Systems, vol. 2. Cambridge: Cambridge University Press.

Marr, D. (1982): Vision. New York: Freeman & Co.

Metzger, W. (1941/1975): Psychologie. Dresden/Darmstadt: Steinkopff (5th ed.).

Metzger, W. (1975): Gesetze des Sehens. Frankfurt/Main: Kramer (3rd ed.).

Nawrot, M. & Sekuler, R. (1990): Assimilation and contrast in motion perception: Explorations in cooperativity. Vision Research 30, 1439.

Neuhaus, W. (1930): Experimentelle Untersuchungen der Scheinbewegung. Archiv für die gesamte Psychologie 75, 315 – 458.

Perkell, J.B. & Klatt, D.H. (Eds.) (1986): Invariance and Variability in Speech Processes. Hillsdale, NJ: Erlbaum.

Pomerantz, J.R. & Kubovy, M. (1986): Theoretical approaches to perceptual organization. In: K.R. Boff, L. Kaufman & J.P. Thomas (Eds.), Handbook of Perception and Human Performance, pp. 36 – 1 to 36 – 46. New York: Wiley-Interscience.

Poston, T. & Stewart, I. (1978): Nonlinear modelling of multistable perception. Behavioral Science 23, 318 – 334.

Ramachandran, V.W. & Anstis, S.M. (1987): Visual inertia in apparent motion. Vision Research 27, 755 – 764.

Repp, B.H. (1984): Categorical perception: Issues, methods, findings. In: N. J. Lass (Ed.), Speech and Language: Advances in Theory and Practice, vol. 10, pp. 243 – 335. Ney York: Academic Press.

Rock, I. (1986): The description and analysis of object and event perception. In: K.R. Boff, L. Kaufman & J.P. Thomas (Eds.), pp. 33 – 1 to 33 – 71. New York: Wiley-Interscience.

Schöner, G. (1989): Learning and recall in a dynamic theory of coordination patterns. Biological Cybernetics 62, 39 – 54.

Schöner, G., Haken, H. & Kelso, J.A.S. (1986): A stochastic theory of phase transitions in human hand movement. Biological Cybernetics 53, 247 – 257.

Schöner, G., Jiang, W.J. & Kelso, J.A.S. (1990): A synergetic theory of quadrupedal gaits and gait transitions. Journal of Theoretical Biology 142, 359 – 391.

Schöner, G. & Kelso, J.A.S. (1988a): Dynamic pattern generation in behavioral and neural systems. Science 239, 1513 – 1520.

Schöner, G. & Kelso, J.A.S. (1988b): A dynamic theory of behavioral change. Journal of Theoretical Biology 135, 501 – 524.

Schöner, G., Zanone, P.G. & Kelso, J.A.S. (1992): Learning as change of coordination dynamics: Theory and experiment. Journal of Motor Behavior 24, 29 – 48.

Ullman, S. (1979): The Interpretation of Visual Motion. Cambridge: MIT-Press.

von Schiller, P. (1933): Stroboskopische Alternativversuche. Psychologische Forschung 17, 179 – 214.

Williams, D., Phillips, G. & Sekuler, R. (1986): Hysteresis in the perception of motion direction as evidence for neural cooperativity. Nature 324, 253 – 255.

Yuille, A. & Grzywacz, N.M. (1988): A computational theory for the perception of coherent visual motion. Nature 333, 71 – 74.

Artificial Neural Networks and Haken's Synergetic Computer: A Hybrid Approach for Solving Bistable Reversible Figures

H.H. Szu[1], F. Lu[2] and J.S. Landa[3]

[1] Naval Surface Warfare Center, Dahlgren Division, Code R44, Silver Spring, MD 20903, USA
[2] The American University, Department of Physics, 4400 Massachusetts Avenue, Washington, DC 20016, USA
[3] Presearch Incorporated, 8500 Executive Park Avenue, Fairfax, VA 22031, USA

Abstract: To explain the multistable perception in "reversible figures", an artificial perceptron neural network model is proposed. The networks are composed by a shifting invariant smart eye preprocessing unit, a depth processing unit and the main brain computing unit. The shift invariant preprocessing unit could be achieved by wavelet transform optical filters (Sheng et al., 1993) or fixed artificial neural networks (Widrow & Winter, 1988). The depths processing unit is constructed by two McCulloch-Pitts neurons with time-dependent threshold, which is updated by modified Haken's time-dependent attention parameters dynamics. The main brain computing network is a revised back error propagation network. Following Haken's (1991) description, we adopt Stadler's connectionist approach. We use a polynomial energy function, e.g., ξ^4 field, for the performance measure replacing the least mean square energy. The training of the computing network follows a standard supervised delta learning rule of interconnected weights. The test of "reversible figures" is subsequently controlled by the phase transition tuning parameter driven bottom up from test image data. The effects of the tuning parameter are illustrated. And the modeling of multistable perception is discussed. We demonstrate that the brain computing networks trained with the new energy function generally perform better in training speed and classification of patterns than the standard back error propagation networks trained by the least mean square energy.

1 Introduction

Recent advances in top down design of artificial neural networks have yielded improvement in recognition of labelled objects belonging to identifiable classes. The reversible figure problem represents a different challenge, where the data cannot be defined in advance. One picture can belong equally to two different classes depending on the perception conditions.

Based on the biological findings in vision science that the eye and brain system actually is composed of different subunits, we ad hoc break up the network system into three units for technical reasons. One will serve as the shift invariant preprocessing unit, one as the depth processing unit, and one as the main brain computing unit. The input pattern will first pass through the shift-invariant unit, which will then generate a pattern that is shift-invariant relative to the input pattern. This pattern combined with the information from the depth processing units will then present to the input layer of the main brain computing networks for pattern recognitions. An ANN based and an optical wavelet transform based translation invariant preprocessing units are discussed. A depth processing unit based on Haken's time dependent attention parameter dynamics is introduced.

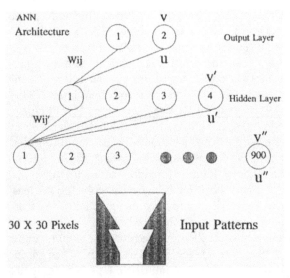

Fig. 1 The main brain computing artificial neural network architecture

We test our brain computing network using reversible figure input patterns and small perturbations. We assume a standard back error propagation network using three layers of sigmoidal nodes. Connections are made from the bottom layer to the hidden layer and then from the hidden layer to the output layer with no local recurrent or intralayer connections (Figure 1). The least mean square (LMS) (Werbos, 1974) energy performance measure is obtained between the desired and actual output of the top layer. This LMS is well suited for determining a best fit for data representation. For cases of classification, Szu and Tefler (1991) have shown an L1 city-block distance in terms of minimum misclassification of error (MME) to be preferable when handling overlapping non-Gaussian distributed data. A quadratic energy function is used by Haken in his Synergetic Computer for pattern recognition. By applying the slaving principle, the synergetic computer can find the nearest attractor to the incoming pattern via solving the time evolving dynamic equation of order parameters. This top down design approach can let one rigorously trace the performance of the network and guarantee the convergence of the network to one attractor, so it has the benefit of not being trapped in a spurious state like traditional neural networks (Haken, 1991). We adapt the single order parameter potential function of Haken's synergetic computer as the artificial neural network energy function, with the attention parameter as the phase tuning parameter to resolve reversible figures. Various networks performances are tested. Analog implementation using summing amplifiers and phototransistors is discussed.

This paper is organized as follows. We start with the main brain computing unit, which is the most important part of the networks. Then we introduce the depth processing unit to model the perception oscillation phenomena. Finally we discuss the preprocessing units for translation, rotation and scaling invariant pattern recognition.

2 Main brain computing unit

The main brain computing unit is the main information processing unit. It is composed by a revised back error propagation artificial neural networks (BPN). It is trainable, and has the memory ability to recognize trained patterns. The basic elements of an artificial neural network (ANN) is the McCulloch and Pitts neurons (Figure 2), which has the following properties.

$$U_i = \Sigma W_{ij} V_j' - \theta_i \tag{1}$$

$$V_i = \sigma(U_i) \tag{2}$$

where W_{ij} are the weights, U_i is the net input value of the neuron, V_i is the output value of the neuron, $\sigma()$ is the sigmoid function and θ is the threshold, which usually could be set to constant. In this paper we will set the threshold equal to zero for the main brain computing net and time dependent for the depth processing unit.

The learing power relies on the updating of network weights W_{ij}. The learning rule of W_{ij} could be the fixed point Hebbian leaning rule :

$$W_{ij} = V_i V_j \tag{3}$$

or the gradient descending learning rule:

$$\frac{\partial W_{ij}}{\partial t} = -\frac{\partial E}{\partial W_{ij}} \tag{4}$$

which will be used in our BPN networks.

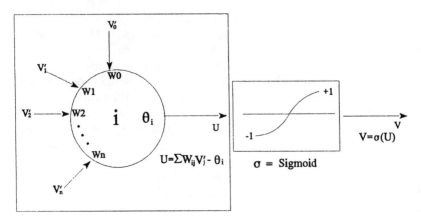

Fig. 2 Artificial neuron, where Θ is the threshold, which is usually set as constant

Our main computing unit is formed by three layer artificial neurons, which assume a standard BPN architecture, but with a new training energy function adapted from Haken's synergetic computers.

2.1 Reversible figures and phase transition

The perception of ambiguous pattern, (Fig. 3) vase/face, has many interesting aspects. Depending on the initial conditions and how we choose to interpret this picture, we could classify it as either a vase or two old men facing each other. Since both interpretations are correct, a network is forced to either converge on a single pattern or separate the pattern into two unrelated classes. A single minimum energy function (Fig. 4) such as LMS, must merge these two interpretations into one class.

Reversible figures are analogous to phase transition phenomena. As the energy of a system changes we notice an unstable point during the phase

Fig. 3 Reversible figure. It can be interpreted either as two old men's faces or a vase

transition where both states exist. The probability that the system is in either state is equal so we cannot classify the system as belonging to state 1 or state 2.

We apply this phenomena by examining the quadratic energy function,

$$V = A\xi^4 + B\xi^3 + C\xi^2 + D\xi + E \qquad (5)$$

using the case of Hopf bifurcation ($A = 1/4$, $B = 0$, $C = -\lambda/2$, $D = 0$, $E = 0$) when $\lambda < 0$ the potential energy V has a single minimum at the origin $\xi = 0$ (Figure 4). When $\lambda > 0$ a double well exists (Figure 5) with the minimum located at $\xi = \pm\sqrt{\lambda}$ and a minimum value of $-\lambda^2/4$. We are now left with

$$V = -\frac{\lambda}{2}\xi^2 + \frac{1}{4}\xi^4 \qquad (6)$$

This energy function is of the same form as the single *order parameter* potential function of Haken's synergetic computer.

2.2 Synergetic computer and BPN networks

To recognize the bistable figure via Haken's top-down approach, we need the attention parameter $\lambda > 0$, and thus an order parameter moving in the potential landscape with two attractors, vase and face respectively.

By solving the time evolving dynamic equation

$$\frac{d\xi_u}{dt} = -\frac{\partial V}{\partial \xi_u} \qquad (7)$$

where ξ_u is the order parameter and V is the potential energy. We can find out, determined by the initial perturbation condition, which attractor the *order parameter* will fall into, so as to recognize the pattern as a vase or a face.

For $\lambda < 0$, there is only one attractor in the energy landscape, which cannot distinguish between patterns, so there won't be any recognition of patterns by synergetic computers. From the order parameter dynamic equation (7) we can have:

$$\xi_j \frac{\partial \xi_i}{\partial t} = -\frac{\partial V}{\partial \xi_i}\xi_j \qquad (8)$$

$$\xi_i \frac{\partial \xi_j}{\partial t} = -\frac{\partial V}{\partial \xi_j}\xi_i, \qquad (9)$$

and adding these two equations together:

$$\frac{\partial(\xi_i\xi_j)}{\partial t} = -(\frac{\partial V}{\partial \xi_i}\xi_j + \frac{\partial V}{\partial \xi_j}\xi_i) \qquad (10)$$

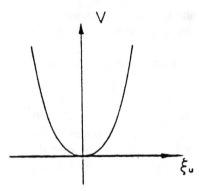

Fig. 4 Single minimum potential well

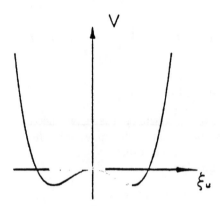

Fig. 5 Double minimum potential well

By adapting Hebbian's learning rule $W_{ij} = \xi_i \xi_j$ we get:

$$\frac{\partial W_{ij}}{\partial t} = -(\frac{\partial V}{\partial W_{ij}} \frac{\partial W_{ij}}{\partial \xi_i} \xi_j + \frac{\partial V}{\partial W_{ij}} \frac{\partial W_{ij}}{\partial \xi_j} \xi_i) \,, \tag{11}$$

which can be further reduced to:

$$\frac{\partial W_{ij}}{\partial t} = -\frac{\partial V}{\partial W_{ij}}(\xi_i \xi_i + \xi_j \xi_j) \tag{12}$$

Finally it can be written as:

$$\frac{\partial W_{ij}}{\partial t} = -(W_{ii} + W_{jj}) \frac{\partial V}{W_{ij}} \tag{13}$$

This formula is very similar to the delta training rule $\partial W_{ij}/\partial t = -\partial E/\partial W_{ij}$ of the back error propagation networks, which reveals the intrinsic relationship between the two dynamics.

Now instead of solving a dynamic equation to find the energy landscape attractor as did in synergetic computer, we use the ANN data driven bottom up approach to reach the attractor.

We begin with a standard feed-forward back error propagation network architecture with the uplink W_{ij} between the output layer $(in, out) = (u, v)$ and the hidden layer $(in, out) = (u', v')$ and lower link W'_{ij} between the hidden layer and the input layer $(in, out) = (u'', v'')$. From the gradient descending learning, we can have the following BPN weights update rule.

For the weights between output layer and hidden layer:

$$W_{ij}(t+1) = W_{ij}(t) + \Delta W_{ij} \tag{14}$$

$$\Delta W_{ij} = \delta_i v'_j \Delta t \tag{15}$$

where

$$\delta_i = -\frac{\partial E}{\partial v_i} V_i(1 - V_i) \tag{16}$$

For the weights between hidden layer and input layer:

$$W'_{ij}(t+1) = W'_{ij}(t) + \Delta W'_{ij} \tag{17}$$

$$\Delta W'_{ij} = (\delta'_i)(v''_j)(\Delta t) \tag{18}$$

where

$$\delta'_i = \frac{dv'_i}{du'_i} \Sigma_j \delta_j W_{ij} \tag{19}$$

We now have a generic delta training rule based on an unspecified energy function. It can be seen that the δ_i is the key element of the six equations. Previously it has been shown (Szu & Telfer, 1991) that least mean square energy function, minimum misclassification error functions, and minimax energy functions are all valid for backpropagation type networks. All these different functions have limitations. The least mean square technique is most common and deeply rooted in our thinking because it is most general. This technique has earned its place in history for its ability to find best fit for a wide variety of data sets. The MME technique helps to separate overlapping features that cause confusion between two distinct classes of objects and are generally more appropriate for automatic target recognition problems. The development of minimax gives us another tool for determination of unknown feature vectors necessary for class separation.

2.3 New training energy learning rule

By adapting Haken's first order parameter potential function as our BPN memory networks training energy, we can have our specific weights update rules. We will use the attention parameter λ as the phase tuning parameter to see its effects on the network performance.

For $\lambda > 0$, we set

$$\xi_{\pm} = \sqrt{\Sigma(T_i - V_i)^2} \pm \sqrt{\lambda}, \tag{20}$$

where T_i is the target output value of output neuron i and V_i is the actual output value of the output neuron. When $T_i = V_i$, $\xi = \pm\sqrt{\lambda}$ are the two attractor positions representing face and vase respectively. Plugging this definition and the new energy function (6) into the arbitrary energy function delta training rule formula (16), we have

$$\delta_i^{\pm} = \xi(\xi \pm \sqrt{\lambda})(T_i - V_i)V_i(1 - V_i) \tag{21}$$

where $\xi = \xi_{\pm}$.

For $\lambda < 0$, we set $\xi = \sqrt{\Sigma(T_i - V_i)^2}$ so

$$\delta_i = (\xi^2 - \lambda)(T_i - V_i)V_i(1 - V_i) \tag{22}$$

Inserting these specific δ_i into the general weights update rule (14–19), the corresponding training rule can be easily obtained.

2.4 Simulations

To test our energy function we break down the reversible figure of face/vase into a two-dimensional binary array. Using a 30 X 30 pixels picture gives us sufficient resolution to identify both figures in the pattern. All the training and testing patterns are listed in Figure 6. We use pattern 1 and pattern 2 as the training patterns for face and vase respectively, pattern 3 through pattern 8 are the perturbed test patterns, pattern 9 through pattern 14 are used to test the connectivity performance of the trained networks. We also tested distorted input patterns, gray scale changed patterns, and varying scaling patterns, etc. The face and vase figures each have 450 pixels, while the perturbation figure apple and earrings each have 60 pixels. To train on this pattern we use 900 input neurons. We vary the number of hidden layer nodes to ensure generality, but limit ourselves to no more than one percent of the input. Our output layer consists of two nodes each one representing the presence of one attractor. Weights are allowed to train until convergence and test patterns are then identified. The performance is then compared to the non-modified LMS networks. We add a momentum term in the training algorithm of both network to accelerate the training speed, with the learning coefficient set to 0.15 and momentum coefficient set to 0.075 the general form of our learning formula is

$$W_{ij}(t+1) = W_{ij}(t) + 0.15\delta_i V_j + 0.075\Delta W_{ij}(t-1) \tag{23}$$

Initial weights are random values between ± 0.015. The network stops training when the condition $Er = \Sigma(T_i - V_i)^2 < 0.01$ is fulfilled.

TEST PATTERNS

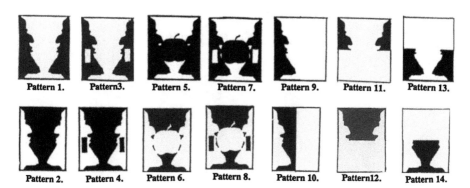

Fig. 6 Training and testing patterns. Patterns 1 & 2 are the two training patterns for the memory networks

2.5 Numerical result

We test the typical LMS energy BPN network converging speed while varying the number of hidden units, The result is shown in Fig. 7. With hidden unit number equal to 1, 2, 3, 5, there is no convergence. When the hidden unit number equals 4, the network converges after 25 iterations, with 6 hidden units the number of iterations drops to 19, but because each iteration takes more time with 6 hidden units than with 4 hidden units, the actual speed may not improve much. As we will see later, all the networks trained poorly when there are 5 hidden units. This suggests that the number of the hidden units should be kept symmetric with input unit number and output unit number. Because here we have even numbers of input and output units, the hidden units should be of even number too.

We then test the training speed of the double well potential energy ($\lambda > 0$) BPN network (Fig. 8). Compare with Fig. 7. Now we can have networks convergence with 3, 4, 5 and 6 hidden units by choosing proper λ values. This reveals the better convergence ability of the double well neural networks. The best overall performance is at 4 hidden units and at λ equal to 14 and 16, where the networks converged only after 10 iterations, that is about 150% faster than the LMS network. Generally, by increasing the λ value, we could have faster convergence. But there is a limit on the increase; after a certain value the network returns to being non-converging. We believe this is because of the general gradient descent training method. With bigger λ, we have steeper gradient, (Figure 9), bigger correction of weights after each iteration. So the network will more quickly reach the energy landscape minimum. But when the step becomes too big, it leads to oscillation within the potential well, the system will be trapped in a middle state and never converge to the minimum. This explanation could be further confirmed by observing the

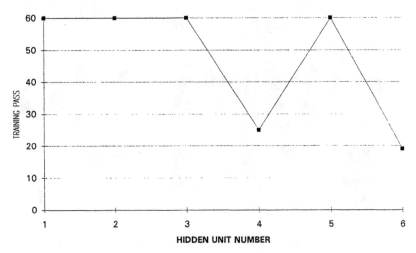

Fig. 7 The effect of different hidden layer neuron numbers on the training speed of a typical LMS back error propagation network. Training pass equals sixty means non-converged training

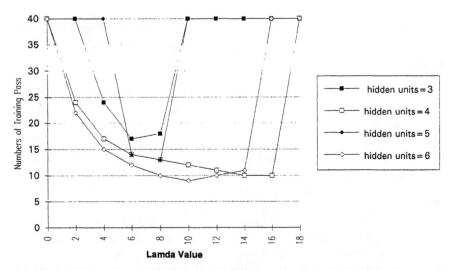

Fig. 8 The effect of different phase tuning parameter λ values and hidden layer unit numbers on the converging speed of double potential well energy trained memory networks

evolution of the Er value after each iteration during the training. For the normal converged training, we observe that the change of the Er value ΔEr is quite big initially, and it becomes smaller and smaller after each iteration, we can see a clear trend that it is moving toward zero until it reaches the preset stopping condition. For the big λ non-converged training, we observe

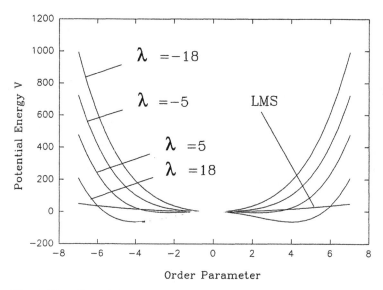

Fig. 9 The effect of the phase tuning parameter on the shape of the energy functions

several very big Er changes initially, and then it drops abruptly to very small changes, and the trend is not toward zero but an intermediate value, e.g. 0.5, which means the system has been trapped in a spurious state.

Figure 10 shows the training speed of the single-well potential BPN network ($\lambda < 0$). Comparing with the double potential well network, we have pretty much the same data structure, but a bigger converged λ value ranges. This could be explained in that the double potential well is separated by a middle point at $V = 0$, so one side of each the double well is rather shallow. With bigger λ value, the training step could easily jump over the middle point sticking in the opposite attractor well. The single potential well has no such tendency, so it can accept much bigger λ values. Another interesting aspect of this single well network is that we can now have a converged two-hidden-unit network.

To test the recognition ability of the three network, we run all fourteen test patterns through the three networks; the result is showed in Figure 11. We could see that generally the network trained by using Haken's ξ^4 potential energy has much better performance in recognizing the perturbed patterns. The standard deviation $\sqrt{(\Sigma(T_i - V_i)^2/2)}$ of all fourteen test patterns is less than 0.06. Because the first two patterns are the actual training patterns, the standard deviation value of all other test patterns should use their value as the reference. Bearing this in mind, we can see that all twelve perturbed patterns are almost perfectly recognized. The recognition ability of the LMS energy trained network is quite different. With four hidden units, it has very big errors in recognizing patterns 5, 7, 13 and 14. With six hidden units, it

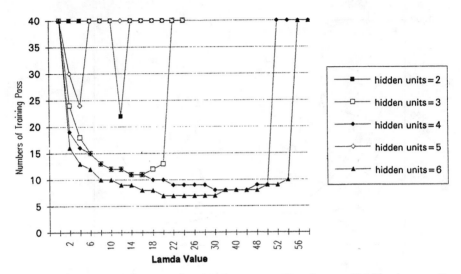

Fig. 10 The effect of different phase tuning parameter values and hidden layer unit numbers on the converging speed of *single* potential well energy trained memory networks

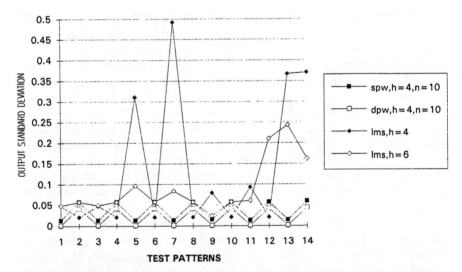

Fig. 11 The comparison of the output standard deviation of four different networks: LMS with four hidden units, LMS with six hidden units, single potential well with four hidden units and $\lambda = 10$, double potential well with four hidden units and $\lambda = 10$. It shows that the two LMS networks have much bigger standard deviations for all test patterns

improves a little, but still has fairly big errors in recognizing patterns 12, 13 and 14. Also we notice that the bad recognition occurred at different patterns with 4 hidden units and 6 hidden units. It shows fluctuation, which we believe could be explained by more profound phase transition phenomenon. This leads us to the next figure.

In Figure 12, we plot the output standard deviation value of test pattern 12 from a six hidden units ξ^4 energy network to the phase tuning parameter λ, when we change λ gradually from negative values passing zero to positive values. We see that as λ approaches the zero point from both positive and negative directions, the value of the standard deviation become bigger and bigger. At zero point, the network doesn't even converge. We interpret this as having even bigger deviation. This finding is similar to the phase transition phenomena. As λ goes from negative to positive, the energy as a function of order parameter ξ^4 changes from a single well to a double well potential. The formerly stable position $\xi = 0$ becomes unstable and is replaced by two new stable positions (Figs. 4, 5). So $\lambda = 0$ is actually a phase transition point. When λ goes from negative toward zero point, the basin of the potential well becomes flatter and flatter, so the system relaxes back to the stable point more and more slowly, which is analogous to the thermal equilibrium phase transition critical slowing down phenomenon. This explains why the training speed gets slower when λ become smaller. Furthermore, because the restoring force (gradient descent) is becoming weaker and weaker, the fluctuation of the order parameter ξ becomes more and more pronounced, thus critical fluctuations occur. At this point, it is very unstable and the system parameters have very high fluctuations. That is why we observe high network output deviation when λ approaches zero. This also explains why LMS energy trained networks have much bigger output value deviation than big λ value ξ^4 energy trained networks. Figure 13 shows the shape of an LMS energy function and a $\lambda = -0.1$ Haken ξ^4 energy function. We can see that at the basin area the LMS function closely resembles the ξ^4 function, and it even has a much flatter bottom. So the LMS function can be emulated by very small λ ξ^4 energy function. In other words, the LMS function is very close to the phase transition point. Therefore we see big output standard deviation at certain test patterns, and the big error output occurred at different test patterns for different number of hidden layer units.

To test the fault tolerance ability of the memory networks, we add noise to the input pattern by randomly reversing the input pixels, then we look at the output error standard deviations. One of the two networks we tested is a single potential well energy trained network, with four hidden units and $\lambda = -12$. Another is a double potential well trained network, with four hidden units and $\lambda = 12$. The test pattern is the face pattern (pattern 1). As shown in Figure 14. Both networks exhibit pretty good performance in recognizing noisy patterns. For up to 40 percent noise, the output standard deviation is under 0.07, which is very good.

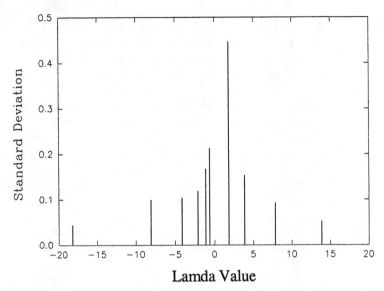

Fig. 12 Phase tuning parameter effect on pattern recognition. The closer the λ to zero (the phase transition point) the higher the standard deviations of the network output values. We use six hidden units and pattern 12 as the test pattern

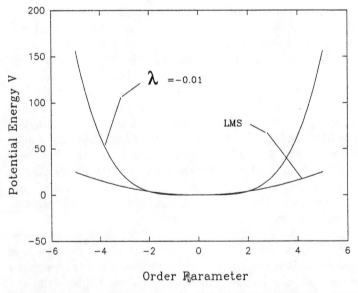

Fig. 13 The shape of the small λ potential well is similar to the LMS energy function shape. The smaller the absolute value of the λ, the closer these two functions are

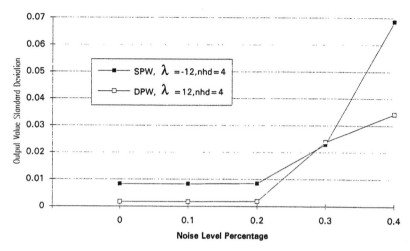

Fig. 14 The performance of the networks under noisy input conditions. The single potential well and the double potential well trained networks both did well for up to forty percent noise level input

We then change the relative gray scale of the input pattern to see if the networks could still recognize the features. For pattern 1, we gradually reduce the face gray scale from 1.0 to 0.1, letting it run through the networks (double potential well, 4 hidden units, $\lambda = 12$). The result is shown in Figure 15. We plot the relative gray scale against the output state that is defined as follows

Output State $= V1 - V2$.

For perfectly trained networks, Output State $= 1$ means it recognizes the face, Output State $= -1$ means it recognizes the vase. We can see from Fig. 13 that there is very little drop in recognizability with the decrease of the relative gray scale. We did the same thing for the test pattern 2. Figure 16 shows even better results.

Figure 17 reveals the ability of the same network to recognize the scaled down face pattern (pattern 1). We gradually reverse the boundary pixels of the face to make it proportionally smaller, then see how the output state value changes. The figure shows that the network exhibit good ability to recognize the feature until the size of the face decreases to about 20% of its original size. After that there is a sudden drop in its performance. So the memory network does have its limits in recognizing scale changed patterns. It is a good practice to implement a separate preprocessing unit to solve this problem rather than relying on the fault tolerance of the main memory net. This concept is further confirmed in Figure 18, which shows the networks performance for shifted input patterns. We gradually shift the left face of the input pattern to the right, the result reveals that the memory net alone is incapable of recognizing the shifted patterns.

326   H.H. Szu et al.

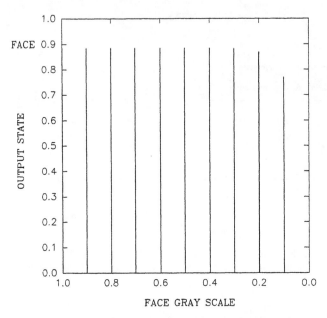

Fig. 15 The recognition by the double potential well energy trained network when presented with pattern 1 of decreased gray scales

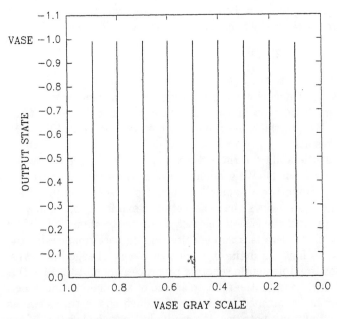

Fig. 16 The recognition by the double potential well energy trained network when presented with pattern 2 of decreased gray scales

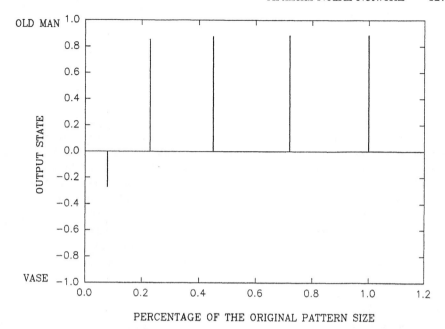

Fig. 17 The recognizability of the double potential well energy trained network when it is presented to scaled down face patterns

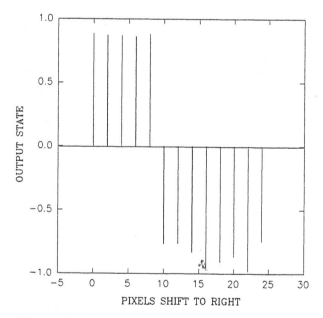

Fig. 18 The recognition by the double potential well trained networks when presented with shifted face patterns. The performance is found to be unacceptable

3 Modeling of perception oscillations and depth processing unit

From the study of ambiguous patterns such as Fig. 3, it is found that when a person looks at such figures, he will experience an oscillation in perception, he will first perceive one feature of the pattern, e.g. the old men's faces, for some seconds, then he will perceive another feature, e.g. the vase, for some seconds, then back again to the first feature.... Some of the major experimental properties of the ambiguous patterns are: (1) There must be two or more possible features of the pattern (alternative perception). (2) When one of the features is perceived, it remains stable for a period of time before it suddenly (or spontaneously) switches to another feature of the pattern. (3) The strength of the two alternatives of perception is not always equal to each other. One is preferred over the other in most cases, but the difference in strength shouldn't be too big to achieve the reversion perception.

These perception oscillation phenomena have been studied by many authors (Köhler and his colleagues, 1940; Kawamoto & Anderson, 1985, Kazuo Sakai, 1993; Haken, 1988; etc.), and a variety of models were reported. The most generally accepted hypothesis to explain these phenomena is the saturation theory proposed by Köhler (1940). He assumes that when a pattern is transmitted from the retina to the visual cortex, it will trigger an electrochemical current which in turn will enhance the resistance of the neural medium so that the neurons become saturated, and eventually force the current to change its direction. Most of the models are based on the saturation assumptions. Kawamoto's model uses a brain-state-in-a-box (BSB) neural network, with time-dependent eigenvalue functions and direct synaptic modifications, and achieved the oscillation between two states. Kazuo Sakai expanded the PDP Schema model by introducing the time dependent bias iteration equations to simulate the perception oscillation of the Necker cube problem. However the theory that fits most closely to the biological findings is Haken's synergetic computer theory, which extends the basic synergetic dynamics by introducing the time dependent attention parameters. The result of this theory has very good agreement with the experiment findings in Künnapas' experiment and Borsellino's experiment. As we look at Fig. 3, the ambiguity actually comes from the inability to distinguish the foreground and the background. If we interpret the black as the foreground then the face is perceived; if we interpret the black as the background then the vase is perceived. Because there is no depth involved, both interpretations are correct. If we look at a real vase or face, there will be no ambiguity, it is easy to distinguish the background from the foreground of a three-dimensional object. So we believe the ambiguity really comes when we look at the drawings on paper. This makes the depth-detection center unable to make solid judgments. The perception oscillation of this drawing comes from the state oscillation of the depth determining network that involves fewer neurons and switches fast. Because the perception switches from one state to another are instantaneous, we believe that they are unlikely to involve massive memory neuron synapse changes. Based on these assumptions and the saturation theory, we extend our network by including a new processing center, the depth processing unit, to distinguish the background and foreground. The new network topology is shown in Fig. 19. The depth processing net (Fig. 16) is composed of two interrelated artificial neurons with self-feedback time-dependent threshold. The structure of the memory net is the same as before, but the sigmoid function is shift down 0.5 units to let the output value be between −0.5 and +0.5 for the purpose of symmetry. The input patterns are also changed to use −0.5 and 0.5 correspondingly. The training process will let the memory net remember the two features in the pattern. The task of the depth processing nets is to pick one of the features from the pattern by distinguishing the foreground from the background. We define Flip = $V1 - V2$ as the indicator of distinguishing. When Flip = 1, it means the input value 0.5 in the input pattern is foreground and −0.5 is background. If Flip = −1, then the input value 0.5 is background and −0.5 is foreground.

We introduce the self-feedback time-dependent threshold dynamics for the two processing neurons:

$$\theta_1(t+1) = \theta_1(t) + \alpha[1 - \theta_1(t) - W_{11}(t)] \qquad (24)$$

$$\theta_2(t+1) = \theta_2(t) + \alpha[1 - \theta_2(t) - W_{22}(t)] \qquad (25)$$

where W_{11}, W_{22} are the self-feedback connection weights, α is a small constant. The corresponding neural output updating rule is:

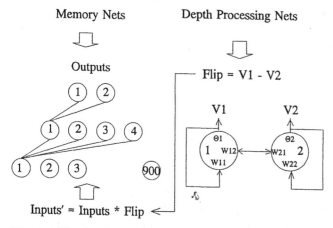

Fig. 19 The depth processing network architecture, where $\Theta 1$ & $\Theta 2$ are time dependent threshold and W_{11} & W_{22} are self feedback connection weights

$$V_1(t+1) = V_1(t) + \beta \left[\theta_1(t)V_1(t) - 4\gamma \sum_{j=1}^{2} W_{1j}(t)V_j(t)\right] \qquad (26)$$

$$V_2(t+1) = V_2(t) + \beta \left[\theta_2(t)V_2(t) - 4\gamma \sum_{j=1}^{2} W_{2j}(t)V_j(t)\right] \qquad (27)$$

where β and γ are constants.

By plugging in the Hebbian learning rule $W_{ij} = V_i V_j$, we see that the four equations have the same form as Haken's four time-dependent attention parameter dynamic equations:

$$\frac{d\lambda_1}{dt} = a - b\lambda_1^2 - c\xi_1^2 \qquad (28)$$

$$\frac{d\lambda_2}{dt} = a - b\lambda_2^2 - c\xi_2^2 \qquad (29)$$

$$\frac{d\xi_1}{dt} = \xi_1(\lambda_1 - A\xi_1^2 - B\xi_2^2) \qquad (30)$$

$$\frac{d\xi_2}{dt} = \xi_2(\lambda_2 - A\xi_2^2 - B\xi_1^2). \qquad (31)$$

For each iteration, the Flip value is calculated, then the modified input value input' = input∗ flip becomes the input of the memory net where the recognition is being made.

Figures 20 and 21 show the numerical simulation result of the model. ($\alpha = 0.02$, $\beta = 0.2$, $\gamma = 0.2$)

Figure 20 shows the output oscillation of the depth processing unit. Figure 21 is the state output of the memory network. We can see that the memory net actually smooths out the wiggles of the output of the depth processing unit and it has very steep jump from one state to another state. This very clearly reveals the biological aspect of spontaneous switching from one state to another.

4 Shift invariant preprocessing unit

Just as the human primary visual cortex (V_1) alone cannot detect all the aspects of the input patterns, the classical back error propagation networks do not work well with shifted patterns. Proper preprocessing units are needed to tackle this problem. This is very similar to the biological function of our visual system. Based on vision science, the eye and brain break up the visual world into various aspects, such as color, form, motion and depth. These pieces of the picture are interpreted in a complex network of processing centers. To form a coherent picture of the world, the eye-brain takes signals from the retinas, relays them through the lateral geniculate bodies. From there the signals

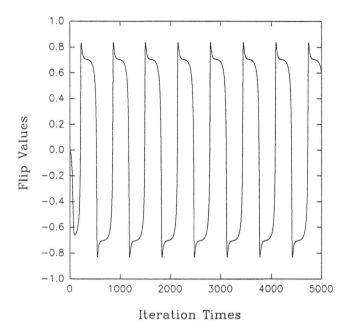

Fig. 20 The depth processing unit output. It shows the time evolution of the Flip values

proceed to a region of the brain at the back of the skull, the primary visual cortex, also known as V_1. They then feed into a second processing area, called V_2, and branch out to a series of other higher centers with each one carrying out a specialized function, such as detecting color, detail, depth, movement, shape or recognizing faces (Grady, 1993). So our visual system is composed of different centers that process a specific aspect of the input information. To use separate units in our artificial neural network system also has the engineering advantages that we can use all kinds of existing sensors (existing technology) to form our smart eye and smart ear preprocessing units. The preprocessing units can also compress the input data and reduce the degree of freedom of the input data to the main computing neural networks, therefore the main nets can become more efficient and more powerful in speed and memory ability.

The key concept of the invariant units is that they will map the input pattern to a new pattern that is not necessarily the same shape as the input pattern but is translation, rotation, and scaling invariant to the input pattern. This invariant new pattern is then sent to the main memory networks for recognition.

There are several ways to construct the preprocessing units to give invariance under translation, rotation, and scaling. Two optical information processing procedure based examples are: using log-polar-Fourier filter

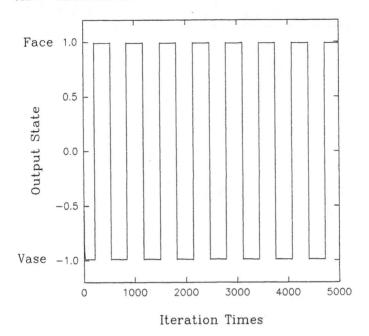

Fig. 21 The main computing unit output. An output state value of 1 means the network is perceiving the pattern as a face. An output state value -1 means the network is perceiving the pattern as a vase

(Casasent & Psaltis, 1976; Cavanagh, 1978, 1984; Szu, 1986) and optical wavelet matched filters (Sheng et al., 1993). The optical wavelet transformation is very powerful in feature extraction and is inherently shift invariant. We can also use a fixed ANN to achieve these invariances, Figure 22 shows an ANN network that is used to implement translation-invariant recognition of patterns (Widrow & Winter, 1988).

Each AD in Figure 22 is an ADALINE network (Figure 23). The weights of the upper left corner ADALINE in each slab are called key weights that are fixed and can be randomly chosen. All other ADALINE weights in this slab have the same weight element as the key weights but with shifted positions. This makes the output value of each slab invariant with respect to the translation of the input patterns. The combination of the output value of each slab will form a new pattern to be sent to the brain computing networks for recognition. By adding many more slabs, the system can be expanded to incorporate rotational and scaling invariance. It is a huge system in terms of the number of neurons involved, but it can always be achieved by using the VLSI electronics implementation. Because the weights are all fixed, once it has been built, it is very fast.

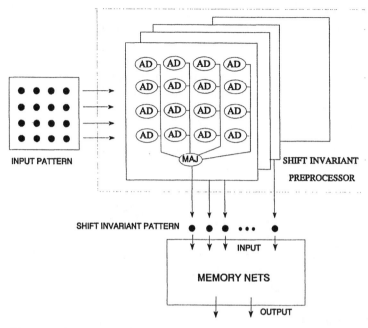

Fig. 22 The shift invariant preprocessing unit. The preprocessor maps the input pattern to a shift invariant new pattern for memory networks to recognize

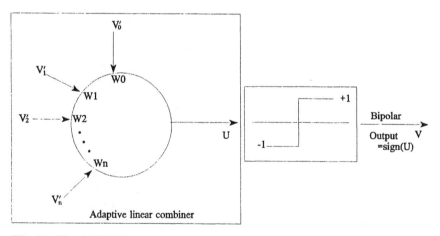

Fig. 23 The ADLINE neuron

5 Hardware implementation

Development of this model using analog electronics and fiber optics depends upon being able to expressly determine the weight matrix and minimization of hidden layer nodes (Rumelhart & MacClelland, 1986). We use simple op-amp summation nodes driven by linear response region of LEDs. Photo-sensitive transistors and thin film optical masks operate as sigmoidal approximations and weight multipliers. We close pack the fiber optic interconnects at a ratio of 10 : 1. This is an additional reason for the previously mentioned limit of hidden nodes to ten percent of input nodes. This processor has the obvious drawback of being case specific since the weight matrix must be determined before any processing.

6 Conclusion

We conclude that by using appropriate optical or artificial neural preprocessor, we can achieve translation, rotation, and scaling invariant pattern recognition. By adopting Haken's single order parameter potential function as the training energy on our back error propagation memory neural network, we get much improved performance both in training speed and recognition of perturbed and distorted patterns. We can manipulate tuning parameters to move attractors further apart or closer together depending on our needs. Moving the tuning parameter toward zero, we will see increased fluctuation, this is believed to be due to the phase transition phenomena. Because the LMS energy function is very close to the phase transition point, we observe increased fluctuation as expected. This leads us to believe that although the LMS energy is the most commonly used training energy function, it is less desirable for recognizing perturbed and noisy patterns. Also using the LMS training energy, the network is slow and poorly convergent. When $\lambda > 0$, we succeed in training the network to converge to both attractors with each representing a pattern. This network shows slightly better accuracy in recognizing perturbed patterns than single well ξ^4 energy trained network, but it has much better performance than LMS energy trained network. When $\lambda < 0$, we get a single potential well that is generally much steeper than the LMS single potential well. To train a BPN network toward its minimum energy point, it is very desirable to have big jumps initially, and then as the system moves close to the minimum point, the jump step becomes smaller and smaller until converging. This cannot be done by adjusting learning rates, because they are fixed in the training. But by using this ξ^4 single potential well, we actually achieve this requirement, as the gradient descent is very big initially, and becomes smaller and smaller when it approaches to the minimum. The LMS potential well, as shown, has a much flatter basin than the ξ^4 single well, so its average gradient descend is much smaller than ξ^4 single well. Therefore we observe much improved performance in training speed and pattern recognition. By apply Haken's time dependent attention parameter dynamics to our depth processing unit, we achieve modeling of the perception oscillation phenomena.

Although today's digital computers are very powerful in many areas, they do have their limitations in certain fields that need some intelligence, such as pattern recognition. The sequential and synchronous computing nature of the von Neumann computer will limit the further increase of its computing speed. To keep all the processing units following the same clock cycle, will generate a lot of cross talk between the units, which is not used in real computing. This overhead will cancel the benefit of further increase of the clock cycle making increasing the computing speed unachievable at a certain point. We refer to this as the second bottleneck of von Neumann computers. The artificial neural network with its limited intelligence and parallel computing power has the potential to form an asynchronous computer to overcome the second bottleneck, and reach a high level of computing speed. With the availability of the ANN VLSI chips today, this technology is becoming more and more appealing.

References

Ditzinger, A., Fuchs, T. & Haken, H. (1992): Synergetic Approach to Phenomena of Perception in Neural and Artificial Systems. Berlin: Springer.

Freeman, J.A. & Skapura, D.M. (1991): Neural Networks: Algorithms, Applications, and Programming Techniques. Addison-Wesley Publishing Company, Massachusetts.

Grady, D. (1993): The vision thing: Mainly in the brain. Discover, 57 – 66, June.

Grossberg, S. & Wyse, L. (1991): A neural network architecture for figure-ground separation of connected scenic figures. Neural Networks 4, 723 – 742.

Haken, H. (1991): Synergetic Computers and Cognition: A Top-Down Approach to Neural Networks. Berlin: Springer.

Intrator, N. & Cooper, L.N. (1992): Objective function formulation of the BCM theory of visual cortical plasticity: Statistical connection, stability conditions. Neural Networks 5, 3 – 17.

Rumelhart, D.R. & MacClelland, J.L., (1986): Parallel Distributed Processing: Exploration in the Microstructure of Cognition. MIT Press.

Sakai, K., Katayama, T., Wada, S. & Oiwa, K. (1993): Perspective reversal caused by chaotic switching in PDP schema model. ICNN.

Sheng, Y., Roberg, D., Szu, H. & Lu, T. (1993): Optical wavelet matched filters for shift-invariant pattern recognition. Optics Letters 18, No. 4.

Szu, H.H. &, Telfer, B. (1991): Minimum misclassification error performance measure for layered networks of artificial fuzzy neurons. IJCNN-91 Seattle.

Wagner, K. & Psaltis, D. (1987): Multiple layer optical learning network. Applied Optics 26, 25061 – 25076.

Werbos, P. (1974): Beyond regression; New tools for prediction and analysis in the behavioral sciences. Ph.D thesis. Harvard University.

Widrow, B. & Winter. R. (1988): Neural nets for adaptive filtering and adaptive pattern recognition. Computer, March.

Multistability in Geometrical Imagination: A Network Study

H.-O. Carmesin

Institute of Theoretical Physics and Center for Cognitive Sciences,
University of Bremen, D-28334 Bremen, Germany

Abstract: A network model for the visual system is studied which obeys 5 *physiological* constraints and explains many *psychological* observations. More generally, the model explains:

how collective neural states are established by discontinuous increases in the number of synapses,

how these collective states establish multistabilities and

how these multistabilities establish the semantic ambiguity that to a given scenario geometric relations out of a huge set of possible geometric relations can be assigned.

1 Introduction

The visual system is essential for humans, so many semantic ambiguities originate from the use of the visual system. Accordingly, the study of the visual sytem and its ambiguities has a long tradition in cognitive psychology (Köhler, 1971; Piaget, 1975; Kruse, Stadler & Wehner, 1986).
There have been many attempts to model the visual system with neural networks (Grossberg, 1988); however, the network investigators only rather occasionally used the rich experimental fundus offered by psychology (Köhler, 1971; Piaget, 1975). In contrast, I started an extensive network study (Carmesin, 1992, 1993) directly from Piaget's concentrated results (1975, 1000 pages). The presented network models are constructed in order to solve the tasks studied by Piaget (1975); whether such tasks are usually performed by the visual system is not investigated. While the full set of tasks is modelled in Carmesin (1992), we concentrate on illustrative examples here; among the full set of tasks are *to find a yellow pearl, to distinguish sequences of pearls, to draw the way home, to redraw pictures, to distinguish knots, to estimate sizes, to relate sizes, to foresee the effect of motions*. Such modelling has recently been used by biologists Crick and Koch (1990).

The models are linked to biology as follows. Five physiological constraints are obeyed, the neuronic dynamics is of the McCulloch and Pitts type (1943), and the coupling dynamics uses a result about the Hebb rule (Hebb, 1949; Carmesin, 1992, 1993; Kreyscher, Carmesin & Schwegler, 1993) in a slightly idealized and analytically efficient manner.

The modelled networks perform active interventions in order to excite desired collective neural states which analyze a presented scenario. The collective states are highly but not uniquely determined by the presented scenario. So, an ambiguity remains. On the neural level, this ambiguity corresponds to the multistability of the collective states. The modelled ambiguities have been observed in psychological experiments (see Figs. 3–5 below or Piaget, 1975), and the corresponding multistabilities will hopefully be observed in future physiological experiments.

2 Neural network

The modelled effects are essentially due to the fact that *there are significantly more neurons in the brain than synapses at one neuron*. For the sake of explicitness, this fact is expressed through the numbers 10^{12} and 10^4 in the following.

Five physiological conditions are obeyed.
PC1 Neuronic constraint: There are roughly 10^{12} neurons.
PC2 Synaptic constraint: There are roughly 10^4 synapses per neuron.
PC3 Time constraint: The synapses are so that a task is performed by the visual system as fast as possible.
PC4 Retinotopic map: The external scenario is represented by the activity of a two dimensional assembly of 10^{12} neurons (one eye and all retinotopic maps together). This assembly is called map, its centre is called fovea.
PC5 Eye movement: The eye can move. If the network does not excite an eye movement, then the eye moves so that something significant is at the fovea which is viewed.

Dynamics of neurons. We consider a network of neurons s_i with the values ± 1. The dynamics is defined through the *neuronic equation* (Caianiello, 1961; van Hemmen & Kühn, 1991)

$$s_i(t+1) = \text{sgn}[\Sigma_j J_{ij} s_j(t) - \lambda_i].$$

Thereby sgn is the signum-function and λ_i is a threshold value. The matrix elements J_{ij} are called coupling constants or short couplings and correspond to the strengths of the synapses. The equation defines how signals are transferred between the neurons, the coupling J_{ij} transfers the signal $J_{ij}s_j$ from s_j to s_i. The *time delays* of the signals are one step of time in the above equation. In the human brain there occur very different delay times. We assume accordingly that couplings with longer delay times are also available, i.e., we generalize the above equation slightly. Another slight but essential generalization is introduced in Sect. 5 and concerns a *time out*.

Dynamics of synapses. The synapses are formed so that a task is performed by the visual system as fast as possible (first priority) and with as few synapses as possible (second priority). The number of synapses is called *complexity*, so the complexity is minimized with second priority. Altogether, the synapses are determined by an *optimization problem*. In order to avoid a conceivable misunderstanding I emphasize that the networks found in nature need not be similar to those found through the optimization problem, however, the natural networks are at least as complex as (or slower than) the analyzed networks.

Analysis of the network.

High resolution limit. The above optimization problem is solved approximately and so that it becomes exact in the following asymptotic network N_∞. The network N_∞ is the same as the above, except that it has α synapses per neuron and α^3 neurons. The limit $\alpha \to \infty$, called *high resolution limit*, is analyzed. In the human cortex $\alpha \approx 10^4$. The complexities are calculated in the same approximation.

Ripening. An increase of the number of synapses by at least α is called a *ripening*, because such an increase is expected to take relatively long time.

Subnetworks. We assign some subtasks to subnetworks. These need not be separated from other networks. Subnetworks that handle geometric relations are called *geometry networks*, short GN.

Simplified periphery. Here, some networks can perform tasks only at the fovea and not peripherically. In nature the difference is gradual; tasks can be performed well at the fovea and worse peripherically (Strasburger, Harvey & Rentschler, 1991).

Biological plausibility of the neural network. The constraints PC1-PC5 are idealizations of established physiological facts. The optimization of complexity emerges roughly from the Hebb rule (Hebb, 1949), if the cognitive system repeats successful operations (Carmesin, 1992, 1993; Kreyschler, Carmesin & Schwegler, 1993). Recently, the physiological conditions under which visual processing might occur within the short term memory have been reviewed by biologists Crick and Koch (1990). I was surprised when I recognized that many such conditions are similar to those modelled here, or obtained for this network model below. In particular, the (important) spatio temporal coincidence obtained below in network IA has been required in Crick and Koch (1990).

3 Foveal multistability

Task 0. *Detection:* The visual system shall detect (with a single neuron) whether a small spot of light is presented.

Network 0. The spot of light can be at any of the 10^{12} neurons (PC1). So, the problem is to transfer from any of these 10^{12} neurons a signal to the detector neuron, though the detector has only 10^4 synapses. A solution of this problem according to the primordial optimization is the following. When the spot occurs, then one neuron fires. One time step after the spot occurs, the signal has been transferred to a second neuron and the second neuron fires. Because one neuron can gather at most 10^4 signals from other neurons (PC2), 10^8 such second neurons are necessary. The signals of these second neurons are gathered by 10^4 third neurons analogously and a third neuron fires two time steps after the spot occurred. One fourth neuron gathers the signals from the third neurons and fires three time steps after the spot occurred.
Altogether, the network looks like a tree, consists of 10^{12} synapses and requires three time steps.

Remark. If this task 0 occurs in a real brain, then it is likely that the couplings are different from those of the network 0. But they are not of lower complexity, as emphasized above.

Complexity - velocity dilemma. If the visual system had neurons with 10^{12} synapses each, then this task could be solved in one time step. We observe that the same visual system but without synaptic constraint could perform several times faster than the actual visual system can do. This fact is typical for usual tasks, see below. *The velocity of the visual system is significantly limited by the number of available synapses.*

Eye-movements. If the visual system expects the spot of light to occur at the centre of the fovea, and if the centre of the fovea has 10^8 neurons only, then the task can be performed in only two time steps (PC5). We observe that under appropriate conditions eye-movements are able to speed up the performance of a task. It will turn out below that the complexer a task, the more it is speeded up by eye-movements. *Eye-movements help compensate the limited number of synapses of the visual system.*

The foveal multistability. If the visual system expects two spots of light to occur at different places, then it locates the fovea towards one of these expected light spots. Each of the two locations is equally stable. So there occurs a bistability. More generally, we obtain the following. *When the visual system performs a task, then at an instant of time there are in general several locations of the fovea equally appropriate. But the visual system realizes only one of these locations. This situation is called the foveal multistability.*

4 Preoperational multistabilities

Task 0A. *Pattern recognition* (van Hemmen & Kühn, 1991): A set of up to 10^4 (storage capacity, see PC1, PC2, van Hemmen & Kühn, 1991) patterns is given in a training phase. For the storage capacity, it makes no essential difference whether symmetric couplings are realized or not (van Hemmen & Kühn, 1991). For simplicity, we consider symmetric couplings; as a consequence, we obtain a potential with basins of attraction. The task is to assign each of the $2^{10^{12}}$ possible patterns (PC1) to a similar one of the 10^4 learned patterns.

Network 0A. The network forms a potential with basins of attraction. According to the definition of the above synaptic dynamics we can neglect technical difficulties like spurious states (van Hemmen & Kühn, 1991). We call the state of lowest (or relatively low) potential in a basin of attraction a *pattern classon*. It is a *collective state* of neurons. The basins of attraction give rise to a multistability, because there are several basins.

Task 0B. *Recognition of objects*: A set of objects with (for the visual system) distinguishable *characteristics* (sufficient perspective independent for the distinction between objects) is given in a training phase. For each object a symbol (by social convention) will be agreed in addition. At one time the visual system is shown exactly one object from any perspective and the system shall name it, i.e., assign the agreed symbol.

Network 0B. Because the complexity is minimized, every object will be recognized by a minimal set of characteristics. The system operates in a similar way as for the recognition of patterns, however not in the space of the patterns but in the space of the characteristics, called *characteristic space*. This space is the state space of corresponding subnetworks, called the *characteristic networks*. In this state space there are basins of attraction, in analogy to the above. An object is recognized as soon as the network has taken the state of lowest potential of a basin of attraction. This collective state is called *object classon*. The basins of attraction give rise to a multistability. The multistabilities of tasks 0A and 0B are purely due to the parallel processing of networks, but not due to an operation like the location of the fovea. For that reason we call these multistabilities *preoperational*.

5 Separation multistability

Task IA. *Object separation:* Several pearls are lying on the table. One of the pearls is yellow. (The yellow colour is allowed to be partially rubbed off.) An action (e.g. to raise single pearls) as aid is impossible (e.g. because the pearls are far away). There are no conspicuous neighbourhoods (e.g. all pearls except one are lying close together). The system shall point to the yellow pearl.

First basic mechanism for IA: Coincidence. Physiologically, the neurons in the map do not fire continuously, but in spikes. At the beginning, spikes are fired in some stochastic manner. Accordingly, we model spikes followed by a time out (this aspect is similar as in Gerstner & van Hemmen, 1993). For the sake of formal simplicity, we introduce *neuronic equations for spikes with time out*: Neurons take values at times 0, 1/2, 1, 3/2, 2, ... $s_i(t+1/2) = minimum[-s_i(t), sgn(\Sigma_j J_{ij} s_j(t) - \lambda_i)]$. As a consequence, whenever a neuron fires at a time t, then it does not fire at a time t+1/2. *It will turn out that this is the essential effect of the spikes for binding signals and for searching for novel signal configurations.* At the time when an object is classified in the characteristic network, the characteristic network receives the characteristics from certain places in the map, called *registering places*. The characteristic network sends signals to all places in the map; these signals excite another signal at the registering places, called *coincidence signal*. So the registering places are marked by the coincidence signals; the latter are used for further processing.

Necessity of a ripening. It seems to be obvious that the unification of the characteristics is formed, i.e., the characteristics of pearls and the characteristic yellow are both sought. But that is too easy, because a yellow dice could lie on the table, but no yellow pearl. Then you would find on the table the characteristics for a pearl and the characteristic yellow, but no yellow pearl. Consequently, it is not enough to search in the characteristic space, the system has rather to make a *spatial relationship* between the characteristic yellow and the characteristics of a pearl. Because an action which could create a spatial relationship is impossible (according to the task), and there are no conspicuous neighbourhoods between the pearls, only the relative position of the pearl to the system remains as the *spatial clue*.

This relative position can only be established in the map. Therefore, the system has to mark the place of the pearl inside the map. This happens without any additional signal in the map (and therefore with a minimum of processing time and complexity) by the fact that the place of the viewed pearl is the fovea (PC5) (first of all any pearl you like). But it is not enough to mark the place of the pearl, because then the whole map would be active and yellow colour could be found at any time by neurons at other places. Therefore, it is necessary to passivate by signals those places which *are not assigned to the found pearl*, i.e., to *fade out* such places. Here a coupling with each place is necessary (to fade out the places they have to get a corresponding signal). Thus, a ripening is necessary (PC1).

Formation of a spatial relationship to characteristics through a spatio-temporal coincidence. For this we proceed from the object classon. Without increasing the complexity we coordinate the delay time of the signals so that the classification of every object needs the same time t_E. We call the neurons of a place i, which are able to send signals to the characteristic network *characteristic neurons* s_i^m. We assign to every characteristic neuron a *coincidence neuron* $\bar{s}_i^m$. A *latency signal* is a signal which is alone too weak to determine the value of the neuron that it enters. The *unordered classon* is +1 as long as no object is classified, otherwise it is −1. The coincidence neuron $\bar{s}_i^m$ gets a latency signal from the corresponding characteristic neuron s_i^m by a coupling with the delay time $t_E + 1$, as well as another latency signal with the coupling −1 from the unordered classon c_{ne} and $\bar{s}_i^m$ takes the value 1 when two positive latency signals come in. This "networking" has the following effect.

At the time when an object is classified in the characteristic network, the characteristic network receives signals from certain characteristic neurons at places i. The coincidence neurons of these places are switched on. These places are called registering places.

Second basic mechanism for IA: Frame. It would be conceivable to leave only registering places unpassivated. An object can be registered admittedly by very local characteristics, without leaving the main part of places of the object unpassivated. This will only be avoidable if the system adds an aspect to the characteristics, namely, a *compactness*. The compactness is made by the formation of a frame by signals, because signals are the only means by which a frame can be established in the map. The frame is necessary to make the persistent localized observation of an object at all possible.

ring-frames constitute the full-frame. It is reasonable, to make the frame approximately isotrope. The formation is possible in three time steps (PC1, PC2), analogously as for network 0. In order to make possible frames of any radius, the frame has to be constructed by approximately concentric circles, we call these *ring-frames*. A ring-frame is established by 10^4 neurons which represent the ring-frame by their activity. The ring-frames establish the frame, also called full-frame, by their activity.

Formation of ring-frames for coincidence neurons. Concentric ring-frames are made with the fovea in the centre. Each ring-frame is made of 10^4 places i. Each place i has a *frame neuron* s_i^u. To activate it you proceed from the active coincidence neurons $\bar{s}_i^m$. One time step after a coincidence neuron is +1, all frame neurons s_j^u of the frame corresponding ring-frame are active: For it every coincidence neuron of a frame sends a signal with the coupling 1 to every frame neuron of the ring-frame. The threshold of every frame neuron is $10^4 - 1$, to balance the negative signals of the passive coincidence neurons.

Formation of a full-frame. To form the full-frame, all those ring-frames have to be activated (internal frames), the radius of which is smaller than that of the activated ring-frame of largest radius, all other ring-frames have to be passivated (external frames). For it all frames are sending signals by couplings of strength +1 to all internal frames, but no signals to external frames. The reaction time is the same for all ring-frames, because there are 10^{12} places, therefore there are 10^8 ring-frames with 10^4 places each. Every ring-frame is able to send 10^8 signals simultaneously, namely, 10^4 from each of its 10^4

frame neurons. Therefore a given ring-frame is able to send simultaneously a signal to every internal ring-frame.

For it a *conjugated frame neuron* $\bar{s}_i^u$ is introduced. It gets the above mentioned signals from the other frames and it is +1 when it gets at least one positive signal (possible by a suitable threshold). The conjugated frame neuron $\bar{s}_i^u$ sends a signal by a coupling of strength +1 to every frame neuron s_i^u of its ring-frame. The threshold will be subsided (see above), so that a frame neuron s_i^u is +1, if it gets a positive signal from at least one of coincidence neurons of its ring-frame or from one of the 10^4 conjugated frame neurons of its frame. All frame neurons s_i^u are sending a positive latency signal to every characteristic neuron of the corresponding place which is so weak that it has normally no influence on the characteristic neuron. But the GN sends latency signals homogeneously to all characteristic neurons so that only characteristic neurons at places with active frame neurons s_i^u can be active in the next time step, and so that these characteristic neurons are active under the same conditions as before. I.e., the spaces which are situated in external ring-frames are faded out.

Network IA. The whole networking guarantees that one time unit after the classon c_j of the classified object is +1, all corresponding coincidence neurons s_i^m are +1. 2 time steps later, all frame neurons s_i^u of the corresponding frames are active, 3 time steps later the corresponding frame neurons $\bar{s}_i^u$ of the inside situated frames are activated and 4 time steps later, the frame is completed. I.e., all frame neurons s_i^u which are inside a circle which surrounds all registering places are active. 5 time steps later $s_i^m = -1$ for all characteristic neurons outside the frame. I.e., *the fade out is completed 5 time steps after the object was classified.*

Optimization. The above is the quickest way to produce the fade out. Because 5 time steps at the earliest after the classon is activated, a coincidence can be found (1 time step) and this can be communicated to 10^{12} neurons (3 time steps) by signals and with the latter one the fade out can be excited (1 time step).

In a ring-frame there are 10^4 coincidence neurons, each of which interacts with the 10^4 frame neurons. For this networking there are therefore 10^8 synapses in each ring-frame. There are 10^8 ring-frames, therefore 10^{16} couplings are used. A lower complexity would enlarge the processing time. This ripening is called *frame ripening*. It makes possible the following:

1. Searching instability. The above networking has an important side-effect: Every state of the persistent finding of an object which is at all classifiable in the system is stable if that object is presented. It will be unstable if no accordant object is presented. The unordered classon c_{ne} is unstable if any object which can be classified is presented. Exactly this *stability profile* is appropriate to a searching process. We call such an instability *searching instability*. The question of how the first finding happens shall be ignored in this case, because a lot of mechanisms are in question, from a fluctuation by chance up to the use of a cleverly thought-out searching strategy.

2. Selective searching instability. We study the seeking of an object out of a chosen class of objects, specified by a set of characteristics. A selective searching instability is achieved by the system in the following way. In the characteristic network, exactly those states which belong to the class relevant objects are made latent by signals. This is made with minimal complexity, so that the characteristics are imitated during the finding of the object, but with appropriate strength so that it is only possible to classify objects which are imitated as well as registered in the map. For that purpose the imitating network sends the signals which it generates into the characteristic network, however, with latent strength.

These imitations shall not be achieved by the characteristic network, because it cannot distinguish between imitated states and states which are caused by the map. And the imitation shall not be caused through the map, because it is impossible to distinguish between signals caused by sensory inputs and signals caused by imitation. That is the reason why the imitation shall be made by another network, the *recognition network*. This recognition network is possible as a network which has a basin of attraction for the chosen set of characteristics.

3. Finding the yellow pearl. The system can find the yellow pearl by using the frame ripening. Depending on the difficulty of the scenario, the GN can use the frame ripening for an appropriate process of finding.

3.1. Finding by imitating the characteristics. The recognition network imitates the characteristics of a pearl and the characteristic yellow. The system can find a yellow pearl. But it can also find a blue pearl with a yellow dice, because the spatial relation established by the coincidence does not fully exclude this possibility in this process.

3.2. Finding by successive imitation of characteristics. The recognition network imitates the characteristics of a pearl first. As a consequence, the system can find a pearl. While the pearl is persistently found, the recognition network imitates the characteristic yellow. As a consequence, the system can continue to find the pearl, if there is yellow colour in the frame. The system can still persistently find a blue pearl, if yellow colour happens to be in the frame.

3.3. Improved finding by contraction. The above ambiguity (in 3.2) can be decreased as follows. After the system found the pearl in the process 3.2, the GN contracts the frame by interventions through signals in such a manner that the pearl is still found (by trial and error). In this manner the frame becomes approximately as large as the pearl which is persistently found. Then the GN proceeds as in process 3.2. It is now quite unlikely that yellow colour is still in the frame but not at the pearl. So it is likely that the pearl is found persistently only if it is yellow.

Remarks: The definition of a yellow pearl is not unambiguous at all. For instance, the colour may be partially rubbed off; or the boundary of the pearl might be ambiguous, if the pearl is made of cloth. Therefore several processes of finding are adequate. More refined constructions of frames, including anisotropic frames, are possible; this is beyond the scope of the present study.

On the ripening complexity. The high complexity of the ripening would be avoidable in two ways: Hierarchical: Places are combined in groups by couplings, these groups are combined in groups of a higher class and so on. This possibility requires too much processing time because each hierarchical level requires at least one processing step. Statistical: The system could arrange a subset of neurons which is distributed in the map and processes signals rapidly and ignores those spaces that do not belong to the subset. But then the reliability and the resolution would be restricted. Such a system might exist additionally, but the main system has a high resolution (von Campenhausen, 1981). Further experiments are desirable to examine whether the resolution is reduced under some conditions.

Separation multistability. In general, the process of searching has several stable final states. Such a (final) collective state of the neurons is called the *separation classon*. The multiplicity of final states gives rise to a multistability, called *separation multistability*.

Neighbour multistability.

Task IB. To find a neighbouring object. A quantity of objects, where an object x will be found persistently is presented. A neighbouring object y has to be found (and named).

Network IB. The task will be solved with minimal complexity, if the existing networks are used. The frame will be enlarged by a GN (if necessary gradually) in such a way that a neighbouring object will finally be inside the frame. For this neighbouring object there is the same searching instability as for the original one. So, it will be finally found. The GN has to be able to activate all frames with one coupling, i.e. 10^8 couplings are necessary. This ripening is called *frame-field ripening*.

Neighbour multistability. As there are in general several neighbours available, but only one found, there occurs a multistability, called *neighbour multistability*. The established collective neural state is called *neighbour classon*.

6 Sequence multistabilities

Task IIA. A rectangular table is given. At two edges there are toy telephones. The telephones have to be connected with telegraph poles and the lines of telegraph poles have to run parallel to the table edges.

Network IIA. The solution of the task demands the transmission of the orientation of the table edge to the orientation of a line of telegraph poles.

That provides the recognition of the orientation of the table edge. In addition, one direction on the straight line has to be kept during the process of construction of the telegraph poles.

Necessity of a ripening. To recognize the orientation of the table edge one point of the table edge will be moved into the fovea (PC5). As the other recognized places of the table edge are situated at any place at all in the map, in general, every place in the map has to be hitched up. Therefore a ripening is necessary.

Kinds of ripening.

(1) Recognition of the orientation. As the places of the table edge could be scattered in a finite angle interval, the smallest sector which contains all places with a $\bar{s}_i^m = 1$ of the table edge should be registered. Such a sector can be registered at every side of the fovea. Here, places which are situated up to 36° apart should be "united", see below for details. This procedure is similar to that with the frame (IA), accordingly the networkings are similar. A number of 10^4 places builds an *elementary sector*, this corresponds to the frame (IA). Five *sector neurons* $s_i^r, s_i^l, s_i^z, s_i^c$, and s_i^R are introduced, they correspond to the frame neuron s_i^u.

The activation begins with a coincidence neuron as follows. If a coincidence neuron $\bar{s}_i^m$ of an elementary sector is active, then the sector neurons s_j^c and s_j^z of the elementary sector are active in the next time step (networking analogous to IA). Each of the 10^4 neurons s_j^z of an elementary sector activates 10^3 neurons s_k^l, that are in the 10^3 neighbouring elementary sectors that are situated to the left of the elementary sector of the coincidence neuron $\bar{s}_i^m$. So, in each of the 10^7 next elementary sectors to the left there is one sector neuron s_k^l activated. If one sector neuron s_k^l is activated, then all 10^4 sector neurons s_k^l of the elementary sector are activated at the following time step. The elementary sectors to the right are activated analogously. As there are 10^8 elementary sectors, the neurons s_n^r and s_k^l in one tenth of the elementary sectors are activated in this manner, i.e., the corresponding neurons are activated in the neighbouring 36° of the primordial coincidence neuron. If the neurons s_i^r and s_i^l of a place i are active, then s_i^c will also be active (Fig. 1). This can be achieved as follows. Let x be the number of coincidence neurons $\bar{s}_i^m$ of the place i. Then the coupling is as follows:
$s_i^c(t+1) = \sigma(\bar{s}^{m_1}(t) + \bar{s}^{m_2}(t) + ... + \bar{s}^{m_x}(t) - x + 0.5 + 0.5s_i^r(t) + 0.5s_i^l(t))$.
Altogether, exactly at the places i of the searched sector, the sector neurons s_i^c are active and the orientation of the table edge has been registered by this activity.

(2) Optimization for the recognition of the orientation. In this manner, the recognition of the searched sectors happens as quickly as possible, because an activity $\bar{s}_i^m$ is transferred to 10^4 neurons s_i^c and s_i^z at the *first* time step, at the *second* time step to 10^7 neurons s_i^r and s_i^l in neighboured elementary sectors, at the *third* time step to 10^{11} neurons s_i^r and s_i^l in the same elementary sectors, at the fourth time step to those s_i^c for which s_i^l and s_i^r are active.

Roughly 10^{16} couplings are used for the quick signal processing. The ripening is named *sector ripening*. The complexity is moderately to the scale 10^{16}, i.e., minimal.

(3) Transmission. After the orientation has been registered, the GN moves the eye saccadically (i.e. quickly and without the possibility of correction) (see e.g. von Campenhausen, 1981) so that the places of the telephones are situated in the fovea. The saccadic eye movement is quicker than the switching time of the neurons, consequently, the orientation is still shown by the s_i^c. After the activity of the s_i^c has decayed, another look to the table edge is necessary.

(4) Keeping the direction. To keep the direction during the construction of the telegraph poles it should be guaranteed that only in one of the two possible sectors are the neurons s_i^c active. The sector in which the most s_i^c are active early should be chosen, for details see the following networking.

Unite 10^4 neighbouring elementary sectors to a *block* and assign a neuron s^B to this block. At the beginning $s^B = 0$. In each elementary sector one s_i^c is assigned which is coupled to the corresponding s^B with strength $+1$. To each neuron s_i there is another one related by a point reflection at the fovea, we denote the latter neuron (also) by $\tilde{s}_i$. The couplings to s^B are as follows $s^B(t+1) = \sigma(\Sigma^{Block}(s_i^c(t) - \tilde{s}_i^c(t)) + 30000 s^B)$. Consequently, s^B is zero, as long as the activities of the corresponding sector neurons s_i^c and $\tilde{s}_i^c$ balance each other, afterwards, s^B takes the value 1 if s_i^c dominates first, and -1 if $\tilde{s}_i^c$ dominates first.

10^4 neurons s^R are introduced and they are zero initially. The couplings to these s^R enter from all those 50 percent of the neurons s^B that are assigned to the neurons s_i^c above and left of the fovea as follows $s^R(t+1) = \sigma(\Sigma^{Blocks\ above\ fovea} s^B(t) + 20000 s^R)$. Consequently, all s^R take the same value. This value is zero, as long as the activities of the corresponding s^B balance each other. Afterwards that value is taken which dominates first.

Analogously, 10^4 neurons $\bar{s}^R$ are introduced and activated that receive signals from those s^B that correspond to neurons *above and right of the fovea*.

If the place i is above and left of the fovea or below and right of the fovea, then the neuron s_i^R is 1, if s^R is 1 *and* s_i^c is one, and -1 otherwise. If the place i is above and right of the fovea or below and left of the fovea, then the neuron s_i^R is 1, if $\bar{s}^R$ is 1 *and* s_i^c is one, and -1 otherwise.

The networking achieves the desired direction instability, i.e., a sector is activated and the sector that is related by a point reflection at the fovea is passivated. The ripening is called *constant-direction ripening*.

(5) Optimization of the constant-direction ripening. The timing is as follows. At time step 1 the s^B are switched, at time step 2 the s^R are switched, at time step 3 the s_i^R are switched. This is as fast as possible, because at each place signals from each place must be taken into account (two time steps

since the elementary sectors are already processed) and the conjunction with s_i^c must be processed at each place.

The complexity is 10^{12} for the couplings of the s^R to the s_i^R, 10^8 for the couplings of the s^B to the s^R and 10^{12} for the couplings of the s_i^c to the s^B. So, the complexity is roughly 10^{12}. This is minimal, because each place needs a signal.

Sequence multistabilities. The process of solving this task with the visual system is a collective state of neurons, called *straight sequence classon*. (This collective state is significantly extended in time and not homogeneous in time.) First, this collective state establishes one orientation out of several orientations. So, there is an *orientation multistability*. Second, this collective state establishes one direction out of two directions. So, there is a *direction bistability*.

Process of following.

Task IIB. A closed pearl necklace is presented. The system has to follow the pearls with the eye and name eventually the colours for the control (Fig. 2). Remark: To make the eye movement the system has to mark the places of the connecting line for some neighbourhood in the map by neuronal activity. Because by the mere searching of neighbourhoods of the network IB, the sequence would neither mark an orientation nor keep the direction.

Network IIB. The GN arranges the finding of a first pearl (IA), the places of which are moved to the fovea by eye movement (PC5). Afterwards the GN arranges the finding of a neighbouring pearl (IB) and then it arranges the activation of a corresponding sector. The GN lets this next pearl move to the fovea by an eye movement. Then the GN excites the searching of another pearl inside the still active sector and neighbouring to the previous pearl. As the pearl necklace does not lie straight, an extension of the sector is necessary. This happens similar to the frame field ripening. Into every elementary sector there enters a coupling of a GN. It can activate a neuron $\bar{s}_i^m$ and as a consequence the whole elementary sector. As there are 10^8 elementary sectors, it requires 10^8 couplings, the ripening is named *sector field ripening* and switches as soon as possible with minimal complexity.

More sequence multistabilities. As there are in general several pearls in a neighbourhood, but only one is found, there occurs a multistability, called *successor multistability*. As there are in general several ambiguous situations in which the line bifurcates approximately, but only one path is followed, there occurs a multistability, called *bifurcation multistability*. The collective state of neurons is called *sequence classon*.

Control of the direction bistability.

Task IIIA. Given are two thin rods on which a set of pearls of different colour is stuck in the same order. Both rods are either presented side by side in the same direction or in the opposite direction. That means either (red yellow blue) (red yellow blue) or (red yellow blue) (blue yellow red). The system has to distinguish reliably whether both orders are equal or not. By that it has to recognize the changed order as unequal; and it has to come to a conclusion only by examination with the eye.

Network IIIA. The direction instability decides the GN by the production of a corresponding field (i.e. by corresponding signals which are sent to relatively many neurons) in one direction. With it it is necessary that the 10^4 neurons s^R can be switched from outside the map. That requires 10^4 couplings. The ripening is called *direction field ripening*, switches as quick as possible in one time step. and has the minimal complexity 10^4.

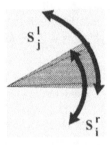
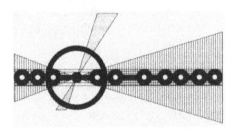

Fig. 1 The sector neurons s_i^c are active in that sector, in which signals are recieved from the left *and* from the right

Fig. 2 Necklace with pearls. A straight necklace is illustrated for simplicity. Those pearls are sought, which are in the neighbourhood (circle) *and* in the sector with active neurons s_i^c (large sector) but not in any other sector (small sector)

7 Multistabilities in comparison

Complexity of multistabilities and development. The studied multistabilities have two significant properties. *Ripening property: The multistabilities become possible only after a huge number of synapses has been implemented, i.e., after a ripening has been established.*

By the complexity of a multistability we understand the complexity of the ripening which makes possible the multistability. In order to visualize these complexities, we use a logarithmic scale (Fig. 6).

Basic property: The multistabilities are presented in a certain order. A multistability presented later requires not only its own ripening, but also the ripenings of the multistabilities presented before.

The multistabilities of the *stages* IA, IB, IIA, IIB and IIIA correspond to the developmental stages studied by Piaget (1975). A more detailed study (Carmesin, 1992) shows that the presented networks explain all those experiments in (Piaget, 1975) that concern two dimensional tasks.

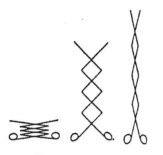

Fig. 3 Nürnberger scissors, closed, opened slightly, opened

Fig. 4 Psychological observation: Stage I. At stage I a child can detect the relation *closed*. As a consequence, it recognizes the rhomb as a closed figure and draws it accordingly

Fig. 5 Stage IIA. At this stage the child detects the relations *straight*, *equally oriented* and *successor*. So, it may recognize a rhomb as the closed sequence straight line, corner, straight line, corner, straight line, corner, straight line, corner (right part). However, for that purpose the child must follow one out of three possible edges, when it reaches the point at which two rhombs are linked. Here, the bifurcation multistability occurs. Presumably, it is easier to follow in a straight manner; in that case the child produces intersections (left part)

Comparison with Piaget's work. Here, a theory is presented which explains the psychological experiments *and* obeys the physiological constraints PC1 – PC5. As a consequence, the observed developmental stages are explained in terms of *discontinuities of complexity* as follows. The performance of a task requires roughly 10 – 100 synapses (in a GN), while a ripening requires more than α synapses; in the limit $\alpha \to \infty$ there occur discontinuities of complexity at the ripenings. As a further consequence, the ripenings establish those multistabilities and the control of those multistabilities by which the cognitive system analyzes a given scenario. Piaget also suggested a developmental stage IV. In the present study we identify this stage as very different from the others, because no ripening corresponds to the stage IV. Instead stage IV corresponds to the learning of detailed knowledge about geometrical rules. These geometrical rules are usually described by axiom systems (Cederberg,

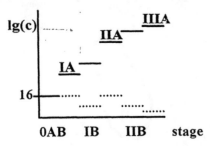

Fig. 6 Complexity of multistabilities. The multistabilities are indicated at the horizontal axis, while their complexities are shown at the vertical axis in logarithmic scale (lower curve). According to the basic property also the sums of the logarithms of the ripenings already performed are indicated (upper curve)

1989, pp. 26, 201). The cognitive system modelled here can learn any such rules for any goemetry. For instance, the presented cognitive system can learn Euclidean geometry as well as hyperbolic geometry.

The following is emphasized. Piaget (1975) studied a set of more than hundred geometrical tasks. He partitioned these tasks into 7 subsets, one for each of the stages IA, IB, IIA, IIB, IIIA, IIIB, IV. The same partitioning has been obtained as a result of the five physiological constraints together with the optimization. For comparison, the number of possible partitions of 100 tasks into 7 subsets is very roughly $7^{100} \approx 10^{80}$.

Multistabilities, their collective states and their invariants. The multistabilities consist of *single stabilities* of several corresponding *collective states*. Each of these single stabilities establishes a stabilitiy of the collective state with respect to certain *irrelevant changes*. In other words, the collective state remains *invariant*, if these irrelevant changes are applied. The invariants are *geometrical relations*, except for the preoperational stages 0A and 0B. The immediately detected geometrical relations enable the system to detect further geometrical relations (Carmesin, 1992). For stage 0A the irrelevant changes are fluctuations, for the stages 0B-IIB the irrelevant changes are noise and perspective. Here the collective states are states of neurons. They are presented together with their invariants in Table 1.

Table 1. Multistabilities at various stages, their collective states with their invariants and geometrical relations.

Stage	classon	invariant	further geometrical relations
0A	pattern	pattern	—
0B	object	object	—
IA	separation	area of object	inside, closed
IB	neighbour	neighbourship	near, overlapping
IIA	straight sequence	successorship	adjacent, equally oriented, straight
IIB	sequence	successorship	long, connected

Psychological example. "Nürnberger scissors" are presented to children and opened slightly. The children are asked to draw the rhombs and how they develop when the scissors are opened further (Peaget, 1975). Typical results are shown for a child at stage I and for another child at stage IIA in Figs. 3–5. The bifurcation multistability (Fig. 5) establishes a *semantic ambiguity* in the geometric relations which the cognitive system assigns to the scissors: The cognitive system relates the edges either so that they form rhombs, or so that they form crossing lines. In contrast, equal orientations are detected without ambiguity for the presented scissors; so, the children draw these without difficulty (Fig. 5).

8 Conclusion

A network model for the visual system is presented which obeys five physiological constraints and solves Piaget's tasks. To solve the tasks, the system establishes multistable collective neural states. These collective states analyze a given scenario by active interventions with neural signals and eye movements.

The collective states establish geometric relations for the scenario. While there is a huge set of possible geometric relations which the cognitve system *can* assign to a given scenario via its collective states, the cognitve system *does* in practice assign only a small subset of geometric relations to the scenario. *The multistability among the collective states establishes the semantic ambiguity that occurs when the cognitive system processes the small subset instead of the huge set of geometric relations.* In this manner, many psychological observations have been explained (Figures 3 – 5) (Carmesin, 1992).

Acknowledgements

I thank Peter Kruse, Helmut Schwegler and Michael Stadler for stimulating discussions.

References

Caianiello, E.R. (1961): Outline of a theory of thought processes and thinking machines. Journal of Theoretical Biology 2, 204 – 235.

Campenhausen, C. von (1981): Die Sinne des Menschen. Stuttgart: Thieme.

Carmesin, H.-O. (1992): A minimization principle emerging from the Hebb rule. Neuronal Network. (to be published).

Carmesin, H.-O. (1992): Netzwerkmodellierung der Entwicklung der geometrischen Anschauung. (unpublished).

Carmesin, H.-O. (1993): A minimization principle emerging from the Hebb rule I: Analysis. In: N. Elsner & D.W. Richter (Eds.), Gen – Gehirn – Verhalten. Stuttgart: Thieme.

Carmesin, H.-O. (1993): Entwicklungsstufen der anschaulichen Bedeutungskonstruktion. In: H. Brügelmann & H. Balhorn (Eds.), Bedeutungen erfinden, im Kopf, mit Schrift und miteinander, pp. 66 – 70. Konstanz: Ekkehard Faude.

Cederberg, J.N. (1989): A Course in Modern Geometry. New York: Springer.

Crick, F. & Koch, C. (1990): Towards a neurobiological theory of consciousness. Seminars in the Neurosciences 2, 263 – 275.

Gerstner, W. & Hemmen, J.L. van (1993): Universality in neural networks: The importance of the mean firing rate. Biological Cybernetics 68, 363 – 374.

Grossberg, S. (1988): Neural Networks and Natural Intelligence. Cambridge: MIT.

Hebb, D.O. (1949): The Organization of Behaviour. New York: Wiley.

Hemmen, J.L. van & Kühn, R. (1991): Collective phenomena in neural networks. In: E. Domany, J.L. van Hemmen & K. Schulten (Eds.), Models of Neural Networks, pp. 1 – 105. Berlin: Springer.

Köhler, W. (1971): Die Aufgabe der Gestaltpsychologie. Berlin: de Gruyter.

Kreyscher, M., Carmesin, H.-O & Schwegler, H. (1993): A minimization principle emerging from the Hebb rule II: Computer simulation. In: N. Elsner & D.W. Richter (Eds.) Gen – Gehirn – Verhalten. Stuttgart: Thieme.

Kruse, P., Stadler, M. & Wehner, T. (1986): Direction and frequency specific processing in the perception of long range apparent movement. Vision Research 26, 327 – 335.

McCulloch, W.S. & Pitts, W.H. (1943): A logical calculus of ideas immanent in the nervous activity. Bull. Math. Biophys 5, 115 – 133.

Piaget, J. (1975): Gesammelte Werke. Stuttgart: Klett.

Strasburger, H., Harvey, L.O. & Rentschler, I. (1991): Contrast thresholds for the identification of numeric characters in direct and eccentric view. Perception & Psychophysics 49, 495 – 508.

Part VI

Brain Processes

A Psychophysiological Interpretation of Theta and Delta Responses in Cognitive Event-Related Potential Paradigms

M. Schürmann[a], C. Başar-Eroglu[b], T. Demiralp[a,1], and E. Başar[a]

[a] Institute of Physiology, Lübeck Medical University, D-23538 Lübeck, Germany
[b] Institute of Psychology and Cognition Research and Center for Cognitive Sciences, University of Bremen, D-28334 Bremen, Germany

Abstract: The present paper combines a review of event-related potentials (ERPs) with empirical data concerning the question: What are the differences between auditory evoked potentials (EPs) and two types of ERPs with respect to their frequency components? In this study auditory EPs were elicited by 1500 Hz tones. The first type of ERP was responses to 3rd attended tones in an omitted stimulus paradigm where every 4th stimulus was omitted. The second type of ERP was responses to rare 1600 Hz tones in an oddball paradigm. The amplitudes of delta and theta components of EPs and ERPs showed significant differences: In responses to 3rd attended tones there was a significant increase in the theta frequency band (frontal and parietal locations; 0–250 ms). In delta frequency band there was no significant change. In contrast a diffuse delta increase occurred in oddball responses and an additional prolongation of theta oscillations was observed (late theta response: 250–500 ms). These results are discussed in the scope of ERPs as induced rhythmicities. The intracranial sources of ERPs, their psychological correlates and the role of theta rhythms in the cortico-hippocampal interaction are reviewed. From these results and from the literature a working hypothesis is derived assuming that delta responses are mainly involved in signal matching, decision making and surprise, whereas theta responses are more related to focused attention and signal detection.

1 Introduction

In the last two decades the investigations showed that event-related potentials (ERPs) consist of 'exogenous' and 'endogenous' components (Picton and Hillyard, 1974; Hillyard and Picton, 1979; Picton and Stuss, 1980; Başar and Stampfer, 1985; Näätänen 1988). The term 'exogenous component' stands for a feature or component of the ERP which correlates with changing physical parameters of the stimuli and 'endogenous component' means a component

---

[1] Present address: Electro-Neuro-Physiology Research and Application Center of Istanbul University, Istanbul Faculty of Medicine, Çapa, Istanbul, Turkey

Springer Series in Synergetics, Vol. 64 **Ambiguity in Mind and Nature**
Editors: P. Kruse, M. Stadler © Springer-Verlag Berlin Heidelberg 1995

which varies only in relation to the given tasks to the subject, which probably modulate the intrinsic brain mechanisms in perceptual processes. During the development of this type of research, the endogenous components have been found to correlate with higher hierarchial levels of information processing like expectation, short and long term memories as well as attention (see e.g. Heinze et al., 1990, for an ERP study of focused visual attention).

A late ERP component following the early exogenous potentials with a latency shift of 250–400 ms – called the P300 wave – is an example of endogenous ERP components. The P300 response obtained by means of the oddball paradigm, has been used successfully for psychological research and clinical diagnostics. Since the earlier applications there have been several excellent reviews concerning methodological, psychological and clinical aspects of this paradigm (see Hillyard and Picton, 1979; Johnson, 1988; Rösler, 1982; Birbaumer et al., 1990; Näätänen, 1988,1990; Regan, 1989; Woods, 1990).

An important goal in the analysis of physiological correlates of the P300-response consists in searching sources of generators giving rise to the P300 component of the event related potential. Accordingly, several investigators used intracranial electrodes to be able to localize the sources in human recordings and animal models (see Halgren et al., 1986; Harrison et al., 1990; Paller et al., 1988; Başar-Eroglu et al., 1991a).

Several of these authors do indicate the existence of multiple generators including sources in hippocampus, parietal, frontal and several other areas of the association cortex (for a review see Knight et al., 1981; Smith et al., 1990; Paller et al., 1988).

In our earlier publications on P300 we used two types of approaches:

- The analysis by means of frequency characteristics of the P300 response to emphasize the role of EEG synchronization and enhancement (Başar et al. 1984; Stampfer and Başar, 1985; Başar and Stampfer, 1985; Başar, 1988; Başar-Eroglu et al. 1992)
- Experiments by means of a passive P300 paradigm with intracranial electrodes in the cat brain (Başar-Eroglu and Başar, 1987, 1991; Başar-Eroglu et al. 1991a,b).

We concluded that the most prominent component of P300 is the so-called 'theta' response in several structures of the cat brain, and the 'delta' response in human brain.

The present study is on one hand an extension of the work by Stampfer and Başar (1985) with the oddball paradigm including results on various locations of the human scalp; on the other hand a physiological approach to P300 by using the natural EEG-frequencies. As Mountcastle (1992) states, the role of EEG as a most important physiological signal of the central nervous system is in reappraisal. It was a frequency domain approach that enabled Demiralp and Başar (1992) to analyze ERP components faster than the P300 wave: As these components overlap with early exogenous components, they cannot be detected in most paradigms. They used frequency domain analysis of ERPs in order to detect endogenous components of ERPs, which may not necessarily differ from the exogenous components in time and space, by means of a method which is sensitive to changes in frequency components of the ERPs. Their paradigm was based on earlier studies: Başar et al. (1989) carried out a series of ERP studies on human subjects by applying a modified form of the omitted stimulus paradigm of Sutton et al. (1967). The paradigm consisted of auditory or visual stimulations with regular interstimulus intervals where some stimuli were omitted in a random or regular order with various degrees of probability. The subject's task was to mark mentally the time of the omitted stimulus. With this type of paradigm especially when the stimulus omission occurred in a regular manner (for example every 4th stimulus was omitted) quasideterministic, reproducible patterns of EEG signal occurred anticipating the omitted stimulus. The subjects reported that they had paid attention to the rhythm of preceding stimuli to be able to fulfill the task. Demiralp and Basar (1992) applied this paradigm to test whether event-related changes occur in responses of different brain areas to the stimuli which precede or follow the omitted stimulus.

They showed that the frequency analysis approach may differentiate the responses to the stimuli which are coupled with a cognitive task from the standard EPs, detecting some specific changes in frequency components, whereas the time domain analysis of the same responses show no prominent differences.

In the present study, frequency domain analysis is used as a tool to compare responses to auditory stimuli of three different types:

- stimuli in a standard EP paradigm,
- stimuli preceding the omitted one in the omitted stimulus paradigm described above ('3rd attended stimuli')
- oddball stimuli in an oddball paradigm.

In this study we tentatively conclude, also in the light of the new extensive analysis and survey by Miller (1991), that theta and delta responses in the hippocampo-cortical system of the brain dominate or control the ERPs and might be interpreted as correlates of some functional states as selective attention and decision making.

A general review of animal P300 experiments and human P300 is also given in this study in order to bridge physiological states and psychological correlates.

2 Methods

2.1 Subjects and environment

Each of the three parts of the experiments (auditory evoked potential, omitted stimulus paradigm, oddball paradigm) was carried out on 10 voluntary, right-handed, healthy subjects, 19–21 years of age. The subjects did not have any known neurological deficit and did not take any drugs which are known to affect the EEG.

The subjects sat in a soundproof and echo-free room which was dimly illuminated. The room was also shielded to attenuate the environmental electromagnetic noise effects. After the electrode placement a few minutes of rest time is given to subjects to get them familiar with the environment.

2.2 Data acquisition and equipment

Electrode placement. The data were derived with Ag-AgCl disc electrodes placed on frontal, vertex, parietal and occipital (F3, Cz, P3, O1) recording sites of the international 10–20 system against the reference of earlobes. All electrode impedances were maintained at less than 5 kOhms. The EOG was also registered to mark eye movement artifacts.

EEG amplification and digitization. All data were amplified by means of a Schwarzer EEG apparatus (time constant: 0.5 s; low pass filter with cut-off frequency at 70 Hz (24 dB/octave) for anti-aliasing; 50 Hz notch filter (36 dB/octave)). 1 s pre- and 1 s post-stimulus EEGs were digitized with a sampling rate of 500 points/s and stored on the hard-disk of the computer. The recording of data and stimulation were controlled by a HP 1000 F computer which was also used for the off-line analysis of the data.

Artifact rejection. For the elimination of artifactious trials, two on-line artifact rejection procedures are applied in addition to the manual off-line selective averaging procedure (see 2.4): (i) An automatic on-line artifact rejection procedure is used for the elimination of global artifactious EEG epochs, i.e., trials with extremely high amplitudes are rejected. (ii) The EEG is monitored and recorded continuously on paper during the experiments and the subjects can be observed via closed circuit TV, so that the technician can mark the trials with artifacts during the recording. It is also possible to pause the recording procedure by a button press, if long lasting artifacts occur in the EEG.

Stimulations. As auditory stimuli 80 dB, 1500 and/or 1600 Hz tones with 0.5 ms rise-time and 800 ms duration were presented binaurally.

2.3 Experimental paradigms

Firstly the spontaneous EEG was registered for a few minutes to determine global characteristics of subjects' spontaneous EEG activity and arousal state at the beginning of the experiments. This period helped also the subjects to become familiar with experimental conditions.

Thereafter on the first group of subjects auditory EPs were recorded and omitted stimulus and oddball paradigms were applied with short resting periods in between.

The auditory EP experiments consisted of the presentation of 1500 Hz tones with interstimulus intervals (ISI) randomly varying between 2.5–4 s with a mean value of 3 s.

In the *oddball paradigm* the tones were presented in a pseudorandom sequence with 1600 Hz tones occurring 20 % of the time and 1500 Hz tones occurring 80 % of the time. The interval between tones varied randomly from 2.5 to 4 s with a mean value of 3 s as in the auditory EP experiments. The subjects were instructed to keep a mental count of the number of 1600 Hz tones (non-frequent target tones).

The *omitted stimulus paradigm* consisted of a series of 1500 Hz tones with a constant interstimulus interval of 3 s, but every 4th stimulus was omitted and this time the subject's task was to mark mentally the virtual onset time of the omitted stimuli. At the end of the experiment they also were required to report about their performance in predicting the onset times of omitted stimuli. These subjective reports were taken into consideration in selecting the valid experiments. If they said that they were not successful, the experiment was either repeated or the subject's data were excluded from further analysis steps (for a more detailed description of the paradigm see Demiralp and Başar, 1992).

2.4 Data analysis

Before describing the method used we want to explain the theoretical basis of the analyses carried out on the data. Resonance is the response that may be expected of underdamped systems when a periodic signal of a characteristic frequency is applied to the system. The response is characterized by a 'surprisingly' large output amplitude for relatively small input amplitudes, i.e. the gain is large.

Resonance phenomena or responses to forced oscillations can be analyzed in the direct empirical way as follows: A sinusoidal signal of a frequency f is applied to the system. After a certain period sufficient for the damping of the transient, only forced oscillations will remain, having the frequency of the signal. Then the amplitude of the applied signal (input), the amplitude of the forced oscillations (output) and the phase difference between input and output will be measured. Gradually increasing the frequency from $f = 0$ to $f = f_0$, the output amplitude relative to the input amplitude and the phase differences will be measured as a function of frequency (amplitude characteristics and phase characteristics, respectively; Solodnikov, 1960).

Although this approach reveals the natural frequencies of a system, only a small number of workers have investigated the behavior of the EEG response using sinusoidally modulated light and sound signals (for details on pioneering experiments, see Van der Tweel, 1961). Difficulties result from the requirement for evoked responses to sinusoidal signals of over at least three decades of stimulation frequencies, evoked responses in each stimulation frequency being averaged using at least 200 stimuli. Another difficulty comes from the frequent changes in brain activity stages: they may change within a few minutes and have a limited duration, which is not sufficient for the application of sinusoidal stimuli of different frequencies.

There is, however, another way of obtaining the frequency characteristics of a system, called transient response frequency characteristics (TRFC) method: according to general systems theory, all information concerning the frequency characteristics of a linear system is contained in the transient response of the system and vice versa. In other words: knowledge of the transient response of the system allows one to predict how this system would react to different stimulation frequencies, if the stimulating signal was sinusoidally modulated. If the step response $c(t)$ of the system – in our case: the sensory evoked potential – is known, the frequency characteristics, $G(j\omega)$ of this system can be obtained with a Laplace transform, i.e. a one-sided Fourier transform:

$$G(j\omega) = \int_0^\infty \frac{d\{c(t)\}}{dt} \exp(-j\omega t)\, dt$$

($\omega = 2\pi f$, where f is the frequency of the input signal).

The frequency characteristics $G(j\omega)$ – including the information of amplitude changes of forced oscillations and the phase angle – is also called the frequency response function. It is a special case of the transfer function and is, in practice, identical with the transfer function (Bendat and Piersol, 1968). The amplitude frequency characteristics $|G(j\omega)|$ and the phase angle $\phi(\omega)$ can be obtained by numerical evaluation – using a Fast Fourier transform – with the help of a digital computer.

Although this transform is valid only for linear systems, it can be applied to nonlinear systems as a first approach (Başar, 1980): the errors due to system nonlinearities are smaller than errors resulting from the length of measurements in sinusoidal stimulation experiments given the rapid transitions of the brain's activity from one stage to another.

Finally, a limitation of this approach has to be mentioned: By application of sensory stimuli, the brain is not directly stimulated with the proper input signal – there are physiological transducers (cochlea, retina, skin) between the input signal and the measured electrical output. Therefore a direct comparison of the input and output signal is impossible; instead, the relative output amplitudes – or the magnitude of the maxima in the amplitude characteristics – are to be compared.

The methodology to evaluate EPs, AFCs and digitally filtered data was previously described (e.g. Başar, 1980, 1983). The essential steps are as follows:

- Recording of EEG-EP epochs: With every stimulus presented a segment of EEG activity preceding and the EP or ERP following the stimulus were digitized and stored on computer disc memory. This operation was repeated about 100–200 times.
- Selective averaging of EPs: The stored raw single EEG-EP or EEG-ERP epochs were selected with specified criteria after the recording session: EEG segments showing movement artifacts, sleep spindles or slow waves were eliminated.
- Amplitude frequency characteristics (AFC) were computed according to the formula given above.
- Digital filtering: EP frequency components were computed using digital filters without phase shift (Başar and Ungan, 1973). The limits of the passband filters used are not arbitrarily chosen. Filters are applied only for selectivity channels, or tuning frequencies indicated by clear peakings in the amplitude frequency characteristics.

2.5 Data reduction and statistical evaluation

Quantification of frequency components of transient EP and ERP. Maximal peak to peak amplitudes of the filtered responses in a predefined time window – i.e. maximal amplitude values of different frequency components of the averaged EP or ERP – are measured and statistically processed. Thus the medians and 95% confidence intervals of the data of all subjects are obtained. Medians and 95% confidence intervals are preferred to other parameters because the normality of the distribution of the data cannot be tested on a sample of 10 to 20 observations.

The values obtained are displayed in histograms sorted according to the experimental conditions and recording sites. The histogram presentation allows a simple visualization and comparison of the amplitudes, together with the scalp distribution of various frequency components, under different experimental conditions.

Statistical evaluation of the results. The maximal peak to peak amplitude values of various frequency components of responses obtained in auditory EPs, omitted stimulus and oddball experiments are tested for the significance of differences by means of a nonparametric test, since they do not appear to be normally distributed. For this purpose the Wilcoxon-Wilcox test (Sachs, 1974) is used. The significance values below 0.05 are given in the tables and figures.

3 Results

From the omitted stimulus paradigm, the response to the stimulus directly preceding the omitted one was included in the analysis because all subjects reported that they had attended to its onset time to be able to mark mentally the virtual onset of the omitted stimulus (see Demiralp and Başar, 1992, for details). From the oddball paradigm only responses to the nonfrequent task-relevant target stimuli were selected for the analysis.

In Fig. 1 the grand averages in time domain and the amplitude frequency characteristics (AFCs) calculated from the grand averages are shown. On the ground that Demiralp and Başar (1992) found consistent changes only in the theta frequency band, we focused our attention mainly to the low frequency components of evoked as well as event-related potentials.

3.1 Differences between AFCs of auditory EPs recorded at different locations

The comparison of AFCs computed from auditory EPs elicited in different recording sites revealed differences of the frequency contents. As described in previous studies, the vertex auditory EP (Cz) showed a peak at 7 Hz with a shoulder at 10 Hz whereas in the parietal region these frequencies were at the same level with an additional side peak occurring at 4 Hz. The frontal response (F3) showed characteristics similar to vertex in alpha range, though it had a more concave form in the sub-alpha band. In the occipital area (O1) a residue of ongoing alpha activity and a smooth theta peak were detectable. Demiralp and Başar (1992) interpreted these differences as possible manifestations of a distributed processing of the stimuli in the brain. Different brain structures might respond in different frequency bands corresponding to the changing quality and function of the neural networks in these structures.

3.2 Differences between time domain grand averages and AFCs of responses in three paradigms

The responses to the 3rd attended stimuli in F3, Cz and P3 locations showed marked increases in the amplitudes of N100-P200 complexes compared with the standard auditory EPs. These increases in the amplitudes were accompanied by increases of theta band (3–6 Hz) amplitudes in AFCs in the frequency domain. In the AFC of the occipital recording there was an increase in the theta band accompanied by a decrease in the alpha peak, though no evident response could be detected in the time domain in this location.

In the oddball experiments the target responses showed the characteristic late P300 complexes in all recording sites including the occipital area where the earlier components were not clearly identifiable. The P300 waves in the time domain were accompanied by additional prominent delta peaks with a

Theta and Delta Responses 365

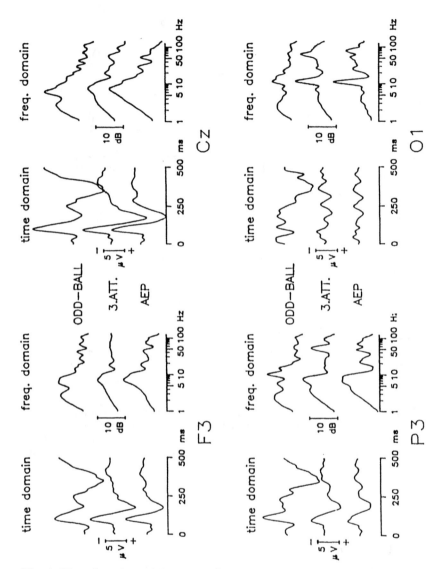

Fig. 1 Time-domain and frequency-domain representations of grand averages of auditory evoked potentials, responses to 3rd attended tones in the omitted stimulus paradigm (3. ATT) and responses to nonfrequent target tones in the oddball paradigm (oddball) obtained in frontal, vertex, parietal and occipital (F3, Cz, P3, O1) recording sites

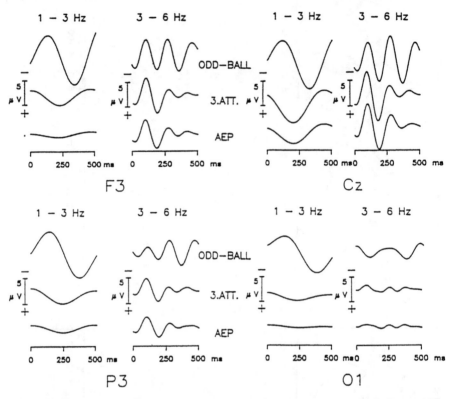

Fig. 2 Delta and theta frequency components of grand averages of auditory evoked potentials (AEP), responses to 3rd attended tones in the omitted stimulus paradigm (3.ATT) and responses to nonfrequent target tones in the oddball paradigm (oddball) obtained in frontal, vertex, parietal and occipital (F3, Cz, P3, O1) recording sites

center frequency of 2 Hz in the frequency domain. A similar change occurred also in the AFC of the occipital area.

The comparison of frequency domain representations of responses obtained in all 4 recording sites to standard auditory stimulation, 3rd attended stimulus and to oddball tones revealed a progressive increase of amplitudes of sub-alpha frequency components and a progressive decrease of dominant center frequency of the activity in this frequency band.

3.3 Adaptive filtering of the responses

For further analysis the signals were filtered by using digital filters, which caused no phase shift. The bandpass limits of the filters were selected according to the peaks in AFCs. We differentiated in the sub-alpha frequency range two frequency bands which were shared commonly by all conditions

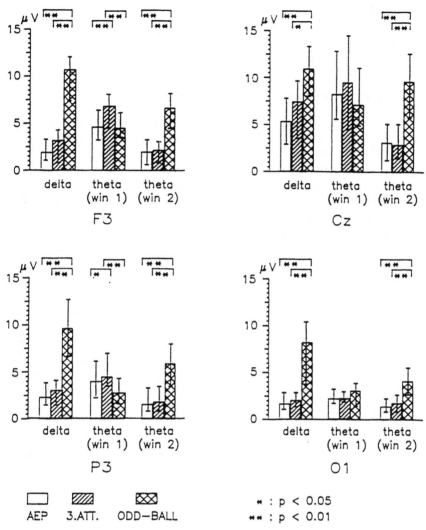

Fig. 3 The medians and 95 % confidence intervals of maximum amplitudes of delta, theta and alpha frequency components of auditory evoked potentials (AEP), responses to 3rd attended stimuli in the omitted stimulus paradigm (3.ATT) and responses to nonfrequent target tones in the oddball paradigm (oddball) obtained in frontal, vertex, parietal and occipital (F3, Cz, P3, O1) recording sites. The statistically significant differences are marked with symbols representing the significance levels

with some changes in center frequencies. They consisted of delta band between 1 and 3 Hz and theta band between 3 and 6 Hz. Fig. 2 shows the frequency components of grand averages obtained in all 3 experiments, filtered in these frequency bands.

3.4 Statistical analysis of filtered waveforms

For statistical testing of the changes of these frequency components under different conditions in cognitive performance, we filtered individual subjects' responses with corresponding bandpass filters. The maximum peak to peak amplitudes of filtered frequency components in relevant time windows were measured and tested for significance of differences between the three experimental paradigms by means of the Wilcoxon-Wilcox test. For the delta frequency band a single time window between 0 and 500 ms has been used, which is approximately equal to the period of a single delta oscillation in the frequency band 1–3 Hz (center frequency at 2 Hz). For theta oscillations two different time windows have been used to be able to identify the prolonged oscillatory activity with smaller damping factors and/or delayed enhancements. Previous studies of our group (Stampfer and Başar, 1985; Başar and Stampfer, 1985) reported prolonged and enhanced theta oscillations in P300 responses. The first time window covered the N100-P200 complex of auditory evoked responses (0–250 ms), and the second time window included the P300 complex (250–500 ms). The medians of the amplitudes of delta, theta and alpha components and 95 % confidence intervals obtained in 3 paradigms are shown in Fig. 3. The significant differences are marked with symbols representing the significance levels (See also Table I).

The responses to the 3rd attended stimuli in the omitted stimulus paradigm depicted in all recording sites nonsignificant increases in delta frequency band compared with the standard auditory EPs. In oddball paradigm, there were further increases in the amplitudes of the delta components, which were statistically significant compared with the responses to 3rd attended stimuli ($p < 0.01$ at F3, P3, O1 and $p < 0.05$ at Cz) and with the auditory EPs ($p < 0.01$ at all locations).

In responses to 3rd attended stimuli in the omitted stimulus paradigm statistically significant increases of theta oscillations occurred in the early part of the ERPs (window 1: 0–250 ms) in frontal (F3) and parietal (P3) locations compared with the standard AEPs ($p < 0.01$ and $p < 0.05$, respectively) and compared with the oddball responses ($p < 0.01$ in both locations). In the late part of ERPs no significant theta change occurred in responses to 3rd attended stimuli. Furthermore, in oddball responses significant theta increases were registered only in the late part of the responses (window 2: 250–500 ms). The theta changes in oddball responses were more widely distributed in comparison to those in responses to 3rd attended tones, which were localized in frontal and parietal regions. The theta increases could be observed in all

Table I: The medians of maximum amplitudes of delta, theta and alpha frequency components of auditory evoked potentials (AEP), responses to 3rd attended stimuli in the omitted stimulus paradigm (3.ATT) and responses to nonfrequent target tones in the oddball paradigm (ODDBALL) obtained in frontal (F3), vertex (Cz), parietal (P3) and occipital (O1) recording sites. The *percent changes of amplitudes* in (3.ATT) and (ODDBALL) conditions as the percent of the standard AEP amplitudes are given in parentheses. Statistically significant differences are marked with symbols representing the significance levels (*: $p < 0.05$, **: $p < 0.01$)

		delta (1-3 Hz)	theta (3 - 6 Hz) window 1 (0-250 ms)	window 2 (250-500 ms)
	AEP	1.8	4.7	2.0
F3	3.ATT	3.0 (66 %)	6.7 (43 %)**	2.0 (0 %)
	P300	10.6 (489 %)**	4.5 (-4 %)	6.5 (225 %)**
	AEP	5.3	8.2	3.0
Cz	3.ATT	7.3 (37 %)	9.5 (16 %)	2.9 (-3 %)
	P300	10.9 (106 %)**	7.2 (-12 %)	9.7 (223 %)**
	AEP	2.4	4.0	1.6
P3	3.ATT	3.2 (33 %)	4.4 (10 %)*	1.8 (13 %)
	P300	9.7 (304 %)**	2.7 (-33 %)	5.8 (263 %)**
	AEP	1.6	2.3	1.3
O1	3.ATT	1.9 (19 %)	2.3 (0 %)	1.5 (15 %)
	P300	8.2 (413 %)**	2.8 (22 %)	3.9 (200 %)**

4 recording sites and were statistically significant in comparison to both the auditory EP and 3rd attended tone responses ($p < 0.01$ at all locations).

4 Review and Discussion

4.1 A comparison of different ERP paradigms focused on delta and theta responses

Demiralp and Başar (1992) described the differences between frequency components of the auditory and visual evoked responses elicited in a standard evoked potential paradigm and those elicited by the stimulus preceding the omitted one in an omitted stimulus paradigm. In this paradigm the subjects had to mark mentally the virtual onset time of the omitted stimulus. The expectancy and increased attention directed towards the 3rd stimulus (predecessor of the omitted stimulus) induced increases in theta frequency components of the evoked responses.

In the present comparative study, we used the auditory oddball responses for comparison. The delta and theta components of ERPs in both paradigms revealed significant differences. In responses to 3rd attended tones there was a significant increase in theta frequency band at frontal and parietal locations only in the early part of the ERP immediately after the stimulation (window 1: 0–250 ms). Similar changes were also observed in visual modality, but with a more pronounced theta increase in parietal region in comparison to the auditory modality (Demiralp and Başar, 1992). In delta frequency band there was no significant change in 3rd attended tone responses.

In oddball responses a prolongation of theta oscillations (a slower damping) was observed. Significant increases in theta frequency band were recorded only in the late part of the ERP (window 2: 250–500 ms). The theta increases in oddball responses were distributed in all locations like the significant delta increases (F3, P3, Cz, O1).

In the following sections, we will
- comment on whether there are 'pure sensory' or 'pure cognitive' paradigms in ERP research,
- discuss the results presented above and those of Demiralp and Başar's study (1992), and
- tentatively develop a working hypothesis for which a survey of ERP measurements including intracranial recordings is necessary.

4.2 'Pure sensory' or 'pure cognitive' paradigms in ERP research?

Differences in frequency contents of EPs obtained from different brain locations suggest that responses recorded from different areas of the brain might originate from different neuron populations with possible different functional meanings.

Başar et al. (1991) assumed the impossibility to design a pure sensory or pure cognitive paradigm in EP research. Even under minimal conditions necessary for sensation, the higher cognitive functions of the brain are not to be totally rejected. On the other hand, a well defined cognitive performance needs to be controlled by certain physical events as an interface between the internally running events in subject's brain and the experimenter. Therefore, it may be awaited that during the standard EPs, besides the sensory processing, various cognitive processes come into play. Recent studies of Posner and Petersen (1990) emphasized the topographical characteristics of cognitive processing. Goldman-Rakic (1988) showed in a neuroanatomical study the parallel distributed networks in primate association cortex. The assumption of Başar et al. (1991) together with the results of the psychological and neuroanatomical studies mentioned above (Posner and Petersen, 1990; Goldman-Rakic, 1988) suggest a distributed sensory-cognitive parallel processing system in the brain. In such a system the primary sensory processes and various associative or cognitive functions might be coactivating in different brain structures during the perception of a physical stimulus. This type of distributed parallel processing could be responsible for the differences of frequency contents of responses obtained in different locations.

4.3 Increased theta-response elicited by stimuli preceding the omitted stimulus

If it is hard or even impossible to design 'pure sensory' or 'pure cognitive' paradigms in ERP research, how can 'exogenous' and 'endogenous' ERP components be differentiated? This question is particularly important for ERP components faster than the P300 wave: they probably cannot be detected in most paradigms used, because they overlap with early exogenous components (Picton and Stuss, 1980; Desmedt et al., 1983; Näätänen, 1988; Başar et al., 1991). Many solutions have been proposed to isolate spatiotemporally overlapping components (for a review see Picton and Stuss, 1980). According to the authors, isolation through experimental manipulation and factor analysis have been the most commonly used methods, which yielded objective results. However these methods lack the physiological interpretability of the isolated components.

Frequency domain analysis may provide important tools to solve this problem, as was demonstrated by Demiralp and Başar (1992): They used two types of ERP paradigms – standard EP paradigms and omitted stimulus paradigms, with both auditory and visual stimuli. They compared responses to stimuli in the standard paradigm and responses to stimuli preceding the omitted one (3rd attended stimuli). They found increased peak-to-peak amplitudes for the main peaks in responses to 3rd attended stimuli: N100-P200 for auditory stimuli and N140-P200 for visual stimuli. Frequency domain analysis showed that this increase in amplitude was mainly due to the selective enhancement of the theta components. The same theta increase is seen in both auditory and visual modalities. This change in low frequency components of ERP is modality independent.

Başar and coworkers (Başar, 1980, 1988; Başar and Özesmi, 1972; Başar et al., 1975a,b,c) have shown that sensory EPs might be considered as a superposition of wave packets in various frequencies with varying degrees of frequency stabilization, enhancement and time locking within conventional frequency bands of the ongoing EEG activity. Furthermore, the authors showed that these phenomena can occur interindependently in various frequency bands in separate applications of the stimulus (sweeps) along an ERP recording: "These variations support the hypothesis that different neural or psychophysiological mechanisms come into operation following stimulation" (Stampfer and Başar, 1985).

In a recent study Başar et al. (1991) emphasized that the response of a primary sensory area consisted mainly of an enhancement in alpha (8–13 Hz) frequency band to adequate stimuli whereas its response to inadequate stimulation was dominated by the theta activity. The authors found a parallelism between these findings and the studies which showed that primary sensory stimuli elicit impulses or volleys converging over thalamic centers to primary sensory areas, whereas the 'sensory stimulation of second order' usually reaches the cortex over association areas (Shepherd, 1988). In this framework the theta dominance during inadequate stimulation of a primary sensory area was interpreted as a possible manifestation of responsiveness of various brain areas in cases of association processes involved in global associative cognitive performance. Mizuki et al. (1980) showed that midline prefrontal region of the cortex generated regular theta rhythms during the performance of simple repetetive mental arithmetic tasks.

The increases in the theta components of the ERPs in comparison to the standard EPs is in agreement with the results of Başar et al. (1991) and Mizuki et al. (1983) mentioned above. Our findings supplement these results in terms of a probable functional assignment to the theta band activity. Considering the theta dominance in responses of primary sensory areas to inadequate stimulation and the appearance of theta rhythms in EEG recorded on association areas during mental tasks together with our results, we incline to think that neural circuits which perform associative functions share a common information channel which operates in the theta frequency range.

Furthermore, the *topographical differences* in weights of increases of theta components in ERPs suggest an association between the investigated cognitive function, the frequency content of the ERP and the topography of the frequency components.

4.3.1 Increase of theta components is highest in frontal recordings

During cognitive performance, Demiralp and Başar (1992) measured the highest – statistically significant – theta increases in frontal and parietal recording sites. In auditory modality, the theta increase was absolutely dominant in the frontal area (44% increase) whereas in visual modality the theta increase in frontal recording site was slightly higher than that in parietal recording site (48% versus 45%). This selectivity has a parallelism with the results of Fuster (1991). Studies of Fuster are based on single unit recordings in prefrontal cortex of monkeys, which showed a high anticipatory activation level of frontal neurons in time delay tasks. Since the cognitive task in our study was also mainly based on anticipation to an expected stimulus, it is not surprising that the greatest changes are in frontal regions.

Our findings showing the strong participation of the frontal cortex in fulfilling a cognitive task are also in accordance with the results of Knight et al. (1981) on patients with frontal cortex lesions. Results of these studies indicated that the frontal lobes exhibited a modulating influence upon the endogenous negativity of ERPs produced in selective attention tasks.

4.3.2 In visual modality the secondary dominant theta increase occurs in the parietal recordings

In Demiralp and Başar's study (1992), the parietal area was found to be the secondary dominant theta center: in visual modality there was a percentual theta increase slightly below that obtained in frontal area (45%) whereas in auditory modality the increase of parietal theta activity was not so prominent (10%). This property of visual ERP is in accordance with the specific functions of parietal cortex in visual information processing, as shown by means of cellular measurements on macaque monkeys (Mountcastle et al., 1975; Lynch et al., 1977). According to Mountcastle, 50% of the investigated neurons in the parietal association area 7 of the inferior parietal lobe are visual fixation cells which are active as the animal looks at visual targets that are linked by a strong motivational drive; the rest of the neuronal elements are light-sensitive with large and bilateral receptive fields (Yin and Mountcastle, 1977; Robinson et al., 1978). Lynch et al. (1977) suggest that the neurons in posterior parietal cortex are involved in selective visual attention processes. Mountcastle et al. (1981, 1984) showed that the enhanced responsiveness of light-sensitive neurons in the inferior parietal lobe does not merely occur with changes in general arousal but is more specifically related to the visual attention directed to the target light. Petersen et al. (1988) showed by means of positron emission tomography similar effects in the parietal cortex of normal humans.

The specific parietal theta increase we observed during visual perception in an attentive state with high expectation supports the view of Başar and Stampfer (1985) that the activity of neuronal populations involved in specific stages or parts of perceptual processes use different frequency channels, and hence these specific activities can be differentiated by means of the frequency analysis applied to the surface recorded evoked responses.

4.4 ERPs as reflections of induced rhythmicities: A review of sources and psychological correlates

4.4.1 A survey of intracranial sources of the ERPs

Despite the interest among psychologists and neurologists, the *neural origin of human event related potentials* is not known. Since P300 is an example of an ERP which is widely investigated, we will first focus our attention to its electrogenesis.

Because of the cognitive correlates of the P300, many researchers have assumed that ERPs are generated in neocortex. *Neocortical generators* postulated for the human P300 have included frontal cortex (Courchesne, 1978; Desmedt and Debecker, 1979; Knight et al., 1981; Wood and McCarthy, 1985). Some other workers described the centroparietal or temporoparietal association cortex as site of generators (Vaughan and Ritter, 1970; Simson et al., 1977; Goff et al., 1978; Pritchard, 1981) since the P300 amplitude is the largest in these areas.

Multiple subcortical sites as generators have been proposed by Wood et al. (1980). The studies of Halgren et al. (1986) suggest involvement of the hippocampal formation and the limbic system in P300 generation. However, according to Wood et al. (1982) and Smith et al. (1990) surface topography of the P300 is unchanged after unilateral temporal lobectomy which includes the hippocampus, and hippocampal P300s may not be volume conducted to the surface. The analysis of Başar-Eroglu et al. (1991b) also excludes the possibility of a volume conduction from CA3 layer of the hippocampus to cortex; however, a facilitation of the neural signal via hippocampo-cortical system is not excluded by the description of these authors.

Many authors postulated that generator systems can be better explored in an *animal P300 model* than in the human brain. Such studies have provided evidence for the neocortical and/or limbic system involvement. The marginal gyrus, suprasylvian gyrus and hippocampus have all been postulated as generators of the P300 in the cat (O'Connor and Starr, 1985; Buchwald and Squires, 1982). Frontal cortex (Boyd et al., 1976) and/or locus coeruleus (Pineda et al., 1987) have been postulated as necessary components for P300 generation in the monkey. A P300-like potential remains after bilateral ablation of the hippocampus. Harrison et al. (1990) have shown that the cat-P300 is present after the ablation of the association cortex. These authors also suggest multiple generators of the cat-P300, such as marginal and suprasylvian gyri as well as the hippocampal region.

Studies of P300 following unilateral anterior lobectomy, which includes anterior regions of the hippocampus in humans (Wood et al., 1982; Stapleton and Halgren, 1987) or bilateral hippocampal lesions in monkey (Paller et al., 1988) have failed to identify any related change in P300 amplitude. On the other hand, physiological events in the hippocampus have been demonstrated to contribute to scalp-recorded activity (Altafullah et al., 1986). Smith et al. (1990) tried to isolate the anatomical locus of neural activity most important for directly generating the scalp P300 and they suggested that the critical locus is the lateral neocortex of the inferior parietal lobule. Further, Smith et al. assume that activity in the hippocampus and frontal lobe may only make minor contributions to scalp recordings. They suggest that the scalp-recordable P300 might best be conceptualized as only the most readily observable aspect of synchronous activity occurring across widely distributed yet highly integrated cognitive activities. The analysis of Katayama et al. (1985) indicated thalamic negativity associated with the endogenous late positivity (or P300) by using conditioning in humans and cats.

Recently, Başar-Eroglu and Başar (1991) and Başar-Eroglu et al. (1991a,b) published a series of results from experiments with freely moving cats and by using a passive P300-paradigm and assumed that P300-like potentials have multiple cortical and subcortical generator sites including reticular formation of the brain stem, hippocampus and auditory cortex. According to these latest results of our group, the P300 potential is most significant, stable and has the largest amplitudes in CA3-layer of the hippocampus of the intact cat brain. Başar-Eroglu et al. (1991b) further showed that the hippocampal P300 manifests an enhancement of the theta activity of the field potentials and/or a type of resonance phenomenon in the theta frequency range. Fig. 4 shows, only for discussion purposes, the comparison of spectral activity of spontaneous field potentials and the P300 response in the hippocampus (compare the hypothesis of Miller, 1991 in section 4.4.2.

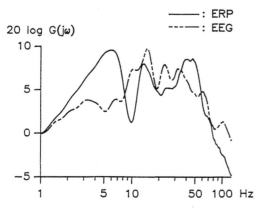

Fig. 4 The comparison of spectral activity of spontaneous field potentials and the P300 response in the hippocampus (from Başar-Eroglu et al., 1991b)

4.4.2 The 'diffuse theta-response system in the brain' and the cortico-hippocampal interplay

According to a number of neurophysiological, psychological and biophysical approaches the oscillatory phenomena in neural tissues merit nowadays considerable attention to understand cognitive functions of the central nervous system (Bullock, 1992; Başar, 1992; Galambos, 1992; Petsche and Rappelsberger, 1992; Pfurtscheller and Klimesch, 1992; Gray et al., 1992; Tononi et al., 1992). According to the results of experiments on the P300 response of the cat brain we will here analyze the responsiveness in the theta frequency range in an extended manner. Tentative interpretations of the results of our group led us recently to postulate the existence of a 'Diffuse alpha-response system' and a 'Diffuse theta-response system' in the brain. Başar et al. (1991) tentatively assumed that the brain theta response (or the theta component of the evoked potentials) and/or slower responses might reflect the responsiveness of various brain areas in processes involved with global associative-cognitive performance.

In spontaneous human EEG, the theta activity is masked by the alpha rhythm and cannot easily be detected. In earlier studies Rémond and Lesèvre (1957) reported on a predominance of the theta-rhythm in the frontal central region whereas Mundy-Castle (1951) described more pronounced theta activity at temporal regions. The recent results of Westphal et al. (1990) showed that the theta amplitude is highest over the anterior midline (Fz- and Cz-locations) and are in accordance with mapping findings of Walter et al. (1984) and Mizuki et al. (1983).

Miller (1991) reviews extensively that theta activity recorded from the hippocampus has been difficult to find in human subjects because of the difficulties of human central electrophysiology. Some recent evidence that the midline prefrontal region of the cortex can generate theta activity was reported by Mizuki et al. (1980) in certain cognitive states: EEG rhythms of 5–5.5 Hz frequency appears with some regularity during the performance of simple repetitive mental arithmetic tasks.

Lang et al. (1987) used also spectral analysis of frontal EEG and showed that theta frequencies (3–7 Hz) were increased during motor or verbal *learning tasks*. Miller (1991) states that the data of groups of Lang and of Mizuki are compatible with the hypothesis that theta activity in frontal regions is associated with theta activity in hippocampus.

Could the prefrontal cortex show theta activity at the same time as the hippocampus? Aleksanov et al. (1986) showed that during the course of alimentary *conditioning*, and at the time of presentation of the conditioned stimulus (CS), the coherence between hippocampus and prefrontal cortex increased. This increase of coherence was observed especially in the delta and theta frequency ranges. There is yet no clear evidence that the more distantly related areas of association cortex generate theta potentials reflecting the hippocampal ones. Miller (1991), after his strong survey of electrophysiological and functional analysis on the cortico-hippocampal interplay concludes: Despite the lack of evidence at presence, the central prediction still made is that there are cortical nodes with consistent phase relations to hippocampal theta activity.

The facts that in the present reports the theta components in frontal and parietal recordings are significantly enhanced and that the hippocampal P300 response of the brain shows also a dominant theta component (Fig. 4) contribute to underline that some special anatomical structures or preferential structures are more involved in generation of theta resonances.

The resonant theta response of the hippocampus to auditory and visual stimuli was explained in details in a report concerning a component analysis of hippocampal EPs (Başar and Ungan, 1973). Recently, the concept of theta resonance has extensively been analyzed by Miller (1991) who describes the cortico-hippocampal interaction as a basic resonance phenomenon in the theta frequency range. In the well documented analysis by Miller (1991) it is envisaged that Hebbian processes of synaptic strengthening select patterns of loops passing from hippocampus to cortex and back to the hippocampus. Miller (1991) states: "The total conduction delay time round each one of these patterns of loops is envisaged to correspond to the theta period. Such resonance between hippocampus and cortex is envisaged to have an important functional role in registration and retrieval of information in the cortex. Each pattern of resonant loops will raise the activation of specific collection of cells which are widely dispersed across the cortical mantle."

Furthermore, Miller (1991) describes the connectional basis of phase locked loops. According to the anatomical and physiological evidence Miller takes the viewpoint that theta modulated signals are likely to influence limbic and prefrontal areas, and also (directly or indirectly) other areas of (mainly association) cortex.

Our recent findings of the cat hippocampus indicating a strong theta enhancement (or resonance as Başar-Eroglu et al. (1991b) emphasize) in the hippocampal P300 and the hypothesis by Miller (1991) give relevant support to emphasize the implication of cognitive components by means of a 'Diffuse theta response system'. The term 'diffuse' is used in order to describe a distributed system in the brain: it is not yet possible to describe it with sharp anatomical boundaries, with directions of signal flow or with neuron-to-neuron tracking. Such a diffuse theta sytem, which was experimentally based on theta enhancement phenomena in various brain structures including the brainstem, shows seemingly considerable strength (or more resonant loops) in the hippocampo-cortical system (see also section 4.4.2. Such theta enhancement could also be shown in the cat reticular formation (Başar-Eroglu et al., 1991a).

Hebb's rule implies that groups of synapses converging on a single neuron and having the tendency to fire together will become strenghtened as a group. This is, as Miller (1991) also describes, a principle of cooperativity:

"When an animal is in a particular environment, a collection of widely dispersed neurons throughout the cortical mantle may be set into a tonic state of elevated activation, as a result of signals reaching the cortex via any of its sensory systems". Miller (1991) further assumes that there is a series of recursive loops, with some temporal divergence/convergence in the pathway in each direction, and there is the possibility of regular oscillation of neural activity imposed at the hippocampal end. "The Hebbian mechanism in this case should favour the strengthening of connections which permit resonance". Başar-Eroglu et al. (1991 b) have shown that the significant theta response at the CA3-pyramidal layer cannot be recorded in the cortex with the important weight due to volume conduction. Therefore, the significant cognitive theta enhancements (see also Miller, 1991; Mizuki et al., 1980; Lang et al., 1987) in frontal and parietal recordings (see sections 3.4 and 4.1 of this study) might occur in the sense of Hebbian cooperation mechanism among the neuronal populations of the frontal cortex, parietal cortex and of the hippocampus.

4.4.3 General remarks on functional correlates of ERPs: Focused attention, signal detection, recognition and decision making

The P300 has been found to correlate with many variables. These include 'task relevance', 'meaningfulness', 'information delivery', 'resolution of uncertainty', and 'decision making'. Regan (1989) criticized that some of these physiological terms overlapped whereas others were not sharply defined, and in itself this led to controversy. The need for psychophysiological models in ERP experiments has also been pointed out by Münte and Künkel (1990) with respect to clinical ERP applications in psychiatry. According to Woods (1990) selective attention refers to the preferential detection, identification, and recognition of selected stimuli in an environment containing multiple sources of information. In P300 experimental designs, the eliciting stimulus is task relevant, that is, attention is paid to it. The contradictory claim that P300 can be recorded in the situation where low-probability stimuli are ignored (Roth, 1973) is suspect, because it assumes that subjects are able to follow instructions to ignore irrelevant stimuli while they just sit with eyes shut (Ritter et al., 1968; Squires et al., 1977).

According to Hillyard and Picton (1979) "the P300 seems to index the operation of adaptive brain systems". A rather different line of thought, derived from autonomic psychophysiology, uses the language of innate stimulus-response systems such as 'orienting', 'startle', and 'defense' reflexes (Donchin et al., 1984).

In general, a P300 will be produced by task-relevant (Chapman, 1973; Chapman and Bragdon, 1964) stimuli that occur somewhat unexpectedly and require a motor response or cognitive decision (Donchin, 1981; Ritter et al., 1968).

4.4.4 Possible neurophysiological correlates of significant increases in delta and theta frequency responses

Woods (1990) claims that studies of the psychological bases of selective attention have little influence on each other. On the one hand, cognitive psychologists have typically studied selective attention in human subjects performing complex tasks and have monitored attention with simple or complex responses. In contrast, cellular neurophysiologists have examined the effects of selective attention on the firing of single neurons in animals responding to simple stimuli. Başar et al. (1991) as well as Demiralp and Başar (1992) tried to bridge the gap between these two different approaches. They aimed at linking both types of experiments, by including the discussion of a neurophysiological signal which is common in both categories of analyses: The changes in theta frequency component (especially of the hippocampus and cortex) during learning, association phenomena, memory retrieval and attentive behavior. We emphasize that during increased attention (the responses to 3rd stimuli) the significant increases in the theta responses were recorded in frontal and parietal locations. Such overall controlling phenomena in the brain were recently surveyed by Başar (1992). The relation of the theta component to single neuron firing is established in a number of relevant studies (see Lopes da Silva, 1992; Buzsaki, 1992).

We aim in this work also, as in a previous study (Demiralp and Başar, 1992), to reduce this type of controversies by discussing as a limited number of measurable neurophysiological correlates i.e. the significant changes in the delta and theta frequency channels with some evident functional correlates. Various workers describe frequency channels of the EEG as innate response of neural networks or intrinsic frequency channels of single neurons, related to function (Freeman, 1992; Gray and Singer, 1992; Llinas and Graves, 1988) and/or induced ryhthmicities (Petsche and Rappelsberger, 1992; Pfurtscheller and Klimesch, 1992) as described in section 4.4.2. In this study we do not explicitly discuss alpha, beta and gamma (40 Hz) frequency bands which were discussed elsewhere. The topographical distribution of the prolonged alpha response will be soon analyzed in a detailed study. A 40 Hz diffuse (distributed) system was also described by Başar-Eroglu and Başar (1991) and Başar et al. (1992). The significant change of the alpha oscillations consisted especially in the results of oddball paradigm as a prolongation of oscillation, but not significant increase in the amplitude (see also Kolev and Schürmann, 1992).

4.5 Focused attention, signal detection, matching and decision making: Possible psychological correlates of increased delta and theta responses?

We now come back to general definitions of 'focused attention', 'signal detection', 'signal recognition' and 'decision making' and tentatively try to combine the results of the theta response and delta response with the psychological context discussed also in section 4.4.3. We aim, therefore, to link physiological results to psychological correlates in terms of natural frequencies of the EEG. For conclusion we summarize the ERP paradigms used, their psychological contents and the correlated changes in EEG rhythms:

- **Oddball paradigm:** The subjects have at least two different categories of tasks to perform:
  - i) Focused attention and signal detection
  - ii) Matching for target recognition and decision making

  Since the target signals were presented in a random order with random interstimulus intervals, the subjects were not able to prepare themselves for signal detection. The signal detection task refers first to the period after the stimulus delivery. In addition, the subject's task to recognize target stimuli between nontarget ones accomplishes matching and decision making processes. (The matching mechanisms have been extensively analyzed in the recent reviews by Näätänen (1988, 1990)).

- **Paradigm with omitted 4th signal and attention to the 3rd signal:** Mainly one of the above-mentioned task categories seems to be in play.
  - i) Focused attention and signal detection

  Since the stimulation is applied repetetively with constant interstimulus intervals, a preparation takes place prior to the target. Attention is focused to the third signal (see Demiralp and Başar, 1992). The signal is anticipated. Therefore, at the time of signal application there is no relevant surprise or novelty. Accordingly, the difficulty which might stem from 'matching for target recognition' and 'decision making' is highly reduced or not at all involved in the procedure.

Common components in ERPs of both paradigms are: Focused attention and signal detection. Accordingly, we tentatively assume that the delta response, most prominent component of the oddball-ERP, is mostly involved with the signal matching and decision making following a novel or unexpected signal and/or partial surprise. The early theta response (window 1) which is most prominent in frontal and parietal locations during the 3rd attended stimulation is probably due to focused attention which refers to preferential detection of selected stimuli, and not to matching for signal recognition or decision making.

Significant changes in theta and delta responses in different locations under both categories of the paradigms and the possible functional correlates of the applied paradigms are summarized in Table II. During the paradigm with *3rd attended stimulus* theta response increase was recorded only immediately following the stimulation. Moreover, such increases were recorded only in frontal and parietal locations. Seemingly, for this less complicated task a theta response increase to sound stimuli in frontoparietal location is representative for the brain mechanisms in play. Considering the *anticipatory (time estimation) component of the task*, the frontal dominance of the theta increase is in accordance with the studies based on the cellular measurements in the prefrontal cortex (Fuster, 1991). Fuster showed a high anticipatory activation level of frontal neurons in time delay tasks. Başar et al. (1989) have shown that during such a paradigm with repetitive stimulation, regular, ample and quasideterministic alpha rhythms were recorded in several cortical locations prior to the target. Moreover, such alpha rhythms are almost time locked to the target for a period of one second prior to the occurrence of the target. It is remarkable that following these regular and time locked alpha rhythms an enhanced theta response is recorded. When the task additionally involves signal matching, decision making and surprise, the changes in the frequency channels of ERP reach a higher degree of complexity:

- In all locations a marked change of the delta response is recorded.
- Increases in late theta responses (window 2) were also recorded significantly in all locations.

The existence of important delta increase suggests that processes of decision making and surprise are reflected in this slowest EEG-response component.

In summary, an increase in the delta response is probably only related to decision making and matching. In contrast, the theta response seems to be involved in several tasks (focused attention, signal detection, anticipation and expectation) appearing either as an early or late component.

More evidence for the participation of delta response during the procedure of decision making was provided by the experiments at the auditory threshold level. In a recent study, Başar et al. (1992) have shown that at the hearing threshold the evoked potentials of the subjects were reduced to an almost pure delta oscillation, also detectable without frequency analysis or filter application. At the hearing threshold subjects are supposed to be involved in decision making. Thus, the results of Başar et al. (1992) and Parnefjord and Başar (1992) – attributing a decision making function at the threshold level to a delta component – are in good accordance with the results of the present study.

In these analyses special emphasis was placed on the increased theta and delta responses during two different paradigms in order to accentuate the role of the theta response system and its interplay with the hippocampo-cortical system. However, due to the approaches used in this study and also due to earlier results concerning prolonged alpha oscillations, preparatory alpha rhythms before the application of repetitive signals as well as the 40 Hz re-

Table II Significant changes of theta and delta responses in different recording sites under both categories of paradigms and the possible functional correlates of the applied paradigms (*: $p < 0.05$, **: $p < 0.01$)

Paradigms	(1) OMITTED STIMULUS (3rd ATTENDED SIGNALS)	(2) ODDBALL (TARGET TONES)
Description of the task	Subject focuses his attention to the 3rd signal presented repetitively: Probability of target occurrence = 100%, high expectancy, no surprise	Subject focuses his attention to rare oddball tones presented randomly. Probability of target occurrence = 20%, low expectancy, surprise, decision making
Functional Correlates	Association, focused attention, signal detection, expectation, anticipation	Association, focused attention, signal detection, matching for target recognition, decision making, surprise

	3rd ATTENDED LIGHT				3rd ATTENDED TONE				ODDBALL (TARGET) TONE			
	F3	Cz	P3	O1	F3	Cz	P3	O1	F3	Cz	P3	O1
Increase of delta-frequency response	-	-	-	-	-	-	-	-	**	**	**	**
Increase of early theta response	**	*	*	*	**	-	*	-	-	-	-	-
Increase of late theta-response	-	-	-	-	-	-	-	-	**	**	**	**

sponse, it might be stated that the EEG-Dynamics is involved in all cognitive processes in various frequency channels. Several new experimental strategies on new psychological paradigms using the frequency analysis approach of this study might help to identify in a clearer manner the functional correlates of theta and delta responses.

5 Conclusion

Based on the results of the present study and on Demiralp and Başar's study (1992), and based on the animal and human ERP studies reviewed here, the following tentative conclusions are conceivable: Comparison of animal experiments suggests a theta activation circuit or a diffuse theta system, in which the hippocampus might play a key role. Part of the hippocampo-cortical interaction might be attributed to cooperativity in the Hebbian sense.

To a certain extent, the different cognitive ERP paradigms used permitted us to investigate focused attention and decision making separately: ERP frequency components appear to be associated with certain aspects of cognitive performance.

The analysis of scalp recorded human ERPs in the frequency domain allows a link between the cellular neurophysiology and cognitive psychology by discussing a neurophysiological signal which is common in both categories.

Acknowledgements

We are grateful to Dipl.-Ing. F. Greitschus for software development, to B. Stier and A. Aufseß for technical assistance and to Dr. E. Rahn for valuable discussion. This work was supported by the Volkswagen-Stiftung grant No. I/67 678, DFG grant No. Ba 831/5-1, and by IBM-Türk Ltd.

References

Aleksanov, S.N., Vainstein, I.I. & Preobrashenskaya, L.A. (1986): Relationship between electrical potentials of the hippocampus, amygdala and neocortex during instrumental conditioned reflexes. Neuroscience and Behavioral Physiology 16, 199–207.

Altafullah, I., Halgren, E., Stapleton, J.M. & Crandall, P. (1986): Interictal spike-wave complexes in the human medial temporal lobe: Typical topography and comparison with cognitive potentials. Electroencephalography and Clinical Neurophysiology 63, 503–516.

Başar, E. (1980): EEG Brain Dynamics. Relation between EEG and Brain Evoked Potentials. Amsterdam: Elsevier.

Başar, E. (1983): Toward a physical approach to integrative physiology. I. Brain dynamics and physical causality. American Journal of Physiology 254, R510–R533.

Başar, E. (1988): EEG-dynamics and evoked potentials in sensory and cognitive processing by the brain. In: E. Başar (Ed.), Dynamics of Sensory and Cognitive Processing by the Brain. Berlin: Springer.

Başar, E. (1992): Brain natural frequencies are causal factors for resonances and induced rhythms. In: E. Başar & T.H. Bullock (Eds.), Induced Rhythms in the Brain, pp. 423–465. Boston: Birkhäuser.

Başar, E., Başar-Eroglu, C., Rosen, B. & Schütt, A. (1984): A new approach to endogenous event-related potentials in man: Relation between EEG and P300-wave. International Journal of Neuroscience 24, 1–21.

Başar, E., Başar-Eroglu, C., Röschke, J. & Schütt, A. (1989): The EEG is a quasi-deterministic signal anticipating sensory-cognitive tasks. In: E. Başar & T.H. Bullock (Eds.), Brain Dynamics, pp. 43–71. Berlin: Springer.

Başar, E., Başar-Eroglu, C., Parnefjord, R., Rahn, E. & Schürmann, M. (1992): Evoked potentials: Ensembles of brain induced rhythmicities in the alpha, theta and gamma ranges. In: E. Başar & T.H. Bullock (Eds.), Induced Rhythms in the Brain, pp. 155–181. Boston: Birkhäuser.

Başar, E., Başar-Eroglu, C., Rahn, E. & Schürmann, M. (1991): Sensory and cognitive components of brain resonance responses: An analysis of responsiveness in human and cat brain upon visual and auditory stimulation. Acta Otolaryngol (Stockholm) Suppl 491, 25–35.

Başar, E., Gönder, A., Özesmi, Ç. & Ungan, P. (1975a): Dynamics of brain rhythmic and evoked potentials. I. Some computer methods for the analysis of electrical signals from the brain. Biological Cybernetics 20, 137–145.

Başar, E., Gönder, A., Özesmi, Ç. & Ungan, P. (1975b): Dynamics of brain rhythmic and evoked potentials. II. Studies in the auditory pathway, reticular formation and hippocampus during the waking stage. Biological Cybernetics 20, 145–160.

Başar, E., Gönder, A., Özesmi, Ç. & Ungan, P. (1975c): Dynamics of brain rhythmic and evoked potentials. III. Studies in the auditory pathway, reticular formation and hippocampus during sleep. Biological Cybernetics 20, 160–169.

Başar, E. & Özesmi, Ç. (1972): The hippocampal EEG-activity and systems-analytical interpretation of averaged evoked potentials of the brain. Kybernetik 12, 45–54.

Başar, E. & Stampfer, H.G. (1985): Important associations among EEG-dynamics, event-related potentials, short-term memory and learning. International Journal of Neuroscience 26, 161–180.

Başar, E. & Ungan, P. (1973): A component analysis and principles derived for the understanding of evoked potentials of the brain: A study in the hippocampus. Kybernetik 12, 133–140.

Başar-Eroglu, C. & Başar, E. (1987): Endogenous components of event-related potentials in hippocampus: An analysis with freely moving cats. In: R. Johnson Jr., J.W. Rohrbaugh & R. Parasuraman (Eds.), Current Trends in Event-related Potential Research. Amsterdam: Elsevier.

Başar-Eroglu, C. & Başar, E. (1991): A compound P300-40 Hz response of the cat hippocampus. International Journal of Neuroscience 60, 227–237.

Başar-Eroglu, C., Başar, E., Demiralp, T. & Schürmann, M. (1992): P300 response: possible psychophysiological correlates in delta and theta frequency channels. International Journal of Psychophysiology 13, 161–179.

Başar-Eroglu, C., Başar, E. & Schmielau, F. (1991a): P300 in freely moving cats with intracranial electrodes. International Journal of Neuroscience 60, 215–226.

Başar-Eroglu, C., Schmielau, F., Schramm, U. & Schult, J. (1991b): P300 response of hippocampus with multielectrodes in cats. International Journal of Neuroscience 60, 239–248.

Bendat, J.S. & Piersol, A.G. (1968): Measurement and Analysis of Random Data. New York: John Wiley.

Birbaumer, N., Elbert, T., Canavan, A.G.M. & Rockstroh, B. (1990): Slow potentials of the cerebral cortex and behavior. Physiological Reviews 70, 1–41.

Boyd, E.H., Boyd, E.S. & Brown, L.E. (1976): Long latency evoked responses in squirrel monkey frontal cortex. Experimental Neurology 515, 22–40.

Buchwald, J.S. & Squires, N.S. (1982): Endogenous auditory potentials in the cat. In: C.D. Woody (Ed.), Conditioning: Representation of Involved Neural Function, pp. 503–515. New York: Plenum Press.

Bullock, T.H. (1992): Introduction to induced rhythms: A widespread, heterogenous class of oscillations. In: E. Başar & T.H. Bullock (Eds.), Induced Rhythms in the Brain, pp. 1–26. Boston: Birkhäuser.

Buzsaki, G. (1992): Network properties of the thalamic clock: Role of oscillatory behavior in mood disorders. In: E. Başar & T.H. Bullock (Eds.), Induced Rhythms in the Brain, pp. 233–248. Boston: Birkhäuser.

Chapman, R.M. (1973): Evoked potentials of the brain related to thinking. In: F.J. McGuigan & R.A. Schoonover (Eds.), Psychophysiology of Thinking: Studies of Covert Processes, pp. 69–108. New York: Academic Press.

Chapman, R.M. & Bragdon, H.R. (1964): Evoked responses to numerical and nonnumerical visual stimuli while problem solving. Nature 203, 1155–1157.

Courchesne, E. (1978): Changes in P3 waves with event repetition: Long-term effects on scalp distribution and amplitude. Electroencephalography and Clinical Neurophysiology 45, 754–766.

Demiralp, T. & Başar, E. (1992): Theta rhythmicities following expected visual and auditory targets. International Journal of Psychophysiology 13, 147–160.

Desmedt, J.E. & Debecker, J. (1979): Wave form and neural mechanism of the decision P350 elicited without pre-stimulus CNV or readiness potential in random sequences of near threshold auditory clicks and finger stimuli. Electroencephalography and Clinical Neurophysiology 47, 648–670.

Desmedt, J.E., Huy, N.T. & Bourget, M. (1983): The cognitive P40, N60 and P100 components of somatosensory evoked potentials and the earliest electrical signs of sensory processing in man. Electroencephalography and Clinical Neurophysiology 56, 272–282.

Donchin, E. (1981): Surprise!...Surprise! Psychophysiology 18, 493–513.

Donchin, E., Heffley, E., Hillyard, S.A., Loveless, N., Maltsman, I., Ohman, A., Rosler, F., Ruchkin, D. & Siddle. D. (1984): Cognition and event-related potentials. II. The orienting reflex and P300. Annals of the New York Academey of Sciences 425, 39–57.

Freeman, W. (1992): Predictions on neocortical dynamics derived from studies in paleocortex. In: E. Başar & T.H. Bullock (Eds.), Induced Rhythms in the Brain, pp. 183–199. Boston: Birkhäuser.

Fuster, J.M. (1991): The Prefrontal Cortex. Anatomy, Physiology and Neuropsychology of the Frontal Lobe. New York: Raven Press.

Galambos, R. (1992) A comparison of certain gamma band (40–Hz) brain rhythms in cat and man. In: E. Başar & T.H. Bullock (Eds.), Induced Rhythms in the Brain, pp. 201–216. Boston: Birkhäuser.

Goff, E.R., Allison, T. & Vaughan, H.G. Jr. (1978): The functional neuroanatomy of event-related potentials. In: E.P. Tuetin & S.H. Koslow (Eds.), Event-Related Potentials in Man, pp. 1–79. New York: Academic Press .

Goldman-Rakic, P.S. (1988): Topography of cognition: Parallel distributed networks in primate association cortex. Annual Review of Neuroscience 11, 37–156.

Gray, C.M., Engel, A.K., König, P. & Singer, W. (1992): Mechanisms underlying the generation of neuronal oscillations in cat visual cortex. In: E. Başar & T.H. Bullock (Eds.), Induced Rhythms in the Brain, pp. 29–45. Boston: Birkhäuser.

Halgren, E., Stapleton, J.M., Smith, M. & Altafullah, I. (1986): Generators of the human scalp P3. In: R.Q. Cracco & I. Bodis-Wollner (Eds.), Evoked Potentials, pp. 269–284. New York: Alan R. Liss.

Harrison, J.B., Dickerson, L.W., Song, S. & Buchwald, J.S. (1990): Cat P300 present after association cortex ablation. Brain Research Bulletin 24, 551–560.

Heinze, H.J., Luck, S.J., Mangun, G.R. & Hillyard, S.A. (1990): Visual event-related potentials index focused attention within bilateral stimulus arrays. I. Evidence for early selection. Electroencephalography and Clinical Neurophysiology 75, 511–527.

Hillyard, S.A. & Picton, T.W. (1979): Event related brain potentials and selective information processing in man. In: J.E. Desmedt (Ed.), Progress in Clinical Neurophysiology, vol. 6, pp. 1–50. Basel: Karger.

Johnson, R. Jr. (1988): Scalp recorded P300 activity in patients following unilateral temporal lobectomy. Brain 111, 1517–1529.

Katayama, Y., Sukiyama, T. & Subokawa, T. (1985): Thalamic negativity associated with the endogenous late positive component of cerebral evoked potentials (P300): Recording using discriminative aversive conditioning in human and cats. Brain Research Bulletin 14, 223–226.

Knight, R.T., Hillyard, S.A., Woods, D.L. & Neville, H.J. (1981): The effects of frontal cortex lesions on event-related potentials during auditory selective attention. Electroencephalography and Clinical Neurophysiology 52, 571–582.

Kolev, V. & Schürmann, M. (1992): Event-related prolongation of induced EEG rhythmicities in experiments with a cognitive task. International Journal of Neuroscience 67, 199–213.

Lang, M., Lang, W., Diekmann, V. & Kornhuber, H.H. (1987): The frontal theta rhythm indicating motor and cognitive learning. In: R. Johnson Jr, J.W. Rohrbaugh & R. Parasuraman (Eds.), Current Trends in Event-related Potential Research. (Electroencephalography and Clinical Neurophysiology, Suppl 40), 322–327.

Llinas, R.R. & Graves, A. (1988): The intrinsic electrophysiological properties of mammalian neurons: insights into central nervous system function. Science 242, 1654–1664.

Lopes da Silva, F.H. (1992): The rhythmic slow activity (theta) of the limbic cortex: An oscillation in search of a function. In: E. Başar & T.H. Bullock (Eds.), Induced Rhythms in the Brain, pp. 83–102. Boston: Birkhäuser.

Lynch, J.C., Mountcastle, V.B., Talbot, W.H. & Yin, T.C.T. (1977): Parietal lobe mechanisms for directed visual attention. Journal of Neurophysiology 40, 362–389.

Miller, R. (1991): Cortico-Hippocampal Interplay and the Representation of Contexts in the Brain. Berlin: Springer.

Mizuki, Y., Masotoshi, T., Isozaki, H., Nishijima, H. & Inanaga, K. (1980): Periodic appearance of theta rhythm in the frontal midline area during performance of a mental task. Electroencephalography and Clinical Neurophysiology 49, 345–351.

Mizuki, Y., Takii, O., Nishijima, H. & Inanaga, K. (1983): The relationship between the appearence of frontal midline theta activity (FmO) and memory function. Electroencephalography and Clinical Neurophysiology 56, 56–56.

Mountcastle, V.B. (1992): Prologue. In: E. Başar & T.H. Bullock (Eds.), Induced Rhythms in the Brain. Boston: Birkhäuser.

Mountcastle, V.B., Lynch, J.C., Georgopoulos, A., Sakata, H. & Acuna, C. (1975): The posterior parietal association cortex of the monkey: Command functions for operations within extrapersonal space. Journal of Neurophysiology 38, 871–908.

Mountcastle, V.B., Andersen, R.A. & Motter, B.C. (1981): The influence of attentive fixation upon the excitability of the light-sensitive neurons of the posterior parietal cortex. Journal of Neuroscience 1, 1218–1235.

Mountcastle, V.B., Motter, B.C., Steinmetz, M.A. & Duffy, C.J. (1984): Looking and seeing: The visual functions of the parietal lobe. In: G.M. Edelman, W.E. Gall & W.M. Cowan (Eds.), Dynamic Aspects of Neocortical Function, pp. 159–193. New York: Wiley.

Mundy-Castle, A.C. (1951): Theta and beta rhythm in the electroencephalograms of normal adults. Electroencephalography and Clinical Neurophysiology 3, 477–486.

Münte, T.F. & Künkel, H. (1990): Ereigniskorrelierte Potentiale in der Psychiatrie – Methodische Grundlagen. Psychiatrie, Neurologie und Medizinische Psychologie (Leipzig) 42, 649–659.

Näätänen, R. (1988): Implications of ERP data for psychological theories of attention. In: B. Renault, M. Kutas, M.G.H. Coles & A.W.K. Gaillard (Eds.),

Event-related Potential Investigations of Cognition, pp. 117–163. Amsterdam: Elsevier.
Näätänen, R. (1990): The role of attention in auditory information processing as revealed by event-related potentials and other brain measures of cognitive function. Behavioral Brain Sciences 13, 201–288.
O'Connor, T.A. & Starr, A.A. (1985): Intracranial potentials correlated with an event related potential, P300, in the cat. Brain Research 339, 27–38.
Paller, K.A., Zola-Morgan, S., Squire, L.R. & Hillyard, S.A. (1988): P3-like brain waves in normal monkeys and monkeys with medial temporal lesions. Behavorial Neuroscience 102, 714–725.
Parnefjord, R. & Başar, E. (1992): Delta-Oscillationen als Korrelat für akustische Wahrnehmungen an der Hörschwelle. 37. Jahrestagung der Deutschen EEG-Gesellschaft, Magdeburg.
Petsche, H. & Rappelsberger, P. (1992): Is there any message hidden in the human EEG? In: E. Başar & T.H. Bullock (Eds.), Induced Rhythms in the Brain, pp. 103–116. Boston: Birkhäuser.
Petersen, S.E., Fox, P.T., Miezin, F.M. & Raichle, M.E. (1988): Modulation of cortical visual responses by direction of spatial attention measured by PET. Association of Research on Vision and Ophthalmology, p. 22 (abstr.).
Pfurtscheller, G. & Klimesch, W. (1992): Event-related synchronization and desynchronization of alpha and beta waves in a cognitive task. In: E. Başar & T.H. Bullock (Eds.), Induced Rhythms in the Brain, pp. 117–128. Boston: Birkhäuser.
Picton, T.W. & Hillyard, S.A. (1974): Human auditory evoked potentials. II: Effects of attention. Electroencephalography and Clinical Neurophysiology 36, 193–199.
Picton, T.W. & Stuss, D.T. (1980): The component structure of the human event-related potentials. In: H.H. Kornhuber & L. Deecke (Eds.), Motivation, Motor and Sensory Processes of the Brain: Electrical Potentials, Behavior and Clinical Use, pp. 7–50. Amsterdam: Elsevier.
Pineda, J.A., Foote, S.L. & Neville, H.J. (1987): Long latency event-related potentials in squirrel monkeys: Further characterization of wave form morphology, topographic and functional properties. Electroencephalography and Clinical Neurophysiology 67, 77–90.
Posner, M.I. & Petersen, S.E. (1990): The attention system of the human brain. Annual Review of Neuroscience 13, 25–42.
Pritchard, W.S. (1981): Psychophysiology of P300. Psychological Bulletin 89, 506–540.
Regan, D. (1989): Human Brain Electrophysiology: Evoked Potentials and Evoked Magnetic Fields in Science and Medicine. Amsterdam: Elsevier.
Rémond, A. & Lesèvre, N. (1957): Remarques sur l'activité cérébrale des sujets normaux. In: H. Fischgold & H. Gastaut (Eds.), Conditionnement et réactivité en électroencéphalographie (Electroencephalography and Clinical Neurophysiology, Suppl 6), pp. 235–255. Amsterdam: Elsevier.
Ritter, W., Vaughan, H.G. & Costa, L.D. (1968): Orienting and habituation to auditory stimuli: A study of short term changes in averaged evoked responses. Electroencephalography and Clinical Neurophysiology 25, 550–556.
Robinson, D.L., Goldberg, M.E. & Staunton, G.B. (1978): Parietal association cortex in the primate: Sensory mechanisms and behavioral modulations. Journal of Neurophysiology 41, 910–932.
Rösler, F. (1982): Hirnelektrische Korrelate kognitiver Prozesse. Berlin: Springer.
Roth, W.T. (1973): Auditory evoked responses to unpredictable stimuli. Psychophysiology 10, 125–137.
Sachs, L. (1974): Angewandte Statistik. Berlin: Springer.
Shepherd, G.M. (1988): Neurobiology. Oxford: Oxford University Press.

Simson, R., Vaughan, H.G. Jr. & Ritter, W. (1977): The scalp topography of potentials in auditory and visual go/no go tasks. Electroencephalography and Clinical Neurophysiology 43, 864–875.
Smith, M.E., Halgren, E., Sokolik, M., Bauden, P., Musolino, A., Liégeois-Chauvel C. & Chauvel, P. (1990): The intracranial topography of the P3 event-related potential elicited during auditory oddball. Electroencephalography and Clinical Neurophysiology 76, 235–248.
Solodovnikov, V.V. (1960): Introduction to the Statistical Dynamics of Automatic Control Systems. New York: Dover.
Squires, N.K., Donchin, E., Squires, K. & Grossberg, S. (1977): Biosensory stimulation: Inferring decision-related processes from P300 component. Journal of Experimental Psychology 3, 299–315.
Stampfer, H.G. & Başar, E. (1985): Does frequency analysis lead to a better understanding of human event-related potentials? International Journal of Neuroscience 26, 181–196.
Stapleton, J.M. & Halgren, E. (1987): Endogenous potentials evoked in simple cognitive tasks: Depth components and task correlates. Electroencephalography and Clinical Neurophysiology 67, 44–52.
Sutton, S., Tueting, P., Zubin, J. & John, E.R. (1967): Information delivery and the sensory evoked potential. Science 155, 1436–1439.
Tononi, G., Sporns, O. & Edelman, G.M. (1992): The problem of neural integration: Induced rhythms and short-term correlations, In: E. Başar & T.H. Bullock (Eds.), Induced Rhythms in the Brain, pp. 365–393. Boston: Birkhäuser.
Van der Tweel, L.H. (1961): Some problems in vision regarded with respect to linearity and frequency response. Annals of the New York Academy of Sciences 89, 829–856.
Vaughan, H.G. Jr. & Ritter, W. (1970): The sources of auditory evoked responses recorded from the human scalp. Electroencephalography and Clinical Neurophysiology 28, 360–367.
Walter, D.O., Etevenon, P., Pidoux, B., Tortrat, S. & Guillou, S. (1984): Computerized topo-EEG spectral maps: Difficulties and perspectives. Neuropsychobiology 11, 264–272.
Westphal, K.P., Grözinger, B., Diekmann, V., Scherb, W., Reeß, J., Leibing, U. & Kornhuber, H.H. (1990): Slower theta activity over the midfrontal cortex in schizophrenic patients. Acta Psychiatrica Scandinavia 81, 132–138.
Wood, C.C. & McCarthy, G. (1985): A possible frontal lobe contribution to scalp P3. Social Neuroscience Abstracts 11, 879.
Wood, C.C., Allison, T., Goff, W.R., Williamson, P.D. & Spencer, D.D. (1980): On the neuronal origin of P300 in man. In: H.H. Kornhuber & L. Deecke (Eds.), Motivation, Motor and Sensory Processes of the Brain: Electrical Potentials, Behaviour and Clinical Use, pp. 51–56. Amsterdam: Elsevier.
Wood, C.C., McCarthy, G., Allison, T., Goff, W., Williamson, P.D. & Spencer, D.D. (1982): Endogenous event-related potentials following temporal lobe excisions in humans. Social Meuroscience Abstracts 8, 976.
Woods, D.L. (1990): The physiological basis of selective attention: Implication of event-related potential studies. In: J.W. Rohrbaugh, R. Parasuraman & R. Johnson Jr. (Eds.), Event-related Brain Potentials, Basic Issues and Applications, pp. 178–209. Oxford: Oxford University Press.
Yin, T.C.T. & Mountcastle, V.B. (1977): Visual input to the visuomotor mechanisms of the monkey's parietal lobe. Science 197, 1381–1383.

Slow Positive Potentials in the EEG During Multistable Visual Perception

C. Başar-Eroglu[1], D. Strüber[1], M. Stadler[1], P. Kruse[1] and F. Greitschus[2]

[1] Institute of Psychology and Cognition Research and Center for Cognitive Sciences, University of Bremen, D-28334 Bremen, Germany
[2] Institute of Physiology, Medical University of Lübeck, D-23538 Lübeck, Germany

Abstract: The present study deals with the kind of EEG activity changes concerning the perceptual switching. Seven subjects observed a multistable pattern (stroboscopic alternative motion: SAM), and were instructed shortly to press the button immediately after perceptual switching with the aim of detecting some neurophysiological parameters of EEG activity. Our results indicate, to our knowledge for the first time, that the EEG changes can be observed during multistable perception. A consistent slow positive wave occured following perceptual switching, therefore we called it **perceptual switching related positivity**. Furthermore the frequency component of this potential has a similarity to the frequency content of stimulus locked P300.

1 Introduction

The phenomenon of multistability in general is characterized by the property of a dynamic system to reach different stable states in a nonlinear self organized manner. Several kinds of such stable states or attractors exist. For instance, when a system relaxes towards an equilibrium state, one may say that it relaxes to a so-called stable fixed point. When it oscillates coherently, one speaks of a limit cycle attractor with a specific frequency.

When certain parameters of a system are changed, the specific attractor on which the system had formerly been, is destabilized and replaced by a new one. For instance, a fixed point attractor may be replaced by two fixed point attractors or several of them. Such bifurcations or nonequilibrium phase transitions are associated with the spontaneous emergence of new macroscopic structures, be they temporal, spatial, or functional (see Haken, 1991).

Such evolution of new stable states or the switching from one stable state into another undergoing a phase of high instability has been described in various disciplines, e.g. in physics, chemistry, and biology. Also in the cognitive sphere there are many processes showing this kind of behaviour, e.g. in memory, thinking, problem solving, perception (see Kruse, 1988 and Stadler & Kruse, 1990 for review).

Especially in the field of visual perception, the fundamental importance of instability as a process characteristic for the emergence of new macroscopic structures becomes evident when viewing so-called reversible figures. Reversible or ambiguous figures make up a well-known class of visual phenomena in which a stimulus pattern alternates spontaneously between different stable percepts or interpretations during continuous viewing. Familiar examples are the Necker cube, the Schröder staircase, the Mach folded card, Rubin's vase/faces, the Maltese cross, Fisher's man/girl, and Boring's young girl/old woman (see Attneave, 1971, for examples of these and other reversible figures). The curious multistable character of ambiguous figures has intrigued scientists for over a century and a half, since an early report of such a phenomenon was published by Necker in 1832. This curiosity concerning multistable perception may be caused by its incompatibility with the everyday experience of a stable perceptual world. The cognitive variability of the spontaneous, i.e. not stimulus driven, perceptual reorganization brings the phenomenon of multistability in an intermediate position between "normal" stimulus driven perception and sensory independent cognitive processes like thinking or imagination. From this point of view ambiguous figures are rare perceptual situations in which we are able to "see the breakdown of a very powerful and pervasive disambiguation mechanism that is quite general in cognition" (Kawamoto & Anderson, 1985). Despite extensive research in the field of multistable perception since Necker's time this phenomenon is still far from being explained. One problem is caused by the heterogeneity of perceptual multistability. There is a great variety of different ambiguous patterns on different levels of complexity explained by different theoretical contributions (see Gräser, 1977; Kruse, 1988). Perhaps the best known hypothesis to explain the dynamics underlying multistable perception is that of satiation or adaptation, proposed by Köhler (1940). Based on this concept of adaptation and the attractor-concept used in the self-organization theory outlined above, a mathematical model of perceptual multistability has been proposed (Ditzinger & Haken, 1989, 1990; Kruse, Stadler & Strüber, 1991). The perceptual appearence of a multistable process can be described as follows: An unchanged stimulus pattern is perceived in a certain orientation or interpretation (stable state A). This state will after some time continuously be destabilized until critical fluctuations occur and the pattern is spontaneously switched to another orientation (stable state B). This stable state begins again to destabilize until the pattern reswitches to stable state A and so on (Figure 1).

A number of interesting questions arise concerning the relation between EEG and multistability: Is it possible to associate certain changes in the EEG with the presumed increase of critical fluctuations preceding the perceptual switching from one stable state to the other? Do differences occur between the alternation of highly unstable and highly stable phases? Does stimulus-induced synchronization of the EEG activity occur during multistable perception? How are the spatial distributions of the synchronized states in the various scalp regions? Are there hemispheric lateralization effects? Much research still has to be done in order to answer all these questions.

In the present study we introduce a psychophysiological paradigm and report experiments to record potential changes in the EEG concerning the perceptual switching (see also Başar-Eroglu et al., 1993).

2 Method

2.1 Subjects

Seven healthy, right-handed volunteers (3 female, 4 male) aged between 22 and 42 years participated in this experiment. All subjects had normal or corrected-to-normal vision and were naive as to the purpose of the experiment.

2.2 Stimulus pattern

The ambiguous stimulus pattern used in this experiment is based on the phenomenon of apparent motion (AM). Dynamic ambiguous AM displays seem to be most adequate to analyze multistable behaviour because their relevant variables can be better controlled than in static pictures and they are very reliable in their perceptual characteristics (Kruse, Stadler & Wehner, 1986). The one used here is the so-called stroboscopic alternative motion (SAM), presented on a computer screen. Four spots are positioned in an upright rectangle. Two diagonally positioned spots are always flashed simultaneously and after a short pause replaced by a flashing of the other two spots (see Figure 1). During fixation of the SAM a horizontal apparent motion alternates clearly and regularly with a vertical apparent motion when the distance of the lights and the time of flashing is properly tuned (Hoeth, 1966). In this study the flashing-frequency of the diagonal point-lights was 2 Hz. The resulting flashing-duration was 165 ms for each diagonal with 80 ms pause between flashing. The horizontal to vertical ratio of the point-lights was 5/8 which as a rule leads to equal probability of perceived horizontal and vertical motion while a quadratic positioning of the spots would favor the vertical motion (Hoeth, 1968). The horizontal distance of the spots was 2.4 cm, the vertical 3.8 cm. At a viewing distance between subject and SAM-display of 150 cm the resulting visual angle for the horizontal direction is 0.92°, and for the vertical direction 1.45°.

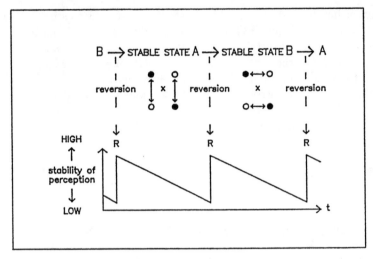

Fig. 1 Schematical representation of the Stroboscopic Alternative Motion (SAM). R: reversion or perceptual switching and x: fixation point. Stable state A corresponds to vertical motion, and stable state B to horizontal motion

2.3 Experimental procedure

The subjects sat in a soundproof and echo-free room which was dimly illuminated. Their right hand was mounted onto a flat switch which could be pressed nearly effortlessly. There was a first session to ensure that each subject was able to see both motion directions. The SAM-display was presented on a computer screen for about ten minutes. The subjects were instructed to look at the fixation point in the middle of the SAM-display for the whole time and to suppress blinks and eye movements as much as possible, especially during perceptual switching. They indicated the changes of the perceived motion direction by shortly and slightly pressing down their right index finger, thus breaking a contact impinging to one channel of the EEG-writer. So the muscle effort for finger movement was minimized.

In addition, we performed the following control experiment: EEG recording during voluntary finger movement in order to estimate the motor potential.

2.4 Electrophysiological recording

The EEG was recorded from Ag-AgCl electrodes at positions F3, F4, C3, Cz, C4, P3, P4, O1, and O2 according to the International 10 - 20 System.

Left earlobe electrode served as a reference. All electrode impedances were maintained at less than 5 KOhms. The EOG was also registered from medial upper and lateral orbital rim of the right eye. EEG activity was amplified by using a band-pass of 0.01 – 70 Hz and digitized on-line at a sampling rate of 250 samples per second and stored on the hard disc of the computer (HP 1000 F) for off-line EEG analysis. All channels were displayed on paper and on-line by monitor scope in order to observe both single trials and averaged signals.

2.5 Data analysis

Selective averaging of EEG data: In an off-line procedure, the on-line recorded, digitized and stored row single EEG signals were selected with specified criteria after the recording session. Movement artifacts and eye blinking were eliminated with double check. Twenty percent of the recorded 120 epochs were eliminated, because of muscle and eye movement artifacts. The remaining epochs were averaged time-locked to the onset of the finger movement. In other words, the start of finger flexions were marked as trigger point for the data analysis. Referred to that trigger point (t=0), the averaging was performed in two time windows. The first window (start 4 s prior to trigger point, duration 8 s) represented either a highly destabilized perceptual state prior to switching and the following stable state or voluntary finger movement respectively. The second window (start 4 s after trigger point, duration 8 s) represented only the stable perceptual states after switching.

The methodology to evaluate the averaged EEG data, Amplitude Frequency Characteristics (AFC) and digitally filtered data was previously described (Başar, 1980; Başar-Eroglu et al., 1992). The method we used here was basically the same except the minor modifications. Therefore, we briefly describe the methods as follows:

Amplitude Frequency Characteristics (AFC): The selectively averaged EEG data were transformed to the frequency domain with the Fast Fourier Transform (FFT) in order to obtain the AFC (not presented in this study).

Digital filtering: For the present study the data were digitally filtered with no phase shift. The limits of the band pass filters are not arbitrarily, but adequately chosen according to the dominant maxima of the AFC.

2.6 Statistical analysis

The maximal peak to peak amplitude value of a consistent wave related to finger movement was measured in the time window -1000 ms and 0 ms with the help of graphics display of the computer. Statistical evaluation was performed as follows: the normality of the distribution of data cannot be tested on samples of nine locations of seven subjects each. Therefore, for inter-electrode comparisons a nonparametric analysis of variance was performed (Friedman's two way analysis by ranks) with post-hoc Wilcoxon-Wilcox tests (Lienert, 1986). Furthermore, we used Wilcoxon matched pairs signed rank test in order to compare the amplitude values of consistent waves obtained in two different experimental conditions.

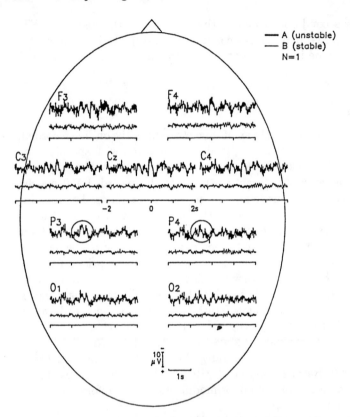

Fig. 2 Averaged EEG from one subject of two different phases in the same experimental session. **A:** the upper curves with thick lines show the unstable state prior to perceptual switching and the stable phase after task related finger movement (t=0). **B:** the lower curves with thin lines present only the stable phase

3 Results

Figure 2 shows two averaged recordings of one representative subject during the same experimental session (multistable perception). The upper curves of all locations present the unstable phase prior to perceptual switching and the stable phase after task related finger movement, whereas the lower curves illustrate only the following stable phase. We observed in all the locations slow potentials during the unstable phase. The amplitude of the EEG was higher in unstable phase than in stable phase. Although we present here only one subject for sake of simplicity, we observed these results in all the subjects.

Starting from this observation, we wondered whether these slow potentials are particularly related to perceptual switching or whether these slow waves occur because of task related finger movement, since the subjects should press the button immediately after perceptual switching. Therefore, we performed

Table I Mean amplitude values (peak to peak, μV) and standard errors of perceptual switching related positivity of frontal, central, parietal and occipital sites from seven subjects. The statistically significant difference is $p < 0.05$, Friedman-Test

Recording site	Mean Value (μV)	Standard Error
F3	5.02	1.03
F4	6.14	0.97
Cz	8.91	1.97
C3	8.08	1.32
C4	8.37	1.02
P3	7.69	1.04
P4	10.11	1.21
O1	6.95	0.83
O2	5.62	1.13

a control experiment, in which the subjects were instructed to move their right index fingers voluntarily in order to estimate the contribution of motor potentials.

Figure 3 presents two averaged recordings of *one subject* in order to compare two different experimental conditions.

A: *Multistable perception*: Potentials during the perception of a multistable pattern (SAM) in various locations with task related finger movement (finger movement onset t=0). As we mentioned above, the task was defined to press the button immediately after perceptual switching.

B: *Control condition*: Voluntary finger movement related potentials without any perceptual task (motor potentials).

We observed in several locations slow compound potentials approximately 1 s prior to finger movement as presented in Figures 2A and 3A (the upper curves of all locations are marked by triangles). This compound **perceptual switching related potential** has an initial negative peak and is usually followed by a positive slow wave at about 500 ms prior to finger movement (t=0), whereas the potential changes recorded during the experiments with the voluntary finger movement do not depict such a slow potential (Figure 3B: lower curves). Inasmuch as our interest is focussed in perception related potentials prior to finger movement, we present but not do discuss the motor potential following the finger movement (0 – 2000 ms) in both cases, to begin with. Since the shape of the perceptual switching related slow potential has a similarity to the time courses of the so called **response-locked positivity** and the same scalp distribution as the stimulus-locked P300 wave of event

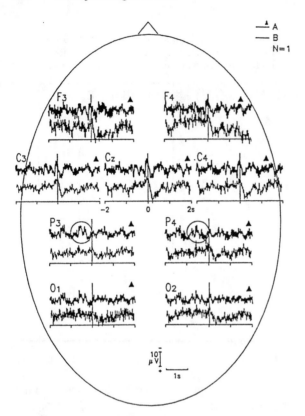

Fig. 3 Recording from the two different experimental conditions of one subject. **A:** Multistable perception; the upper curves with thick lines present perceptual switching related potentials, which were marked by triangle. **B:** Control condition; the lower curves with thin lines show voluntary finger movement related potentials. Vertical lines (t=0 ms) indicate movement onset. Ordinates: amplitudes in μV, abscissa: time in s. Negativity is up

related potentials (ERPs), we focussed our analysis to the recordings from the parietal site and not to the spatial distribution of these potentials (see Luck & Hillyard, 1990, and Hillyard, 1992, personal communication). As already mentioned above, we use the term **perceptual switching related positivity** for our study instead of response-locked activity for description of the slow positive potentials during multistable perception about 500 ms prior to finger movement.

The perceptual switching related positivity in the parietal location has a greater amplitude in comparison to other scalp-sites (Figure 2A, in circle). It is also noteworthy that the parietal location does show smaller response to finger movement in comparison to those recorded from pre- and central areas. As can be easily seen, frontal and especially central sites (C3, Cz, C4) show a slow negativity prior to finger movement (Figure 2B). This slow negative potential is well defined as the readiness potential (Kornhuber & Deecke, 1965).

Table I presents the mean value amplitudes of positive wave for the seven subjects from all the scalp locations, evaluated in the time window -500 ms to 0 ms. As can be seen, the amplitude of positive wave is largest at the parietal location (P4) and smallest at the frontal location (F3). The measured maximal amplitude values were tested for significance of differences between scalp locations by means of Friedman test, ($p < 0.05$). The P4 amplitude is significantly larger than F3 and O2 ($p < 0.1$, Wilcoxon-Wilcox). A similar, although nonsignificant, trend is seen for all the electrode locations. Besides that, we observed in all the subjects that the lateralization of positivity is different, however, more subjects are needed to test significant differences.

In Figure 4 we present the slow compound potential recorded (during multistable perception) from right parietal sites for the 7 subjects under study, because this location shows the largest amplitude of slow positive wave. All the subjects presented show a slow perceptual switching related positivity approx. 500 ms prior to finger movement. We want to mention here that although in all the subjects the perceptual switching related positivity has a similar shape, an exact time course cannot be defined. It exists only a global similarity. It is also known that reaction times of individuals vary between 300 and 700 ms. Accordingly, fluctuations at the time axis of the starting of potential changes were to be expected.

Figure 5 presents grand average curves of the two different experimental conditions for parietal site (P4) evaluated from all subjects. The grand average curve of Figure 5A were evaluated from the data of Figure 4 and it refers to the first experimental condition (A: multistable perception). This curve shows a marked compound slow positive wave between -500 ms and 0 ms (t=0: onset of finger movement) and is also in good accordance with the results of single curves presented in Figure 4. Figure 5B illustrates the grand average curve from the second experimental condition (B: control condition), in which the subjects pressed the button voluntarily. Both of the curves were presented versus mean value of EOG. As can be seen, the positive deflection occurred after perceptual switching with an amplitude value of about 5 μV and is thus, distinguished from the other curve. We compared the amplitudes of the positive waves between -500 ms and 0 ms of the two experimental conditions for all subjects. The amplitudes of perception related positivity were significantly higher in task A ($p < 0.05$, Wilcoxon Test).

Figure 6B and C illustrate the filtered mean value curves of Figure 6A (also presented in Figure 5A) in two different frequency bands (0.1 – 5 Hz and 0.1 – 12 Hz). This is the best straightforward argumentation of frequency composition of the perceptual switching related positivity. The comparison of the filtered curves in two different frequency ranges shows that most of the energy of the perceptual potential is in the frequency band between 0.1 and 12 Hz. This result has a meaningful similarity to the frequency content

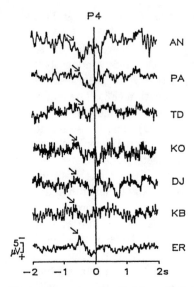

Fig. 4 Perceptual switching related potentials from seven subjects averaged from parietal site (P4). The arrows indicate the beginning of perceptual switching related positivity. Each curve labelled with the initial letter of name of the respective subject

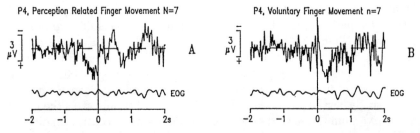

Fig. 5 (**A**) Grand average curves over seven subjects computed from Figure 4. (**B**) Grand average curve over seven subjects during voluntary finger movement. Both grand average curves were presented versus grand averages of EOG. Vertical lines (t=0 ms) indicate movement onset

of stimulus-locked P300 wave. However, it is interesting to note that a 10 Hz wave train of about -500 ms is superimposed to the recorded potential when the filter is a broader one and also contains the alpha band (Figure 6C). Until now we could not ensure whether single experiments and single trials from individual experiments are in accordance with this global finding seen in the mean value curves.

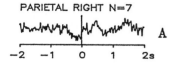
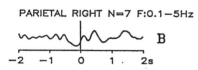
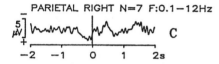

Fig. 6 (A) Unfiltered perceptual switching related potentials from Figure 5A were presented with filtered curves (**B:** 0.1 – 5 Hz) and (**C:** 0.1 – 12 Hz) respectively

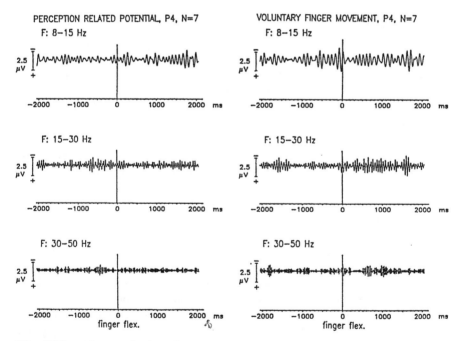

Fig. 7 Filtered curves in three frequency ranges of perception related potentials (left column) and motor potentials (right column). Filter ranges: (from top to bottom) 8 – 15 Hz, 15 – 30 Hz, 30 – 50 Hz respectively

It is well known that the motor potentials are strongly related with the high frequencies of EEG. Therefore we extended our analysis to the high frequency components of EEG and compared two different experimental conditions in the same frequency bands. Figure 7 illustrates the comparison between perception related potentials and motor potentials. We used for this purpose the conventional frequency ranges of EEG. The amplitude of alpha band (8 – 15 Hz) was decreased after perceptual switching (prior to movement onset) whereas the amplitudes of beta (15 – 30 Hz) and gamma band (30 – 50 Hz) were increased. The opposite changes were observed in motor potentials. The amplitudes of beta and gamma bands of the motor potentials were decreased prior to voluntary finger movement and increased after movement onset as was expected, whereas alpha response increased prior to movement. In other words, in the potentials occuring during multistable perception, the frequency patterns are opposite to those potentials related to voluntary finger movement. According to these results, we can state that the EEG changes after perceptual switching are closely related to multistable visual perception.

4 Discussion

4.1 The phenomenon of multistability

In the case of visual perception, any stimulus situation except a completely homogeneous visual field can be regarded as ambiguous. The figure-ground segmentation and the boundary function of discontinuity or contour is always a performance of the perceiver and not an inherent property of the stimulus (Kruse & Stadler, 1990). Therefore, investigating the dynamics underlying the mechanism of multistability will be essential for understanding perception in general. Basically, the same idea was already mentioned by the gestalt psychologist Wolfgang Köhler in 1940.

Since the techniques of human electrophysiology gained an important weight in the last decade to analyze the dynamics of sensory and cognitive processing, we applied a series of methods in our study containing advanced techniques including the selective averaging and the adaptive digital filtering.

A central problem of perception is still the time scale of cognitive order formation. The emergence and change of stable percepts is far too quick to measure the presumed nonequilibrium phase transitions themselves. One possible way out of this methodological dilemma is to look for neurophysiological parameters which are fast enough to allow a measurement of the neuronal macrodynamics correlated with the reversion process (perceptual switching).

Accordingly, the results obtained in the present study may open new insights into the understanding of these phenomena from the neurophysiological point of view. The relevance of many different macrodynamic neuronal processes for cognitive phenomena has been claimed in the last fifty years of brain research.

Recently, there was a certain revival in the recognition of the importance of the EEG-rhythms for cognitive processes (Mountcastle, 1992). Particularly the pioneering work of Freeman (1975, 1990) on the rabbit's olfactory bulb sheds new light on the understanding of how perceptual processes are connected with global neural mass-activity recorded from EEG. Also Petsche et al. (1992) showed that the task involving thinking with images may provoke detectable changes in background EEG.

Taking these results into account and extending them with the results of the present analysis, EEG-recordings seem to be a promising tool to study neuronal correlates of phase transitions between different perceptual states. In the case of perceptual multistability EEG-correlates are of a special kind, because no external trigger exists. The real "stimulus" is the switching, i.e. an endogeneous process.

4.2 Multistability and EEG

The electrophysiological analysis performed in the present study shows for the first time that the phenomenon of perceptual switching by viewing multistable patterns is correlated with changes in the EEG (slow positive compound potentials). Accordingly, the present study may open new insights into the understanding of multistable visual perception from the neurophysiological point.

However, we have to mention a word of caution about the contribution of motor potentials. It has been shown that motor potentials, especially Bereitschaftspotentials (readiness potential), not only depend on physical factors of movement, but also on psychological aspects of movement such as attention, motivation, and intentional movement (Deecke et al., 1969, 1984, Freude & Ullsperger, 1987) and are due to the complexity of the movement pattern (Demiralp et al., 1990).

4.3 Contributions of motor potentials to perception

In the paradigm used here, one of the major problems is to extract the pure perception related signals from the motor potentials (MPs) of the cortex. The MPs have their origin mainly in two different types of movements:

A: The eye blinking and eye movement, B: The task related finger movement immediately following the perceptual switching (EMG Onset).

In our analysis, the eye blinking movement was rejected with a double check: (i) by means of selective averaging (off-line) and (ii) automatically artifact rejection (on-line). All signals recorded in the EOG were digitized and averaged in order to detect any contaminations with presented data. This type of motor potential, even one single eye blinking, can strongly distort the compound potentials. Therefore, in our procedure only sweeps which do not contain any eye movement artifact have been taken in the averaged curves. It should be emphasized here that without the use of selective averaging procedure, i.e. without strong elimination of single trials containing eye movement artifacts, it is not possible to obtain clean compound perception potentials.

4.4 What is the contribution of premovement potentials during the time period between −1000 ms and 0 ms?

It is known that the readiness potentials consist of a slow increasing negativity about 1 s prior to voluntary finger movement (Kornhuber & Deecke, 1965; for a review see Regan, 1989). Accordingly, an interaction with BP and presented slow compound potential which lies in the time period between −1000 and 0 ms was expected. In an early study, Deecke et al. (1984) described the premotion positivity (PMP), which begins about 90 ms prior to onset of the movement and occurs ipsilateral precentrally over both parietal regions, especially over the midline. They also emphasized the existence of a conditioned PMP, which anticipates the change in direction of "random V-Figures". It is maximal at FCz location (overlying paralimbic supplementary motor area: SMA).

In our study, we observed a maximal perception related positivity at parietal site (P4). Besides that, in the presented results (cf. Figure 3A) neither a negative slow slope nor an increase of the potential to the negative direction was observed. On the contrary, a positive slow wave occurred instead of a negative slow activity which should be elicited prior to finger movement, whereas voluntary finger movement induces clearly negative potentials (cf. Figure 3B) as expected. It is plausible that the positivity of perceptual switching related potentials superimposes and strongly reduces the negativity of the readiness potential and enhances PMP.

If the consideration above is valid, it can be stated that the perception related potentials should have a much higher positivity in the time axis between −200 and 0 ms. The last assumption is a theoretical one and somewhat speculative. This comparison, however, enables us to strongly state that the measured compound potentials mainly reflect the brain electrical activity related to perceptual switching and not movement.

4.5 Comparison of the perception related positivity and response locked positivity of ERP

ERPs during visual perception were recorded by many authors in order to evaluate parallel and serial models of visual processing in the context of Treisman's feature integration theory (Luck & Hillyard, 1990) or for recording of slow cortical potentials (Elbert, Hommel & Lutzenberger, 1985). The former authors have emphasized that a well-defined response-locked positive wave (Rösler, 1983) occurred prior to EMG onset. This positive wave has the same scalp distribution with a large positivity in parietal sites as the stimulus-locked P3 wave (300 ms after task relevant stimulus) and is therefore be considered the same component.

In our study, we observed also the large positivity of perception related potential in parietal sites. Although the amplitude and latency of perception related positive waves varied between subjects, this wave in parietal location occured without exception. These results are in accordance with the interpretation of Rockstroh and coworkers (Rockstroh et al., 1982, Birbaumer et al., 1990) however, our paradigms and measurements are different. They stated that the negative shifts are a sign of preparation, while any cerebral performance adds positivity. Also Elbert et al. (1985) have shown during visual performance of Necker cube reversal that the positive potentials diminish the negative BP at parieotoccipital sites.

In general, a P300 wave is produced by low probability of the task relevant stimuli that occur somewhat unexpectedly and require a motor response or cognitive decision. In our paradigm it has been observed during the experiments with a perceptual task that the proportion of reversal to nonreversal phase was 1:3 respectively. Furthermore the task was unpredictable and the subjects reported the need for great concentration and attention. Therefore, we assume that the positive wave occurred immediately after perceptual switching and that it has a similarity to the eliciting of stimulus locked P300 waves of event related potentials.

O'Donnell et al. (1988) found a late positive component associated with illusory and physical reversals of the Necker cube. Their findings suggest that illusory reversals are more difficult to discriminate than physical reversals, and require additional cognitive resources for evaluation. Because the Necker cube is tachistoscopically presented for 700 ms with a 3.3-s ISI the positivity found by O'Donnell et al. cannot be elicited during the perceptual switching, as it is the case in our study.

One of the important results of this study is the fact that it is possible to record multistable perception in EEG and ERPs that were elicited during perceptual switching. By using the digital filters we were able to show that most of the energy of the perceptual switching related positivity is in the theta and alpha frequency band. This result has a meaningful similarity to the frequency content of stimulus locked P300 (Başar-Eroglu et al., 1992).

5 Conclusion

Our results indicate that a consistent slow positive EEG wave occurred while perceiving a multistable pattern. We assume that it occurs after perceptual switching, therefore we call it "Perceptual Switching Related Positivity". Accordingly, EEG recordings seem to be a promising tool to study the perception of multistable patterns.

Acknowledgement

This research was supported by DFG grant; RO 481/11-1, Project 6. We are thankful to Dr. Th. Bullock and Dr. A.S. Hillyard, Dr. E. Rahn and Dr. M. Schürmann for valuable discussion of results. Our thanks to Dipl.-Ing. M. Braasch for preparation of a part of experimental setup, to A. Aufseß for technical assistance.

References

Attneave, F. (1971): Multistability in perception. Scientific American 225, 62 – 71.
Başar, E. (1980): EEG-brain Dynamics. Relation between EEG and Brain Evoked Potentials. Amsterdam: Elsevier/North-Holland.
Başar-Eroglu, C., Başar, E., Demiralp, T. & Schürmann, M. (1992): P300-response: possible psychophysiological correlates in delta and theta frequency channels. A review. International Journal of Psychophysiology 13, 161 – 179.
Başar-Eroglu, C., Strüber, D., Stadler, M., Kruse, P. & Başar, E. (1993): Multistable visual perception induces a slow positive EEG wave. International Journal of Neuroscience (in Press).
Birbaumer, N., Elbert, T., Canavan, A.G.M. & Rockstroh, B. (1990): Slow potentials of the cerebral cortex and behavior. Physiological Reviews 70, 1 – 41.
Deecke, L., Bashore, T., Cornelis, H.M.B., Grünewald-Zuberbier, E., Grünewald, G. & Kristeva, R. (1984): Movement-associated potentials and motor control. Report of the EPIC VI motor panel. In : R. Karrer, J. Cohen & P. Tueting (Eds.), Brain and Information: Event Related Potentials. Annals of the New York Academy of Sciences 425, 398 – 428.
Deecke, L., Scheid, P. & Kornhuber, H.H. (1969): Distribution of readiness potential, premotion positivity and motor potential of the human cerebral cortex preceding voluntary finger movements. Experimental Brain Research 7, 158 – 168.
Demiralp, T., Karamürsel, S., Karakullukcu, E. & Gökhan, N. (1990): Movement-related cortical potentials: their relationship to the laterality, complexity and learning of a movement. International Journal of Neuroscience 51, 153 – 162.
Ditzinger, T. & Haken, H. (1989): Oscillations in the perception of ambiguous patterns. Biological Cybernetics 61, 279 – 287.
Ditzinger, T. & Haken, H. (1990): The Impact of fluctuation on the recognition of ambiguous patterns. Biological Cybernetics 63, 453 – 456.
Elbert, T., Hommel, J. & Lutzenberger, W. (1985): The perception of Necker cube reversal interacts with the Bereitschaftspotential. International Journal of Psychophysiology 3, 5 – 12.
Freeman, W.J. (1975): Mass Action in the Nervous System. New York: Academic Press.
Freeman, W.J. (1990): Nonlinear neural dynamics in olfaction as a model for cognition. In: E. Başar (Ed.), Chaos in Brain Function, pp. 63 – 73. Berlin: Springer.
Freude, G. & Ullsperger, P. (1987): Changes in Bereitschaftspotential during fatiguing and nonfatiguing hand movements. European Journal of Applied Physiology 56, 105 – 108.
Gräser, H. (1977). Spontane Reversionsprozesse in der Figuralwahrnehmung. Trier: Dissertation.

Haken, H. (1991): Synergetics - can it help physiology? In: H. Haken & H.P. Koepchen (Eds.), Rhythms in Physiological Systems, pp. 21 – 31. Berlin: Springer.
Hoeth, F. (1966). Gesetzlichkeit bei stroboskopischen Alternativbewegungen. Frankfurt/Main: Kramer.
Hoeth, F. (1968): Bevorzugte Richtungen bei stroboskopischen Alternativbewegungen. Psychologische Beiträge 10, 494 – 527.
Kawamoto, A.H. & Anderson, J.A. (1985): A neural network model of multistable perception. Acta Psychologica 59, 35 – 65.
Kornhuber, H., & Deecke, L. (1965): Hirnpotentialänderung bei Willkürbewegungen und passiven Bewegungen des Menschen: Bereitschaftspotential und reafferente Potentiale. Pflüger's Archiv 284, 1 – 17.
Köhler, W. (1940): Dynamics in psychology. New York: Liveright.
Kruse, P. (1988): Stabilität, Instabilität, Multistabilität, Selbstorganisation und Selbstreferentialität in kognitiven Systemen. Delfin 11, 35 – 58.
Kruse, P. & Stadler, M. (1990): Stability and instability in cognitive systems: Multistability, suggestion, and psychosomatic interaction. In: H. Haken & M. Stadler (Eds.), Synergetics of Cognition, pp. 201 – 215. Berlin: Springer.
Kruse, P., Stadler, M. & Strüber, D. (1991): Psychological modification and synergetic modelling of perceptual oscillations. In: H. Haken & H.P. Koepchen (Eds.), Rhythms in Physiological Systems, pp. 299 – 311. Berlin: Springer.
Kruse, P., Stadler, M. & Wehner, T. (1986): Direction and frequency specific processing in the perception of long-range apparent movement. Vision Research 26, 327 – 335.
Lienert, G.A. (1986). Verteilungsfreie Methoden in der Biostatistik. Bd.1. Meisenheim/Glan: Hain.
Luck, J.S. & Hillyard, A.S. (1990): Electrophysiological evidence for parallel and serial processing during visual search. Perception and Psychophysics 48, 603 – 617.
Mountcastle, V.P. (1992): Preface to E. Başar & T.H. Bullock (Eds.), Induced Rhythms in the Brain. Boston: Birkhäuser.
Necker, L.A. (1832): Observations on some remarkable phenomenon which occurs on viewing a figure of a crystal or geometrical solid. The London and Edinburgh Philosophical Magazine and Journal of Science 3, 329 – 337.
O'Donnell, B.F., Hendler, T. & Squires, N.K. (1988): Visual evoked potentials to illusory reversals of the Necker cube. Psychophysiology 25, 137 – 143
Petsche, H., Lacroix, D., Lindner, K., Rappelsberger, P. & Schmidt-Heinrich, E. (1992): Thinking with images or thinking with language: a pilot EEG probability mapping study. International Journal of Psychophysiology 12, 31 – 39.
Regan, D. (1989). Human Brain Electrophysiology: Evoked Potentials and Evoked Magnetic Fields in Science and Medicine. Amsterdam: Elsevier.
Rockstroh, B., Elbert, T., Birbaumer, N. & Lutzenberger, W. (1982): Slow Brain Potentials and Behavior. Baltimore, MD: Urban & Schwarzenberg.
Rösler, F. (1983): Endogeneous ERPs and cognition: Probes, prospects, and pitfalls in matching pieces of the mind-body puzzle. In: A.W.K. Gailard & W. Ritter (Eds.), Advances in Psychology. Tutorials in ERP Research: Endogenous Components, pp. 9 – 35. Amsterdam: North Holland.
Stadler, M. & Kruse, P. (1990). The self-organization perspective in cognition research: Historical remarks and new experimental approaches. In: H. Haken & M. Stadler (Eds.), Synergetics of Cognition, pp. 32 – 52. Berlin: Springer.

Brain Electric Microstates, and Cognitive and Perceptual Modes

D. Lehmann

Brain Mapping Laboratory, Department of Neurology, University Hospital, CH-8091 Zürich, Switzerland

Abstract: The brain's global functional state is an emergent property of very many local subprocesses, and is influenced by many factors including newly arriving internal (memory) or external (surround) information. Synergetics (Haken) has modelled the cooperative coordination of multiple subprocesses which in turn constrains the individual subprocesses. This is consistent with the state-dependency of information processing in the brain.

The adequate level for studying brain mechanisms of conscious experiences which emerge out of cooperative sub-processes must be a comparably high level of physical measurements: The indivisible and rapidly changing brain electric field appears appropriate. This field offers a "macroscopic" view (Haken) with various temporal resolutions. Brain global functional states up to several seconds ("macrostates") are defined by frequency domain properties; states in the sub-second-to-second-range ("microstates") are defined by spatial properties (field landscapes). Different brain field landscapes (microstates) are likely to represent different brain functions. Quasi-steady landscapes (microstates) persist for varying durations, and are concatenated by very brief transitions. Different brain microstates were found to be associated with different conscious mind states and with different steps in information processing.

Assuming identity of brain states and mind states, we suggest that the formally and functionally identifiable global microstates are the building blocks of conscious experiences, the "atoms of thought". Incorporation of one of these sub-second building blocks into the stream of consciousness might depend on microstate duration and on re-entry mechanisms which permit mirroring of the information in the system within a crucial time window.

1 The brain's functional state

The brain can be said to be in a particular global, functional state at each moment in time. This momentary, global functional state is the momentary sum of all parallel and/or interacting processes at all brain regions, i.e. the global state consists of a mosaic of local states. Sooner or later, the functional state will change because new processes become initiated representing subsequent steps in processing the same information or first treatment steps of newly received information, or because of clock advances.

The brain functional state is re-adjusted at each moment in time within the constraints of the developmental stage, by the momentary metabolic/hormonal condition including disease conditions, by the time of the internal clocks, by past system history as implemented in available factual and procedural memories, and most important, by the momentarily newly arriving information from exterior or interior sources (Koukkou & Lehmann, 1990). Within the constraints set by the present functional state, the state is continuously re-adjusted depending on the demands raised by the processing of newly arriving information. Typically, two major parts of information treatment are distinguished: A first epoch lasting about 50 – 100 msec during which pre-attentive processing involving automatic (non-conscious) decisions are made with reflexive speed but relatively small adaptability, and a later epoch of cognitive processing during which the option arises that the information is treated in the capacity-restricted consciousness channel. Conscious treatment of the information offers a wide range of selectable strategies and access to contextual information, but involves relatively slow decision making in comparison with the automatic mode.

The momentary functional state of the brain is measurable on very many parameters with varying sensitivity. The present paper is centered around measurements of the brain electric field. This parameter is particularly suited for the assessment of human brain mechanisms of conscious and non- conscious perception and cognition because it offers the possibility for non-invasive, continuous recording combined with extreme sensitivity to changes of state and with the high time resolution down to the millisecond range which is necessary for perceptual and cognitive processes. Different aspects of perceptual and cognitive functions concern coarser or finer grained time resolution (Newell, 1992), but the shortest possible units of perceptual-cognitive activity certainly are well in the sub-second range as evidenced by the time between information input and major, complex decisions in terms of overt behavior even when using the consciousness channel, for example in street traffic situations.

2 Changes of functional state in living systems

Living creatures are dynamic, adaptive systems that show incessant changes of their state. Many changes of state result from the interactions with the external and internal environment, many others are driven by innate clocks and basically display tendencies to periodicities. Most obvious fluctuations of the innate type are sleep - wake changes which are easy to observe in motor activity and, as private experience, in the presence or absence of reflective self-awareness. It is of interest here to note that the transitions between wakefulness and sleep are extremely short compared with the sustained durations of either state. (During a boring lecture, one might catch oneself "having dozed off" for a moment, of course after one has come to again; the transitions subjectively appear to be either/or breaks, without intermediate stages being perceived).

In the context of multistability of perception, the topic of this volume, it is necessary to be aware that many different parameters which can be measured in living systems show fluctuations not only within days, hours and minutes but also within seconds or milliseconds. Spontaneous fluctuations tend to periodicities which, however, are strongly influenced by the surrounding (external informations) and by the motivational condition of the organism. On the other hand, the periodicity of a particular fluctuation typically varies within a wide or narrow hysteresis even without clear external influences.

It is of interest that these quasi-periodic fluctuations exist also during complete isolation from external influences (such as new stimuli and "Zeitgeber"), i.e. that they are inherent, self-generated properties of the system. (Brains that were completely isolated from all sensory input continue to demonstrate spontaneous changes of their functional state as measured through properties of their brain electric fields (Villablanca, 1965)). Synergetics (Haken, 1983) has successfully modelled the spontaneous formation of emergent temporal structures (fluctuations of "macroscopic state") in activated systems which consist of very many basic components. Internal clocks can accordingly be seen as autonomous order formations within neuronal self-organisation, and hence, the search for a control center might be unnecessary.

The fluctuations of the system state are important, because all actions and responses of living systems are state-dependent. Identical informations (stimuli) which reach an organism in different functional states are followed by different reactions, for example, during sleep and wakefulness. One might contract this into the statement that the system state determines the fate of the currently treated information.

3 Waveform recordings of the brain electric field

The brain electric field can be recorded using a convenient and non-invasive technique via small electrodes attached to the scalp with a conductive paste. Traditionally, the potential differences between recording locations were viewed as waveforms (electroencephalogram [EEG] and event-related potentials [ERP]), and the analysis strategies concentrated on waveform assessment. The spatial aspect of the brain field was largely neglected, and little attention was payed to the fact that potential differences between two recording locations does not imply unique information about either location. Thus, over the past decades, a multitude of findings has been reported based on recorded waveshapes which obviously depend on the chosen electrode combinations, and are not unique for a given scalp point unless data transformations such as removal of spatial DC offset, or gradient- or current density-computations are employed. This criticism applies where the conventional strategies of data analysis are used for conclusions about the intracerebral locations of model sources which often are assumed to be located perpendicularly under the point of maximal (minimal) potential or power density of the recorded waveshapes. Here, two errors are combined, the assumption of perpendicularity of the source orientation and the assumption that voltage waveforms are independent of the reference electrode.

Nevertheless, brain field analysis using EEG and ERP waveforms has produced a body of results that has contributed considerably to our understanding of mechanisms of brain information processing, where knowledge about the conditions for and sequences of events in the brain, and about strength and type of the events in the brain, are of interest, beyond the also desirable knowledge about the locations of the events.

4 Brain functional macrostates

4.1 "Macrostates" defined by frequencies of field polarity reversal

Examination of EEG waveforms has shown that gross differences of periodicities of polarity reversal of the brain electric field (EEG) are commonly observed during development: With increasing maturational age, the dominant frequencies become systematically faster, changing from about 6 Hz around 4 years of age to about 9.5 Hz ("alpha frequency range") in adults. Likewise, within a given maturational age, increases of vigilance (arousal level) are associated with increased dominant frequency of the EEG; dominant 2 – 5 Hz rhythms during deep sleep are replaced by rhythms of 8.5 – 12 Hz during relaxed wakefulness in most adults. The dimensionality (the correlation dimension) of the EEG waveshapes likewise has been shown to increase systematically with increasing vigilance, from sleep to relaxed wakefulness to conditions of attention.

Within the conceptual framework of self-organization of multi-element systems it is of interest that the periodic polarity reversals of the brain electric field persist not only after interruption of all neural information input to the brain (Villablanca, 1965) but also in isolated slabs of cerebral cortex (Kristiansen & Courtois, 1949).

A closer inspection of the EEG waveshapes which occur "spontaneously" while a subject relaxes with closed eyes in an isolation chamber reveals that there are fluctuations of the frequency and amplitude characteristics of the waveshapes over time. A very typical characteristic of spontaneous EEG waveshape patterns in many adults is the so-called "alpha spindling", an iterative waxing and waning of the amplitude of the dominant alpha frequency rhythm; other spontaneous patterns include intermittent replacements of the alpha rhythm by other dominant frequencies. Segmentation procedures using the autocorrelation function have demonstrated that homogeneous epochs in the range of seconds can be recognized in the spontaneous EEG (Creutzfeldt, Bodenstein & Barlow, 1985). Such frequency-defined brain states are called "macrostates" in the present paper.

4.2 Brain electric macrostates and fluctuations of perceptual threshold ("fluctuations of attention")

At the turn of this century, the school of experimental psychology around Wundt at Leipzig investigated the spontaneous fluctuations of perceptual threshold. The basic observation was that continuous, threshold-near visual or auditory signals (continuous light, continuous sound) perceptually appear to wax and wane within seconds (see Woodworth & Schlosberg, 1954). Wundt and his students called the observed phenomenon "fluctuations of attention", referring to a hypothesized, underlying brain mechanism which is generally conceived of to be under (at least partial) voluntary control. Physiological correlates of the phenomenon were searched in the respiratory cycle and blood pressure waves of the third order. Since the respiratory cycle (including willful inspirations) is known to influence EEG frequency patterns as well as blood pressure and perceptual threshold (Hildebrandt & Engel, 1963; Katayama et al., 1977; Lehmann & Knauss, 1976; Poole, 1961), various cross-influences between these parameters are to be expected. Hypotheses about the location of the mechanisms controlling the fluctuations of perception included peripheral sites, but evidence has been presented suggesting more central parts of the system (e.g., Stadler & Erke, 1968).

A special case of fluctuation of perception are the changes of the subjective percept of a continuous visual input that is presented as a "stabilized image" on the retina (the signal-transducing neuronal layers of the eye). Under threshold-near conditions, the percept of the stabilized retinal image spontaneously fluctuates, alternating between disappearance ("fading") and reappearance within a few seconds. The question arose whether these fluctuations of perception are caused by retinal (peripheral receptors) or central (brain) mechanisms. In two independent studies (Lehmann, Beeler & Fender, 1965; Keesey & Nichols, 1966), the subjects EEG's were continuously recorded while the subjects reported the subjective appearance or disappearance of the stabilized visual target by pressing or releasing a microswitch. The period duration between reappearances of the percept was 4.8 seconds on the average in our study. Both studies concluded that EEG alpha frequency patterns appeared before the disappearance of the percept, and waned before the reappearance of the percept, indicative of shifts of macrostate. Motor reaction time was taken into account in these experiments, using control runs where the target intensity was decreased and increased by the experimenter and motor reaction time was measured.

These observations indicate that the perceived fluctuations of the continuously present signal are caused by state shifts of central (brain) mechanisms, reflected in the changes of EEG patterns between disappearance and reappearance of alpha EEG. It is well known that a disappearance of alpha EEG is observed about 200 – 250 msec after the presentation of a sensory stimulus, in particular after the presentation of a light signal. With increasing time after the onset of the stimulus presentation, the alpha-type EEG activity returns. In other words, a state change from relaxation to what might be called automatically produced attention (Pavlov's "orienting response" to new or important information) is associated with the disappearance of alpha EEG patterns. In the studies of image stabilization, the central (EEG) state change preceded the reported perceptual change (taking into account the reaction time). Thus, one might speculate that the basis for the observed fluctuations of perception are spontaneous changes of brain macrostate which to some extent are driven by inherent periodicities, and which influence brain mechanisms that are also drawn upon when voluntary acts of attention are executed.

4.3 Brain macrostates and multistable percepts

The "Necker Cube", a classical case of multistable, oscillating perception (see Kruse, Stadler & Strüber, 1991), offers an experimental paradigm where the temporal sequence of two very different percepts (the so-called reversal of the cube) based on constant visual information can be studied. Attention and memory processes were proposed among various others as determinants of the perceived reversals (e.g. Reisberg & O'Shaughness, 1984). We agree with the suggestion (Stadler & Erke, 1968) that that the above-discussed fluctuations of perception of threshold-near steady sensory stimuli might offer a clue to the brain mechanisms of perceived reversal. The perceptual fluctuations evidently are centrally controlled. The moments of disappearance and reappearance of the stabilized image might reveal "prefabricated" break points in the sequence of the continuously renewed brain operations that underlie the perception of a continuous percept. At these moments, set by the interaction of internal clock cycles and external influences including some voluntarily controlled events such as willed inspiration or expiration, the system might periodically be in a condition where minor additional factors cause a state shift which leads to a new interpretation of the incoming information. It appears reasonable that such state shifts that lead to dis- or reappearances of the percept might possibly act as carriers of revised interpretations of the percept. Under otherwise comparable conditions, the reappearance frequency in fluctuations of perception (e.g., Lehmann et al., 1965, 4.8 sec; Clarke & Belcher, 1962, 6.4 sec) should then roughly correspond to one half of the reversal frequency of bistable figures which is in some agreement with the reported frequencies (Borsellino et al., 1972, about 2.5 sec). Applying principles of Synergetics, Ditzinger and Haken (this volume) successfully modelled the observed time series.

5 Brain functional microstates

5.1 Brain electric microstates and their functions

A space-oriented approach to brain electric field analysis views the recorded data as series of momentary maps of the brain potential distributions measured on the scalp. When examining such series of momentary maps it becomes evident that their spatial configurations ("landscapes") tend to persist for a certain epoch, then change relatively quickly into a different landscape which again stays quasi-stable for some time. It is obvious that the geometry of the activated neuronal elements must be different if the generated potential distribution is different. If the geometry of activated neurons changes, it might reasonably be assumed that this implies that a function is performed which differs from the previous function, i.e., that a different mode, step or content of information processing takes place. Thus, it appears that in order to identify putative "atoms of thought", the task is to recognize the different epochs of stable brain field configurations, and to identify the functions performed during these epochs.

We have developed segmentation strategies (Brandeis & Lehmann, 1989; Lehmann, 1987; Lehmann, Ozaki & Pal, 1987; Lehmann & Skrandies, 1986; Wackermann et al., 1993) which partition the series of momentary brain electric field maps into epochs of quasi-stable map landscapes but of varying durations; such segmentation into microstates can be done during "spontaneous" activity ("EEG potential fields") and immediately after presentation of information ("event-related potential maps"). Assuming that different brain functions are performed during a different geometry of neuronal activation, and considering that the theoretically available time resolution is as high as the sampling rate, we called these steady segments of brain field map sequences "microstates".

As in all classification strategies, the recognition of borders of the segmentation (the assignment to the same or another class) depends on the chosen window size. Using data-driven strategies to determine the adequate window size for the tolerated variance of the spatial descriptors of the potential fields within a microstate, the mean duration of the microstates was 144 msec in normal subjects during relaxed, task-free wakefulness (Strik & Lehmann, 1993).

Summing up, the hypothesis is that global brain electric field properties change over time not in a continuous but in a step-wise fashion at intervals in the sub-second range, indicative of a concatenation of a potentially identifiable repertoire of global microstates of the brain. Each of these microstates persists for a certain, brief duration and is suggested to represent a particular function, to be followed by another global functional microstate which represents a different function. It is clear that each global microstate must consists of a complex mosaic of local states representing all parallel processes in their various momentary conditions: However, as evident in the experience of consciousness, high-level brain operations result in a homogeneous end product, even though myriads of constituting lower-level processes must operate correctly in order to give rise to the highest-level phenomenon as an undivisible whole. The brain's electric-magnetic field as the sum of all momentary processes appears to be the adequate candidate for the externally measurable representation of consciousness.

5.2 Brain electric microstates as "atoms of conscious thought": Spontaneous visual imagery vs abstract thought

The study of brain field representations of cognitive and perceptual acts has to cope with the problem of possible confoundation of "pure" perception or cognition with representations of input information or task execution routines or response selection routines incorporated in the experimental paradigm. We are using a no-input, no-task, no-response paradigm to avoid these confoundations.

In a study designed to identify the functional (perceptual, cognitive) significance of spatially different brain field microstates within the spontaneous stream of consciousness (Lehmann, Henggeler & Strik, in preparation), we had male volunteers sit comfortably in a sound and light shielded recording room while their brain electric fields were continuously recorded from 20 electrodes on the head. The subjects were instructed to report briefly, without further questioning, "what just has gone through your mind just before the prompt" whenever a gentle prompt tone was sounded. The subject was free not to report any content if he wished to do so, but in such a case had to state whether there was no recall of content or whether he chose not to report it. There were 30 prompts for each of the 12 subjects within about 90 minutes. The brain field data of the second epoch preceding each prompt were screened for artifact. 312 epochs with their reports were eventually available.

The reports were transcribed and classed by two raters into mentation reports dominated by visual imagery (on the average over subjects, 40 %) or dominated by abstract thought (26 %) or neither (34 %).

The brain field data were segmented into microstates characterized by quasi-steady landscapes of the electric field. The potential distribution map of the last microstate before each prompt and of the microstate 2 seconds before the prompt were determined. The brain field map landscape of each microstate was assessed numerically by the (anterior-posterior and left-right) locations of the centroids of the areas of positive and negative voltage referred to the average of all recorded voltages in the map. This procedure invokes a planar dipole model as map descriptor. For each map we computed two descriptors: (a) the angle between the line formed by the two centroid locations and the line along the left-right axis of the scalp map, and (b) the distance between the centroid locations.

For each subject, the descriptors were averaged for either microstate (early and last microstate) and either mentation class (visual or abstract type). Statistics showed that the map landscape descriptor "angle" of the last microstate before the prompt report was, over subjects, significantly different for mentations classed as visual imagery compared vs those classed as abstract thought: Visual imagery cases showed an angle that was significantly smaller than that for abstract thought cases; visual imagery cases deviated on the average 86 degrees from the left-right zero line, abstract cases deviated 100 degrees ($p < 0.01$, $N = 13$). This difference, however, was not found for the early microstate (88 vs 85 degrees, no significant difference) that existed 2 seconds before the report prompt. Thus, the last microstate (with the closest possible temporal proximity to the report given by the subject) demonstrated an electric signature of the cognitive mode of the report, whereas the early microstate (separated by about 1900 msec from the report) did not reflect the verbal report mode.

The observations reported above support the proposal that the brain electric microstates might indeed be appropriate candidates for the "atoms of thought", constituting functional entities that incorporate the brain electric field properties which are available to subjective, private experience as modes of conscious mentation. In the studied case these modes were visual imagery and abstract thought. The relevance of the finding is underscored by the observed temporal gradient, i.e. the fact that a functional specificity could be found between mode of reported mentation and the microstate immediately preceding the report, but not between the microstate about two seconds earlier. Whether the particular current microstate participates as a building block in the stream of consciousness might well depend on its duration (Libet, 1982). However, as the brain's functional organization is known to use multiply repetitive information handling areas, it might well be that additional qualifications for access to consciousness are set by re-entry procedures (shift register type procedures) where the same functional state is re-visited repeatedly within a time window.

6 "Spontaneous" global functional microstates and macrostates

The global microstates are defined by the spatial charactereristics of the brain field, and typically exist in the sub-second range. Global macrostates are defined by their temporal characteristics as discussed earlier; they extend in the range of seconds. Accordingly, a sequence of microstates is embedded in a macrostate. It is conceivable that the functional significance of spatially similar microstates differs if they occur in grossly different macrostates such as wakefulness and sleep, or such as normal and psychotic wakefulness that show different EEG frequency characteristics as well as EEG dimensionality (correlation dimension: Koukkou et al., 1993).

Obviously, further studies are needed to work out the functional significance of the different microstate classes. It can be expected that in experiments where input information and/or tasks are part of the design, identifications of additional individual microstate classes might be obtained. An "alphabet" thus worked out might then be applied to the sequence of observed spontaneous microstates, and the deciphering of its syntax would be the next task. Preliminary results indicate that the transition matrix between states is strongly asymmetric (Wackermann et al., 1993).

7 Brain electric microstates during visual percepts: Illusory "Kanizsa" figures and complete figures

The well-known visual figures with illusory contours (Kanizsa, 1980) have offered an opportunity to study high-level rules of perceptual processes. Brain electric phenomena (a late positive wave of the event-related potential [ERP], the so-called P300) were reported to be associated when the reversal occurred (O'Donnel, Hendler & Squires, 1988; see also Başar-Eroglu et al., 1994); this might be the "receipt message" for a somewhat surprising change of the percept, in the line of the classical ERP studies where rare stimuli in a series of frequent stimuli cause the late positive wave P300. The cause of the change of percept is to be expected in the stimulus-antecedent time, before the crucial event occurs. On the other hand, brain electric signatures following the crucial event can cast light on the brain mechanisms of subsequent cognitive interpretations.

In our work aiming at the identification of brain electric microstates with steps or types or modes of brain information processing, we have investigated the event-related brain fields during the viewing of illusory triangles

(Kanizsa, 1980) and during visual "noise" consisting of the parts of the illusory triangle arranged in a meaningless configuration, in either case while the viewers voluntarily (1) payed attention to the target or (2) paid no attention to the target (Brandeis & Lehmann, 1989). Using the strategy of segmentation of the event-related potential map series into epochs of steady brain field landscapes, microstates could be identified which had similar spatial configurations (landscapes) of their brain potential field maps when voluntary attention was payed to the non-triangle ("noise") picture and when the illusory triangle figure was viewed without voluntary attention directed to it. The earliest microstate of this type was found at 168 – 200msec post-stimulus.

These results suggest that at least some of the brain mechanisms that are automatically activated during figure perception might draw on resources that are also used in voluntary attention. In an earlier study on brain electric potential representations of figure-ground distinctions, a similar conclusion was reached (Landis et al., 1984). Reversely, if attention mechanisms are activated, non-competitive figure perception is expected to become facilitated, and competitive perception inhibited.

8 Brain functional states in brain electric fields, in brain information processing, and in conscious experience

The brain's global functional state is an emergent property of the myriad of local subprocesses that are influenced by a large array of factors; a very important factor is newly arriving information from internal (memory) or external (surround) sources. Synergetics (Haken, 1983) has formalized and modelled the observation that, at each moment in time, multiple subprocesses within a system adapt themselves into a cooperative coordination which in turn acts as "order parameter" that constrains the behavior of the subprocesses. This model is consistent with the observations concerning state dependent information processing in the brain (Koukkou & Lehmann, 1983, 1990). When examining information about the brain electric field, a "macroscopic view" of the system (Haken, 1983) can be applied using various resolutions on the time scale. In the present paper, we considered brain functional states ranging up to several seconds ("macrostates") which are characterized by the properties of the brain field in the frequency domain, and states in the subsecond to second range ("microstates") which are characterized by the brain electric field's spatial distribution, i.e., by its landscape. Both types of states are macroscopic views in Haken's terminology.

We have argued that the adequate level for the study of brain mechanisms of conscious experiences which emerge out of the cooperation of all brain processes must be a comparably high level of physical brain measurements, and that the indivisible and rapidly changing brain electric field appears to be the appropriate candidate.

We have further argued that different landscapes (microstates) of the brain electric field most probably represent different brain functions. Evidence has been reviewed showing that different quasi-steady landscapes of the brain field (microstates) persist for varying durations, and are concatenated by very brief transition periods. Experimental results have demonstrated that different brain electric microstates are associated with different conscious mind states, or with different, identifiable steps in information processing. Accepting the viewpoint of an identity between brain states and mind states, we suggest that the formally and functionally identifiable global microstates of the brain are the building blocks of conscious experiences, the "atoms of thought". Incorporation of one of the basic building blocks into the stream of consciousness might not only depend on state durations but also on re-entry mechanisms which permit mirroring of the information in the system within a crucial time window.

Acknowledgement

The work was partly supported by the Swiss National Science Foundation, the EMDO Foundation (Zürich) and the Hartmann-Mueller Foundation (Zürich). B.H. had received a part-time fellowship from the Gertrud-Ruegg-Foundation (Zürich).

References

Borsellino, A., DeMarco, A., Allazetta, A., Rinesi, S. & Bartolini, B. (1972): Reversal time distribution in the perception of visual ambiguous stimuli. Kybernetik 10, 139 – 144.

Başar-Eroglu, C., Strüber, D., Stadler, M., Kruse, P. & Greitschus, F. (1994): Slow positive potentials in the EEG during multistable visual perception. In this volume.

Brandeis, D. & Lehmann, D. (1989): Segments of ERP map series reveal landscape changes with visual attention and subjective contours. Electroencephalography and Clinicaal Neurophysiology 73, 507 – 519.

Clarke, F.J.J. & Belcher, S.J. (1962): On the localization of Troxler's effect in the visual pathway. Vision Research 2, 53 – 68.

Creutzfeldt, O.D., Bodenstein, G. & Barlow, J.S. (1985): Computerized EEG pattern classification by adaptive segmentation and probability density function classification. Electroencephalography and Clinical Neurophysiology 60, 373 – 393.

Ditzinger, T. & Haken, H. (1994): A synergetic model of multistability in perception. In this volume.

Haken, H. (1983): Synergetics. An Introduction. Berlin: Springer (3rd ed.).

Hildebrandt, G. & Engel, P. (1963): Der Einfluß des Atemrhythms auf die Reaktionszeit. Pflüger's Archiv für Physiologie 278, 113 – 129.

Kanizsa, G. (1980): Grammatica del Vedere. Bologna: Mulino.
Katayama, S., Tsunashima, Y., Yokoyama, S. & Kimura, T. (1977): Respiratory correlates of human sleep spindles. Electroencephalography and Clinical Neurophysiology 43, 491.
Keesey, U.T. & Nichols, D.J. (1967): Fluctuations in target visibility as related to the occurrence of the alpha component of the encephalogram. Vision Res. 6 235 – 244.
Koukkou, M. & Lehmann, D. (1983): Dreaming: the functional state shift hypothesis, a neuropsychophysiolocal model. British Journal of Psychiatry 142, 221 – 231.
Koukkou, M. & Lehmann, D. (1990): Brain states of visual imagery and dream generation. In: R.G. Kunzendorf & A.A. Sheikh (Eds.), The Psychophysiology of Imagery, pp. 109 – 131. New York: Baywood.
Koukkou, M., Lehmann, D., Wackermann, J., Dvorak, I. & Henggeler, B. (1993): Dimensional complexity of EEG brain mechanisms in untreated schizophrenia. Biological Psychiatry 33, 397 – 407.
Kristiansen, K. & Courtois, G. (1949): Rhythmic electrical activity from isolated cerebral cortx. Electroencephalography and Clinical Neurophysiology 1, 265 – 272.
Kruse, P., Stadler, M. & Strüber, D. (1991): Psychological modification and synergetic modelling of perceptual oscillations. In: H. Haken & H.P. Koepchen (Eds.), Rhythms in Physiological Systems, pp. 299 – 311. Berlin: Springer.
Landis, T., Lehmann, D., Mita, T. & Skrandies, W. (1984): Evoked potential correlates of figure and ground. International Journal Psychophysiology 1, 345 – 348.
Lehmann, D. (1987): Principles of the spatial analysis. In: A.S. Gevins & A. Remond (Eds.), Handbook of Electroencephalography and Clinical Neurophysiology, Rev. Ser. vol. 1: Methods of Analysis of Brain Electrical Magnetic Signals, pp. 309 – 354. Amsterdam: Elsevier.
Lehmann, D., Beeler, G.W. & Fender, D.H. (1965): Changes in patterns of the human electroencephalogram during fluctuations of perception of stabilized retinal images. Electroencephalography and Clinical Neurophysiology 19, 336 – 343.
Lehmann, D. & Knauss, T.A. (1976): Respiratory cycle and EEG in man and cat. Electroencephalography and Clinical Neurophysiology 40, 187.
Lehmann, D., Ozaki, H. & Pal, I. (1987): EEG alpha map series: brain micro-states by space-oriented adaptive segmentation. Electroencephalography and Clinical Neurophysiology 67, 271 – 288.
Lehmann, D. & Skrandies, W. (1986): Segmentation of evoked potentials based on spatial field configuration in multichannel recordings. Electroencephalography and Clinical Neurophysiology, Suppl. 38, 27 – 29.
Libet, B. (1982): Brain stimulation in the study of neuronal functions for conscious experience. Human Neurobiology 76, 271 – 288.
O'Donnel B.F., Hendler, T. & Squires, N.K. (1988): Visual evoked potentials to illusory reversals of the Necker cube. Psychophysiology 25, 137 – 143.
Poole, E.W. (1961): Nervous activity in relation to the respiratory cycle. Nature (Lond) 189, 579 – 581.
Newell, A. (1992): Precis of unified theories of cognition. Behavioral and Brain Science 15, 425 – 492.
Reisberg, D. & O'Shaughness, M. (1984): Diverting subjects' concentration slows figure reversals. Perception 13, 461 – 468.
Stadler, M. & Erke, H. (1968): Über einige periodische Vorgänge in der Figurwahrnehmung. Vision Research 8, 1081 – 1092.
Strik, W.K. & Lehmann, D. (1993): Data-determined window size and space-oriented segmentation of spontaneous EEG map series. Electroencephalography and Clinical Neurophysiology (in press).

Villablanca, J. (1965): Electroencephalogram in the permanently isolated forebrain of the cat. Science 138, 44 – 46.

Wackermann, J., Lehmann, D., Michel, C.M. & Strik, W.K. (1993): Adaptive segmentation of spontaneous EEG map series into spatially defined microstates. Internation Journal of Psychophysiology 14, 269 – 283.

Woodworth, R.S. & Schlosberg, H. (1954): Experimental Psychology. New York: Holt.

The Creation of Perceptual Meanings in Cortex Through Chaotic Itinerancy and Sequential State Transitions Induced by Sensory Stimuli

W.J. Freeman

Department of Molecular and Cell Biology, University of California at Berkeley, CA 94720, USA

Abstract: Recent advances in mathematics and EEG research have opened a new field in neuropsychology – modeling brain dynamics, as distinct from modeling behavior. The brain as a dynamic system is unlike any other, so it must be described in its own terms, not by analogy or metaphor. The description should have the forms that describe other dynamic systems, namely sets of differential equations or their equivalents, which describe and predict the transformations that systems apply to their inputs to give their outputs. Brain dynamics that is based on measuring and modeling brain activity reveals that memory stores are not invariant, as learning, recall and recognition proceed. The results give the picture that perceptual transactions with the environment are initiated within the limbic system through emergent patterns that are continually constructed by an evolutionary dynamics. Sensory input that is sought out by the limbic system constrains its dynamics into pre-formed basins of attraction, which provide for the construction of new patterns in an ordered sequence of behavior that is in harmony with the environment.

1 Introduction to dynamics

Zeno's paradox in ancient times held that for an arrow to reach its target it had to cover half the distance first, and half of that first, and so forth. By infinite regression it was inferred never to be able to start at all. This paradox was resolved in the 17th century with the invention by Leibniz and Newton of the calculus of infinitesimals. As the distance of travel decreased so also did the time needed to cross it. In the limit, as the distance and time increments both approached zero, their ratio approached a finite value, the velocity. Thereupon the modern view of dynamics came into being, replacing the ancient world of static physics.

The acceptance and success of this new way of thinking was not based on philosophical considerations. It was solidly grounded in the new mathematics of precise description and prediction. The modeling of the solar system and the tides was followed by a host of applications, first in military hardware and mechanical devices which supported the industrial revolution, then in thermodynamics, electrical power networks, electromagnetic and radiological devices, nuclear power systems, aircraft, computers, and so on. Descriptive metaphors for the study of change also found their way into many other fields, such as drama, literature and psychodynamics, but seldom with the support of predictive equations and then with unconvincing results (e.g. Hull, 1943). Such reasonings have largely been analogies with little disprovability and have led to the formation of conflicting schools of thought that seem more like sects of a religion than domains of research.

Early on the brain was seen as a physical dynamic system, and the new concepts were almost immediately applied in attempts to explain its processes, beginning with Descartes, who conceived the pineal as a kind of valve used by the soul to regulate the pumping of spiritual fluid into the muscles. This was quickly disproven by Borelli using a plethysmograph. There followed the analogies of clockworks, telegraph and telephone systems, thermodynamics which underlay Freud's use of the principle of conservation of energy in his theory of hysteria (Freud, 1893), digital computers, and holographs.

None of these likenesses is adequately supported by use of measurements of brain activity made during its normal functioning, nor is the brain "like" any man-made machine. If anything, it is "like" a natural, self-organizing process such as a star or a hurricane. With the guidance of genes it creates and maintains its structure, and it exchanges matter and energy with its surround. Yet it is unique, in that it moves itself through its surround and incorporates facets of its environment for its own purposes, to flourish and prevail. Its dynamic processes are in time scales of fractions of seconds, and its distance scales are in millimeters. It is these intrinsic processes that we must describe mathematically, which means to construct, solve, and evaluate sets of differential equations. Only then can we say that we understand brain dynamics.

2 Evidence required for dynamic models of the brain

As a dynamic system seen from the outside, the brain takes inputs in the form of stimuli and gives outputs in the form of responses. We should not begin with the whole but with the smallest part that will suffice. For many purposes that is the neuron. We measure its input, such as a volley of afferent action potentials, and its output, such as a postsynaptic potential (PSP), and we use the ratio of output to input to specify an input-output (I-O) pair. We repeat this test under varying conditions, until there are no more surprises. The collection of I-O pairs constitutes the experimental data base.

We examine these pairs and devise a model, which consists of a differential equation that, when solved for the input and the initial conditions at the start of the input, yields a curve that can be fitted to the observed output. We can call this equation an operator that serves to transform the input into the output. For example, a nerve impulse is transformed by an axodendritic synapse into an exponentially decaying PSP, and a model consisting of a differential equation suffices to describe this operation. From this simple beginning, which forms a main foundation of modern cellular neurophysiology and more recently of the new field of neural networks, we generalize to arrays of interconnected neurons and to trains of impulses in an unlimited variety of network configurations.

Herein lies a kind of pathology of modeling. In the minds of modelers these networks grow in complexity and in fascination, independently of their original intent to explain the brain. In order to keep on the track of that goal, it is essential that we refer constantly to I-O pairs from brains that are active in the pursuit of their own goals during the time of recording and measurement. Models of behavior, such as the classics of Freud, Jung, Pavlov, more recently Köhler (1940), Hull (1943), Skinner (1953), Masserman (1961, 1964), Grossberg, (1974) and others, are not brain models, because they are not derived from and immediately tested against the brain activity that underlies and accompanies goal-directed behavior.

My intent in this essay is not only to demonstrate an experimental pathway to the realization of dynamic brain models, which are testable with electrophysiological and neuropharmacological tools, but also to show that the use of behavioral tools from Pavlov, Skinner, Masserman and others is essential in order to elicit, control, and measure the behaviors that must take place while the data for brain modeling are being acquired. Certainly I could not have worked without these tools.

For reasons that I will discuss, data from electroencephalograms (EEGs) are especially important but in ways that have not been generally foreseen. A Task Force of the American Psychiatric Association, for example, has recently completed a list and assessment of the current and potential uses of quantitative EEG ("qEEG") in psychiatry (Luchins, 1990). The list includes temporal spectral analysis, visual imaging of color coded brain activity maps, and multiple discriminant analysis of EEGs and evoked potentials from groups of subjects. Regrettably the Task Force failed to perceive the uses of "qEEGs" as data bases for deriving and testing models of brain dynamics, which supports my main conclusion here, that perception depends dominantly on expectation and marginally on sensory input.

3 Spatial patterns from the olfactory system

My own work has been focussed on the olfactory system in small laboratory animals, because this is the simplest and phylogenetically oldest sensory system, and because it is the most important of the senses for cats, rabbits and rats. It is also arguably the prototype for all other perceptual modalities, owing to its precedence in the evolution of brain function.

The crucial experiments are simple in conception, although it took over a decade to do them successfully for the first time. Rabbits were implanted with an array of 64 electrodes in an 8 x 8 grid placed onto the surface of the olfactory bulb, which provided a "window" for observing spatial patterns of sensory-evoked activity. The rabbits were trained to respond by licking to one odorant, such as amyl acetate accompanied by a reward (water), and merely to sniff in response to another odorant (such as butyric acid) as an unrewarded stimulus. On each trial a set of 64 EEG traces was recorded for 6 sec, that included a control period and a test period. Brief EEG segments lasting 0.1 sec were selected during the times of inhalation of the background, control air or either of the two odorants presented on randomly interspersed trials. Sets of several hundred of these EEG segments were analyzed for each animal. We knew which of the three states each segment came from. The crucial question was, what aspect or aspects of the segments would enable us to classify the segments correctly in respect to the antecedent odorant conditions?

The hypotheses were that, between the time of inhalation and the performance of a correct response, odorant information existed in the bulb as a pattern of neural activity, on the basis of which the animal made the discrimination, and that this information would be detectable in some as yet to be determined properties of the EEGs. Our results showed that the information we sought was indeed manifested in the EEGs. It was identified as a spatial pattern of oscillation in the high frequency range of 40–80 Hz that we termed the "gamma" range, in analogy to the high end of the x-ray spectrum. A common wave form was found in each segment, and the spatial pattern of its amplitude tended to converge toward a reproducible shape each time that the background or an odorant was present. In principle this form of information is quite simple. It is like a frame in a black-and-white movie, in which the carrier wave is the light, and the image is formed by the highs and lows of the intensity of the light.

The finding that the central "code" for olfaction is spatial is not surprising. This was predicted by Adrian (1950) on the basis of his pioneering studies in the hedgehog. The olfactory receptors form a sheet of neurons in the nasal cavity. When a chemical is inhaled, it spreads onto the sheet and stimulates a spatial pattern of action potentials (Kent & Mozell, 1992), much as a retinal pattern of light is transformed to a pattern of ganglion cell activity. This spatial pattern of receptor activity is transmitted to the bulb by unbranched axons that have some degree of topographic order. Adrian found evidence that a new spatial pattern could be expected in the bulb for each new odorant. Additional evidence for spatial coding has been presented by several groups using metabolic labeling techniques (e.g. Lancet et al., 1982). Moreover, it is not surprising that odor information should exist in the induced burst of activity accompanying each inhalation, which rises above the background activity during inhalation, because it is well known that detection of odors takes place on inhalation. But there are two aspects that were very surprising.

4 Surprising properties of the spatial patterns

First, the information was uniformly distributed over the entire olfactory bulb for every odorant. By inference every neuron participated in every discrimination, even if and perhaps especially if it did not fire, because a spatial pattern required both "on" and "off". This fact was demonstrated by repeating the classification test of EEG patterns with respect to odorants, while deleting randomly selected groups of channels. No channel was any more or less important than any other channel in effecting correct classifications. Furthermore, we and many other investigators who have attempted to demonstrate odorant specificity in the discharge rates of action potentials from single neurons have never succeeded. In view of the facts that the minimal number of channels for correct classification of EEG segments was 16, and that each channel reflected the activity of many thousands of neurons, we concluded that the information relating to odorants was carried by assemblies of millions of neurons, and it was not detectable in the activity of the handful of neurons that could be simultaneously recorded with microelectrodes.

This finding established the significance of the EEG as an experimental variable. These electrical potentials resulted from the neural synaptic currents that flowed between the cells, where they summed over the contributions of large numbers of neurons (Freeman, 1975, 1991). Thereby the EEGs gave direct access to the amplitudes of synaptic activity of populations of neurons. Populations constitute the hierarchical level, I believe, at which goal-oriented behavior is elaborated, not the single neurons. Further, this conclusion fully supports the views of Wolfgang Köhler (1940): "Our present knowledge of human perception leaves no doubt as to the general form of any theory which is to do justice to such knowledge: a theory of perception must be a field theory. By this we mean that the neural functions and processes with which the perceptual facts are associated in each case are located in a continuous medium." This medium is not the extracellular electric fields as Köhler envisioned, but the synaptically interconnected neuron populations in the cortex.

Second, we found that the spatial pattern with each learned class of stimulus was not invariant. With the addition of each new conditioned stimulus leading to the formation of a new pattern the pre-existing patterns changed slightly but undeniably. Upon re-introduction of an old stimulus after serial conditioning to new stimuli, a slightly different pattern appeared, not the old one. We trained rabbits serially to odorants A, B, C, D, and then back to A, and found that the original pattern did not recur, but a new one appeared (Freeman, 1991). When the contingency of reinforcement was changed, the stimuli were the same, but their patterns differed. The simplest way was to switch a reward between two odorants, so that the previously rewarded stimulus was no longer reinforced and vice versa. Both spatial patterns changed, though the two odorants did not. So also, in each case, the pattern for the control input "no odorant" changed, even though the background mix of odorants was unchanged. Yet the conditioned responses to the unchanged stimuli did not change, provided the environment itself had not changed, implying that stimulus-response invariance was not maintained by an invariant "memory store" in the brain, but by the environment, and that with each evolutionary change with learning in sensory systems there was a change in the motor system to maintain adaptation to the environment.

This lack of inner stimulus-response invariance is an essential property for a true associative memory system, because the store for each new class of stimuli must be tied in some way to the stores for all of the pre-existing classes in order to establish the fabric of associations. The advantage of this system is that stability of the store is maintained as long as the environment is stable, and, as the environment evolves, the brain changes accordingly. A disadvantage is an inability of the brain to perceive gradual change as change, so that horrendous environmental change may be accepted by the brain without cavil in the process known as "normalization". But if a pattern does not relate to a stimulus, what does it signify? I conclude that it signifies the meaning of the stimulus in the subjective context of the stimulus for the subject, the history of associations, the present background input, current states of arousal and attention, and current goals, in short, the intentionality of the subject (Searle, 1983). However, the question must be answered, where does each specific pattern come from with each specific stimulus? If cortical activity patterns are not derived by filtering, transforming, and storing stimuli and then crosscorrelating the retrieved patterns with new stimuli, how are they constructed, stored, shaped, and used to generate behavior?

5 Evidence for chaotic dynamics in olfaction

EEG waves are characterized as oscillations in each of a set of frequency bands, commonly known as "alpha" (10 Hz), "theta" (3–7 Hz), "beta" (15–30 Hz), and "gamma" (30–80 Hz). When such periodic waves exist, the activity can be extrapolated into the future, so that it is predictable. However, for the gamma activity of the olfactory system the carrier wave was aperiodic. That is, it did not show oscillations at single frequencies, but instead had highly irregular wave forms even over short time segments. No two wave forms were ever precisely the same in successive segments, irrespective of odorant input. Yet the same wave form appeared simultaneously on all the channels, although at different amplitudes, a property that is called "spatial coherence". The commonality of wave form led to spatial patterns of amplitude modulation that were predictable across trials in each day of training.

The possibility that this seemingly random wave form could occur by chance almost simultaneously on all channels was vanishingly small. We sought a mechanism by which a common activity might be imposed in this gamma spectral range onto the whole bulb, either by receptors or by centrifugal pathways from the forebrain and brainstem. We found firm evidence that no external driving could produce the observed commonality of wave form (Freeman & Viana di Prisco, 1986). Indeed, the aperiodic basal activity persisted after the bulb and olfactory cortex were surgically isolated as a unit from the rest of the brain.

These findings led to the insight, already reached independently by other investigators (e.g. Babloyantz & Destexhe, 1986), that brain activity is characteristically chaotic. More specifically, many parts of the brain are capable of generating controlled but locally unpredictable activity that looks like noise but is not. This is significant in several respects. The lesser is that it directs us to search for carrier wave forms that are aperiodic and not for oscillations at specific frequencies such as the alpha or the "40 Hz". A greater significance attaches to the property of chaotic systems in their ability to jump suddenly and completely from one global activity pattern to another, just as, for example, we jump from one word to the next in the rapid flow of speech and from one gait to another in walking, jogging and running. Our EEGs are showing us sequences of patterns, each of which is carried by a chaotic wave form and not by a wave at a single frequency, as previous EEG studies had led us to expect. The greatest significance is that deterministic chaotic systems have the capacity to create new patterns that have never before existed (Shaw, 1984). The forms of the patterns are dependent on the initial conditions, such as are provided in sensory cortex by the concomitant inputs from conditioned and unconditioned stimuli. Deterministic systems operating in steady state and periodic modes cannot do this. I propose that the capacity of sensory systems to operate in a deterministic chaotic mode is the key to understanding the origins of neural activity patterns in the context of perception and intentionality.

The chaotic activity is not an average over the random activity of selected cortical neurons. It is a global property of long-distance interactions between masses of neurons in the whole olfactory system. We found that a surgical or pharmacological, block of the main pathway from the bulb to the olfactory cortex causes the cortical activity to go silent both in action potentials and EEGs, whereas the bulb displays a limit cycle oscillation with each inhalation (Freeman, 1975; Gray & Skinner, 1988). These findings demonstrate that the mechanism for generating the chaos includes interaction between the bulb and the olfactory cortex. We know that activity is transmitted not only from the bulb to the cortex by surface pathways but in the reverse direction by deep pathways. Moreover, the background state between inhalations is not an equilibrium that is perturbed by noise. It is a chaotic state that holds the system in constant readiness to jump to any desired perceptual state at any time.

The chaotic dynamics of the olfactory system seems autonomous, in the sense that the system continues to generate its random-seeming wave forms after isolation from the rest of the brain, but it is actually under close chemical control by other parts of the brain through centrifugal pathways into the olfactory system. These pathways do not carry detailed information in the form of highly specific and structured activity patterns. Instead, they carry global modulations that constitute generic commands, such as "turn on", "turn off", "imprint", "habituate", "attend", "learn", and so forth. The connections to and from the forebrain into the olfactory system are essential for its operation, and the controls appear to be implemented by well known transmitters and neuromodulators, including the brain amines and neuropeptides.

These findings are important for understanding the perceptual ambiguities that occur when stimuli that are unchanging are perceived differently owing to swings in mood and attention. The constructions that are made by the chaotic sensory systems are triggered by stimuli, but their forms arise from previous experience embedded in synaptic changes, and through modulation of the dynamics by neurochemicals. Much needs to be done to understand the details, but the salient lesson is the importance of chaotic neurodynamics for its perceptual constructs.

6 Modeling brain dynamics from the data

These surprising results pose major challenges for development of descriptive models, but they also give us several key insights into the dynamic processes to be modeled. First, we are assured that the proper element for our model is the local population and not the single neuron. This provides an enormous simplification, because the details of neurons in each local neighborhood are submerged as we view the mass action as a local mean field quantity. Its amplitude is expressed as a density of pulses, and the relationship to integrated input is not shown by a straight line (linear) as it is for single neurons (Freeman, 1975). Instead it has the form of an S-shaped curve between threshold and maximal firing density for the population (Freeman, 1987b). Most of the complexities of the single neuron remain at the hierarchical level of the neuron and are not needed for brain modeling.

Oscillations arise when an excitatory population is coupled into a loop with an inhibitory population to form a negative feedback loop (Freeman,

1987b). The oscillations are periodic in a model of the bulb, as we find to be the case in the experimentally isolated bulb. Chaos in the model arises when the bulbar loop is interconnected with the olfactory cortical loop, so that each excites the other. Each has its characteristic frequency. As with two strong willed partners in marital discord, they can. neither agree nor escape each other, so that sustained aperiodic oscillation results. If the two parts are disconnected, the chaos in the model disappears, as it does in the experimental animals, when the bulb and cortex are disconnected.

The bulb and the cortex are each simulated by an array of local oscillators, that are interconnected between the excitatory elements by simulated synapses, so that they excite each other. The inhibitory elements are also interconnected, so that they inhibit each other. These connections couple the oscillators into a layer, and they ensure that the whole array oscillates with a common wave form but with local amplitude differences just as in the bulb of animals. There are large numbers of input and output connections, one pair for each local oscillator. The mutually excitatory synapses between the oscillators are selectively strengthened during learning to "identify" an "odorant" by presentation of examples of a class of stimuli, leading to the formation of a Hebbian nerve cell assembly (Hebb, 1949) consisting of groups of local oscillators that are co-excited by the stimuli. A global command is required in the model to "form an assembly", which simulates the release of a neuromodulator in response to reinforcement by an unconditioned stimulus. Two or more classes of stimuli are formed in sequential "training" sessions, leading to formation of multiple "nerve cell assemblies" that are separated by simulated "habituation".

The end result is that the array of coupled oscillators generates a reproducible spatial pattern of amplitude of a common chaotic wave form whenever an example of a learned class of stimuli is presented to the model. That pattern serves to classify the stimulus presented. We say that for each learned class of stimuli the model has a chaotic attractor with a "basin" of attraction. In dynamic parlance the basin means that the olfactory system (or its model) has certain preferred patterns of activity, to any one of which it naturally goes if given the opportunity. That opportunity is provided by the presence of any stimulus in a class that the system (or its model) has learned to respond to, whether or not the example was included in the original training set. A stimulus of this class provides the input and starting conditions that are needed to place the system into its basin of attraction, so named in analogy to a bowl in which a ball will roll to the bottom and stay there. Our data indicate that the bulb maintains a global attractor, that there is a learned wing and attendant basin of attraction for each class of odorant that a subject can discriminate, and that its basin defines the range of generalization for identification of a stimulus of that class.

Our dynamic model seems to classify spatial patterns in the form of images of randomly oriented objects in the same way that the bulb classifies odorants

(Freeman, Yao & Burke, 1988; Yao et al., 1991). We present it with a few examples of each class of object that it is to identify and train it by increasing the synaptic weights between pairs of local oscillators that are co-activated by the inputs. This forms the equivalent of nerve cell assemblies. Thereafter, with each test input the system jumps from a basal chaotic state to the basin of an appropriate wing of the global attractor, and the output is expressed as a spatial pattern of amplitude of a common chaotic wave form, in the same way that the olfactory bulb jumps to a distinctive spatial pattern of chaotic activity when the nasal receptors are presented with a familiar odorant.

Evidence for this dynamic process of perception has now been accumulated for the olfactory system in rabbits, rats and cats; for the visual system in monkey (Freeman & van Dijk, 1987), cat (Eckhorn et al., 1988; Gray et al., 1989) and human (Schippers, 1990); and quite tentatively for the somatosensory system in human (Freeman & Maurer, 1989). Our chaotic dynamic model implemented according to these principles has been shown to solve classification problems that other neural networks cannot.

Learning to identify a new class of stimulus requires the formation of a nerve cell assembly during training through the modification of synaptic strengths. This is simulated in models by increasing the values of the excitatory feedback gains between nerve cells that are co-activated by a stimulus under reinforcement. Learning takes place by a structural change in the system, such that its behavior in the future is dependent on past experience. In this respect the learning process is similar to the induction of epilepsy (Freeman, 1986) and to changes between sleeping and waking states, which are done by modifications of the parameters or the connections with the model. These changes lead to sustained changes in the dynamics that constitute what are called bifurcations. They are independent of further input.

The transitions back and forth between basal and burst states with respiration do not involve a parameter change but depend on input. Therefore, this change is not a bifurcation, but it is a state change in the same sense that water changes its properties when it boils or freezes. The many wings of the olfactory attractor, one for each class of input that the animal can discriminate, are attached to a common stalk that is revealed during exhalation by the unpatterned interburst activity. The system is forced out of the basal state by receptor input with each inhalation, and if the inspired air contains a known odorant, the system is constrained to oscillate in one of the wings. Only one wing can be visualized with each inhalation and then only briefly for as long as the inhalation lasts. A sequence of inhalations induces a series of state transitions from each wing to the next with brief transit through the basal state. Tsuda (1991) has proposed the term "chaotic itinerancy" to denote this migration through state space along a trajectory that is in part determined by successive input and in part by input from other parts of the brain. The landscape of the dynamic space, itself subject to modification by learning and bifurcation, is the main determinant of the spatial patterns of activity that constitute the output of the system. As with an itinerant peddler no two successive visits to any one town are identical, but there is a recognizable degree of constancy. This view offers another perspective on the chaotic linking process described by Tsuda (1992) for the formation of organized sequences of patterns, something that cannot be achieved by the strict application of basin-attractor theory.

7 Some implications of the findings for philosophy

These experimental data and descriptive dynamic models have profound consequences for understanding how sensory cortexes work as the interface between the rest of the brain and the outside world. The crucial point is made by tracing the course of a stimulus into the receptor layer, where it is transduced into a pattern of action potentials, and then into the cerebral cortex, through the thalamus for other systems but directly for olfactory input. What happens in the bulb is a sudden destabilization of the bulbar system, so that it makes an explosive jump from a pre-existing state, expressed in a spatial pattern of activity, to a new state that is expressed in a different spatial pattern. The pattern is selected by the stimulus, but its shape is determined by prior experience with this class of stimulus. The pattern expresses the nature of the class and its meaning for the subject rather than the particulars of the particular stimulus. The identity of the receptors that are activated is irrelevant and is not retained, because the activated receptors belong in an equivalent class. The output does not express the identity of a chemical material but a collection of experiences that the subject has had with the material. The sensory data serve to trigger the construction of the action pattern that then replaces them.

The need for this process can be comprehended by noting that the olfactory environment is indefinitely rich in odorant substances, only a small portion of which ever come to the attention of a subject or form the basis for action. That portion is different for every subject, because it depends in part on genetic determinants of the system but mostly on previous experience with selected odorants and the contexts of reinforcement. In a word, the rate flow of information from the environment is infinite. Any system for information processing must reduce such a flow to a finite rate, lest it become confused or overwhelmed. Man-made systems do this by means of filters, which are designed by observers to accept the portions that are desired by the observers. Brains have no homunculi to specify their goals and desired inputs, and they rely instead on chaotic processes to generate activity patterns that are finite in dimension. The term "self-organizing" connotes the fact that chaotic systems can create information as well as destroy it. Chaotic neurodynamics of the kind we have modeled may originate novel behavior as diverse as learning by insight, trial and error, invention, recollection whether accurate and faulty, and confabulation. In brief, perception and recall are a flexible dynamic process by which meaning is created.

Because each chaotic pattern is created from within and not imposed from outside, I infer that there is no instance in which raw sense data are incorporated and stored in the cortex as whole patterns or episodic representations. There is only the modification of synaptic weights among populations of neurons, such that after some sequence of experiences there is an appurtenant pattern of neural activity and of behavior in follow up to the presentation, by stimulation or recall, of an example of a learned class of stimuli. The experimentally documented process of "mapping" from each input to each output of the olfactory bulb invokes a fundamental epistemological issue of the relationship between an event or object and the perception of it. The perceptual process that I have sketched here allows the brain to "know" only its own experience of the object and not the "reality".

If we can generalize to the other senses, as I think we should from the data that we already have on hand from other areas, then all of our individual knowledge is created and organized by chaotic processes, that are shaped by receptor input but cannot ever capture the actualities of the matter and energy that impinge on the receptors. We cannot know the infinity of the universe we inhabit, only the self-organized patterns that each of us makes within it.

This conclusion is not new, and it is perhaps not even surprising to many philosophers and psychologists, who are already aware that perception is in the mind of the beholder, and that memory is a creative process rather than a look-up table of imprinted data. But the critical question I am addressing here is: what new evidence can experimental brain science bring to support this conclusion? The discovery of meaningful spatial patterns in EEGs is not in itself convincing, because two kinds of spatially patterned activity co-exist in the bulb with each inhalation concurrently. One consists of the global spatial pattern of cooperative bulbar activity that expresses the meaning of a stimulus. The other kind is stimulus-evoked activity, which has the differing form of spatial patterns of pulse firings in sparse networks of interconnected neurons, and which expresses the properties of the stimulus. Both kinds of patterns can be observed and measured only in part, but they can be inferred to exist in their entirety, the global pattern by spatial ensemble averaging and the pulse pattern by time ensemble averaging (Freeman, 1987a). How can we know which of these patterns is accepted and acted upon by the olfactory cortex to which the bulb transmits?

It is immediately clear that the global pattern that is transmitted from the bulb is accepted by the cortex, because the EEGs of the olfactory cortex are usually highly correlated with those of the bulb in the gamma range of frequencies at which the bulb is driving the cortex (Bressler, 1988). But what happens to the stimulus-evoked activity pattern after it is transmitted from the bulb?

My answer is based on the structure of the transmission pathway and the functional properties of the cortical neurons that receive the bulbar output. Each bulbar axon branches and distributes its output broadly over the cortex. Conversely each cortical neuron receives input from many thousands of widely distributed bulbar neurons, and its dendrites sum that input continually over time. I have demonstrated both experimentally and mathematically that the only portion of the input from the bulb to the cortex that survives this operation of space-time integration is the common wave form, which is the product of the global cooperativity between the bulb and cortex. The stimulus-evoked pulse pattern is localized spatially, and it is poorly coordinated temporally, so that even though it is transmitted to the cortex, as recordings from the pathway have demonstrated, it is expunged by the smoothing process of spatial integration in the receiving neurons. The model asserts that their traces are removed by a "laundering" operation in the second stage of the synthesizing processes of perception. The model asserts that the traces continue to exist in the bulbar messages (Freeman, 1991), so if some aspect of an odor is made significant by reinforcement, the bulb and cortex can flexibly modify their sensitivities to accept that aspect, although not the "whole thing" by any sequence of steps. This would appears to be how higher discriminations are achieved, particularly by humans using language (Rabin & Cain, 1984), but that is another story, because language involves temporal strings of input, whereas my experimental data thus far deal only with serial impressions.

8 Some implications of the findings for cognitive science

There are two main extrapolations from this epistemological lesson to be dealt with. First, brain dynamics is largely a study of the self-organization of patterns along an evolving trajectory, and, second, changes take place by sudden jumps from each pattern to the next. What I have modeled thus far are merely the first two steps in the perceptual pathway for olfaction, and then only for one frame at a time in what is clearly a life-long sequence of frames or steps. The models thus far constitute the barest threshold of brain dynamics. Yet despite the enormity of the task ahead, it is possible to speculate on what might be some of the principles in accordance with which the whole brain takes short steps and creates its own path into the future.

I will focus on the perceptual step that follows identification and classification of a stimulus, which is the estimation of its intensity in the goal-directed behavior of searching. In regard to a scent of predator or prey, its meaning resides not merely in its class but also in its concentration, most importantly with respect to the recent past: is it getting stronger or weaker? But this sequence from two or more sniffs has no value unless the subject can remember where it was and what it did in conjunction with each previous frame, so as to plot its next move toward or away from the inferred source. Three kinds of information are required for assembly of a concentration estimate: the sequence of perceptual samples from the bulb; a record of the sequence of motor commands that instigated the taking of samples, which consists of reafferent messages to the sensory cortex; and proprioceptive data from the musculoskeletal system on whether and how the intended actions were taken. These diverse kinds of neural data must be stored and combined over time periods ranging upward from fractions of a second to the limits of the attention span for the species.

We have no models for how this is done. However, we can say with assurance that it must be done for each sensory modality. Either there is a separate short term memory for synthesizing exteroceptive and proprioceptive inputs for each modality, or the inputs from the several modalities are first combined, including the proprioceptive, and there is a single storage. Brain anatomy clearly indicates that the brain is organized around the latter principle, because all sensory systems feed into the entorhinal cortex, and this is the primary source of input to the hippocampus. In turn the largest projection of the hippocampus is recursively to the entorhinal cortex, which itself projects back to all of the sensory cortexes that it received from.

I propose that an act of perception may arise in the limbic system and be expressed through the entorhinal cortex as reafferent messages to the sensory cortexes and as motor commands to look, listen and sniff. Our EEG evidence from the olfactory system shows that a wave of excitation sweeps over the bulb with the resulting inhalation, and that a state change occurs yielding a burst of oscillation in the gamma range. The burst has a spatial pattern of amplitude that constitutes the desired perceptual message, which is transmitted to the entorhinal cortex as well as other places. From my own and others' observations on other sensory cortical EEGs, I infer that similar explosive jumps take place in those cortexes, with transmission by successive stages to the entorhinal cortex of perceptual messages, perhaps in a time frame coinciding with that for the olfactory message. When combined and transmitted to the hippocampus, this gestalt may shape the activity there, which reconstitutes the entorhinal activity pattern in accordance with the recent past. The basis is thereby laid for a next step in a cycle of command, reafference, sampling of the environment, and integration, that forms a directed sequence of behavior along a search trajectory.

It is obvious that higher order trajectories involving selection among competing and time-varying goals must involve the frontal cortex, and indeed the entire forebrain and the neuromodulatory systems of the brainstem as well. However, in modeling dynamic systems it is important not to go beyond the data, and at present we do not know with absolute assurance whether the several primary sensory neocortices share a common central "code" with each other and with the paleocortex or have radically different forms for the expression of meanings. Furthermore, I reiterate that my speculation is unsupported by models, that is, by descriptive sets of coupled differential equations that have been evaluated by fitting the curves comprising their solutions to data points from the measurements of brain activity during goal-directed behavior. But in principle it could be tested and supported or disproven, if the data and the models were in hand. To achieve that end we need additional large arrays of EEG data from each of these cortical structures along with measurements of the concomitant behavior.

9 Conclusion: A new science of brain dynamics

Surely the brain is no longer interpretable as a hierarchy of reflexes or instincts, nor as a thermodynamic or chemical engine. It is a self-organization of neurons and neuronal pools that is dominated by synaptic interactions and modulations. Only in the past decade have the necessary mathematical and technical tools become available for precise description and prediction of the macroscopic properties of the assemblage that constitutes a brain, and in particular for detailed study of its spatial patterns of activity in domains of a few square millimeters and centimeters.

The development of "biologically inspired" digital computers and artificial neural networks has become a world-wide cottage industry, in parallel with "mentationally inspired" artificial intelligence and cognitive psychology. The products of these systems and devices have been useful for the solution of major problems involving information processing, but the ways in which they operate deviate markedly from the ways in which our brains and those of animals perform.

The penultimate message I wish to convey is that the properties of brain dynamics found at the level of perception in animals are much closer to those we perceive by introspection than the properties of computers, artificial neural networks and artificial intelligence. Correlations between the performance of sets of coupled nonlinear equations modeling EEG waves and our consciousness of the flow of ideas in our mental lives can in no way constitute proof of brain models. We are far from understanding either our brains or our mentation, but I take comfort from correspondence of our models with reports by myself and many others on the properties of subjective experience, and I expect that this validation will encourage philosophers and psychologists to break away from the constraints that have been imposed by the tenets of logic, rule- driven manipulation of symbols, and information theory on the enterprise of coming to understand our organ of the episteme.

References

Adrian, E.D. (1950): The electrical activity of the mammalian olfactory bulb. EEG Clinical Neurophysiology 2, 377– 388.

Babloyantz, A. & Destexhe, A. (1986): Low-dimensional chaos in an instance of epilepsy. Proceedings of the National Academy of Science USA 83, 3513–3517.

Basar, E. (1980): EEG Brain Dynamics. Relation between EEG and Brain Evoked Potentials. Amsterdam: Elsevier.

Basar, E. & T.H. Bullock (Eds.) (1989): Brain Dynamics. Berlin: Springer.

Bressler, S.L. (1988): Changes in electrical activity of rabbit olfactory bulb and cortex to conditioned odor stimulation. Journal of Neurophysiology 102, 740–747.

Eckhorn, R., Bauer, R., Jordan, W., Brosch, M., Kruse, W., Munk, M. & Reitboeck, H.J. (1988): Coherent oscillations: A mechanism of feature linking in visual cortex? Biological Cybernetics 60, 121–130.

Freeman, W.J. (1975): Mass Action in the Nervous System. New York: Academic Press.

Freeman, W.J. (1986): Petit mal seizure spikes in olfactory bulb and cortex are caused by runaway inhibition after exhaustion of excitation. Brain Research Reviews 11, 259–284.

Freeman, W.J. (1987a): Techniques used in the search for the physiological basis of the EEG. In: A. Gevins & A. Remond (Eds.), Handbook of EEG and Clinical Neurophysiology vol 3A. Amsterdam: Elsevier.

Freeman, W.J. (1987b): Simulation of chaotic EEG patterns with a dynamic model of the olfactory system. Biological Cybernetics 56, 139–150.

Freeman, W.J. (1991): The physiology of perception. Scientific American 264, 78–87.

Freeman, W.J. & Maurer, K. (1989): Advances in brain theory give new directions to the use of the technologies of brain mapping in behavioral studies. In: K. Maurer (Ed.), Proceedings of the Conference on Topographic Brain Mapping, pp. 118–126. Berlin: Springer.

Freeman, W.J. & Van Dijk, B. (1987): Spatial patterns of visual cortical fast EEG during conditioned reflex in a rhesus monkey. Brain Research 422, 267–276.

Freeman, W.J. & Viana Di Prisco, G. (1986): EEG spatial pattern differences with discriminated odors manifest chaotic and limit cycle attractors in olfactory bulb of rabbits. In: G. Palm & A. Aertsen (Eds.), Brain Theory, pp. 97–119. Berlin: Springer.

Freeman, W.J., Yao, Y. & Burke, B. (1988): Central pattern generating and recognizing in olfactory bulb: A correlation learning rule. Neural Networks 1, 277–288.

Freud, S. (1895/1954): The project of a scientific psychology. In: M. Bonaparte, Freud, A. & E. Kris (Eds.), The Origins of Psychoanalysis. New York: Basic Books.

Gonzales-Estrada, T.M. & Freeman, W.J. (1980): Effects of carnosine on olfactory bulb EEG, evoked potentials, and D.C. potentials. Brain Research 202, 373–386.

Gray, C.M., Freeman, W.J. & Skinner, J.E. (1986): Chemical dependencies of learning in the rabbit olfactory bulb: Acquisition of the transient spatial-pattern change depends on norepinephrine. Behavioral Neuroscience 100, 585–596.

Gray C.M., Koenig, P., Engel, K.A. & Singer, W. (1989): Oscillatory responses in cat visual cortex exhibit intercolumnar synchronization which reflects global stimulus properties. Nature 338, 334–337.

Gray, C.M. & Skinner, J.E. (1988): Centrifugal regulation of neuronal activity in the olfactory bulb of the waking rabbit as revealed by reversible cryognenic blockade. Experimental Brain Research 69, 378–386.

Grossberg, S. (1974): Classical and instrumental learning by neural networks. Progress in Theoretical Biology 3, 51–141.

Hebb, D.O. (1949): The Organization of Behavior. New York: Wiley.

Hollandsworth, J.G. (1990): The Pysiology of Psychological Disorder: Schizophrenia, Depression, Anxiety, and Substance Abuse. New York: Plenum Press.

Hull, C.L. (1943): Principles of Behavior. New York: Appleton-Century.

Kent, P.F. & Mozell, M.M. (1992): The recording of odorant-induced mucosal activity pattern with a voltage-sensitive dye. Journal of Neurophysiology 68, 1804–1819.

Köhler, W. (1940): Dynamics in Psychology. New York: Grove Press.

Lancet, D., Greer, C.A., Kauer, J.S. & Shepherd, G.M. (1982): Mapping of odor-related neuronal activity in the olfactory bulb by high-resolution 2-deoxyglucose autoradiography. Proceedings of the National Academy of Sciences USA 79, 670–674.

Luchins, D.J. (1990): Task force on quantitative electrophysiological assessment. Psychiatric Research Report 5, 3–13,

Rabin, M.D. & Cain, W.S. (1984): Odor recognition: Familiarity, identifiability and coding consistency. Journal of Experimental Psychology 10, 316–325.

Searle, J.R. (1983): Intentionality: An Essay on the Philosophy of Mind. Cambridge, MA: Cambridge University Press.

Shaw, R. (1984): The Dripping Faucet as a Model Chaotic System. Santa Cruz, CA: Aerial.

Skinner, B.F. (1953): Science and Human Behavior. New York: Appleton-Century-Crofts.

Tsuda, I. (1991): Chaotic itinerancy as a dynamical basis of hermeneutics in brain and mind. World Futures 32, 167–184.

Tsuda, I. (1992): Dynamic link of memory – Chaotic memory map in nonequilibrium neural networks. Neural Networks 5, 313–326.

Yao, Y., Freeman, W.J., Burke, B. & Yang, Q. (1991): Pattern recognition by a distributed neural network: An industrial application. Neural Networks 4, 103–121.

Part VII

Symmetry Breaking in Nature

Multistability in Molecules and Reactions

P.J. Plath[1] and C. Stadler[2]

[1] Institute of Applied and Physical Chemistry
[2] Institute of Organic Chemistry
University of Bremen, D-28334 Bremen, Germany

Abstract: Regarding multistability in chemical systems, the level of individual molecules and the level of the kinetic/thermodynamic description have to be distinguished. The difference in the order of magnitude of these two chemical descriptions is about the Loschmidt number. Therefore multistability in the molecular system has to be discussed in terms of algebraic quantum mechanics, whereas multistability on the kinetic/thermodynamic level is fully described in terms of classical non-linear dynamic systems. Taking two well-known chemical examples – the mesomery of benzene and the catalytic oxidation of alcohols – we discuss multistability with respect to both of the descriptions.

1 Introduction

Atoms, valencies and molecules are the basic ideas of chemistry. A molecule is a connected ensemble of atoms, which represents a chemical substance. It is one of the main tasks of chemistry to find the links between the atoms which form the molecule. In the first place this is a topological question. To answer the question if two atoms are linked to each other, the chemist has to execute a reaction. If the linkage between two atoms is established by means of a reaction, it is called a valency. The graphical representation of the valency is a straight line, whereas the atoms are represented by letters or points. With respect to the reaction of interest the minimal set of all atoms which are connected by valencies together with the sets of these valencies form the molecule. Therefore one may say that the chemical reaction generates the molecules. I. Ugi created a more general idea of a molecule by putting together the minimal set of all molecules taking part in the considered chemical reaction (Ugi et al., 1970). Although in ordinary chemistry the geometrical structure of the molecules is often neglected, a chemical molecule possesses a gestalt which is at least a topological structure.

2 Bistability of the topological gestalt of molecules

In order to illustrate the consequences of this concept we will now focus on the well known example of the benzene molecule C_6H_6. There are two graphical representations of the molecule, the so called Kekulé-formulae (see Figure 1). They result from the chemical structure theory. The two Kekulé-formulae are regarded as extreme cases and the "reality" is much better expressed by the idea of a mesomeric state correlating both Kekulé-formulae, which is graphically represented by a two-headed arrow (see Figure 1), and means that a permanent change between the two extreme cases takes place. But this change is nothing else than a chemical reaction, since valencies are changed constantly. In other words, the mesomeric state is a unique description of the underlying process and is often described by a special symbol which represents its uniqueness (see Figure 2).

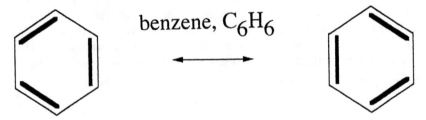

Fig. 1 Kekulé-formula of benzene

Fig. 2 Special symbol which indicates the mesomeric state of benzene

On the other hand there is no doubt about the fact that the benzene molecule is a quantum mechanical system which consists of nuclei and electrons. A quantum mechanical system possesses a holistic character, which under certain conditions exhibits non-holistic classical properties (Amman, 1992; Primas, 1981). The possibility to extract the molecule from the substance via the chemical reaction as well as its topological gestalt are two of these classical features of the quantum mechanical system. But regarding the benzene molecule the holistic character obviously never vanishes completely. Chemists may distinguish two classical molecules – the Kekulé structures

– which differ partially in their topological gestalt, but there remains part of the holistic character of the quantum mechanical systems which cannot be neglected – the mesomery. In order to describe the benzene molecule we need the two classical topological Kekulé-formulae as well as the non-classical holistic description of the so called mesomeric effect.

Following the arguments worked out by A. Amman (1992) there are certain similarities between the chemical description of such molecules and the recognition of pictures. Figure 3 provides an example for the creation of different patterns, based on the extent of holistic perception of the two pictures. While both pictures are a reproduction of one and the same photography of a courtyard, being rotated by 90°, two pictures with almost equally realistic character may be recognized. In the case of the creation of realistic pictures the recognition depends very much on the whole picture surpressing some details, but it is always the holistic character, which gives rise to the recognition of a special realistic view; diminishing the portion of the holistic character by our recognition system the different pictures are created.

Regarding the benzene molecule we can now formulate that the construction of the Kekulé-formula results from the chemical way of diminishing the holistic character of the molecular system and both formulae become realistic just because of the remaining holistic mesomeric effect.

The pictorial drawings of a chemical formula are an interpretation of the topological gestalt of the molecule which can be mathematically expressed using the graph theoretical formalism (Haas & Plath, 1982; Plath, 1988; Plath, Hass & Kramer, 1992). In these terms the formula of a molecule becomes a connected graph $G = (V, E)$ which is an ordered set of a set V of vertices and a set E of edges. One can equate the edges $e_i \epsilon E$ of the graph G with the valencies of the molecule and the vertices $v_i \epsilon V$ with its atoms. This interpreted graph G is indeed the discrete topological structure of the molecule and it is essentially a non-holistic description of the original quantum mechanical system. But as we have mentioned above, one needs at least two Kekulé-formulae and in addition the mesomery effect in order to describe the benzene molecule. That means one needs two classical individual systems with different topological gestalt. If we unify the graphs G_E and G_P of both systems by set theoretical unification of their labelled edges, we will get a non-observable gestalt which represents the reduced holistic object in consideration. Because of its holistic character there is no unique separation of this object. If we choose the usual chemical separation, the remaining holistic character has also to be taken into account, which means that the two structures cannot be fixed. But if we hold the idea that one can describe the molecule using two topological structures, the remaining holistic behaviour can be described by a permanent switching between these two structures. In other words we can state that the binary character of the gestalt of the benzene molecule is a consequence of our goal to diminish the holistic character of this quantum system in order to get a non-holistic gestalt. The concept

Fig. 3a Photography of a californian courtyard

of mesomery is as well a consequence of just this special procedure. At this point of discussion we cannot say if the two gestalts of the benzene molecule are stable and, consequently, whether or not the system is bistable.

If we look for a time-independent description of this constant change between the two classical structures, we have to reduce the system in consideration to its essential variations. That means to look only at those valencies which are changed in the transition from one topological structure to the other one. This can be done using simple algebraic and set theoretical operations working on the graphs G_E and G_P. In this way one gets a set of graphs which represents all structural prepositions about the reaction. The set of edge sets of the graphs S, D_E and D_P (see Figure 4) forms the set of elements on which the discrete topology is defined as a system of subsets. Introducing the set theoretical inclusion relation one gets a partially ordered

Fig. 3b Rotated by the angle of 90°

set of graphs which is a Boolean lattice (see Figure 4) (Plath & Hass, 1983). This is a direct consequence of the fact that these graphs form a discrete topology isomorphic to that of the power set $P(M)$ with $M = \{1, 2, 3\}$. This lattice is known as "reaction lattice". Since it is Boolean, it is atomic and uniquely complemented.

Not only the mesomery of benzene but each pericyclic reaction can be described by such a "three-atomic" Boolean reaction lattice. The particular pericyclic reactions are distinguished by the edge sets of the set theoretical atoms S, D_E and D_P. The diagrams of all these lattices have the structure of a cube (see Figure 4b). The lattices of types of pericyclic reactions and of the mesomeric benzene contain the same abstract dynamic sublattice; it is Boolean as well and its diagram is isomorphic to that of a square or two-dimensional cube (see Figure 5).

This dynamic sublattice LD describes only those valencies which are changed during the reaction. Regarding the set of edges in $E_E \subset D_E$, then the set of edges $E_P \subset D_P$ is its negation $E_P = \neg E_E$ and $E_D \subset D$ represents the formation of E_P from E_E and vice versa. But if the edge set E_E is transformed to E_P during the reaction, there should exist an infinite number of other edge sets of differently weighted graphs. The set of all edge sets forms an orthomodular lattice of which the center – the edge sets of $\emptyset$ and $D : \{E_\emptyset, E_D\}$ forms a one-atomic Boolean sublattice. This means that only elements of the Boolean center but not of the total orthomodular lattice are observable, since the orthomodular lattice is a holistic description of the

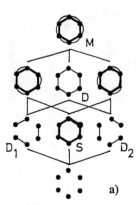 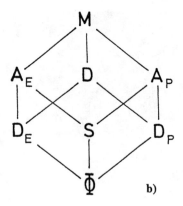

Fig. 4 (a) Three-dimensional Boolean lattice of the edge sets of the graphs of the mesomeric "reaction" of benzene (b) Lattice-theoretical generalization of mechanistic pericyclic reactions as shown in (a)

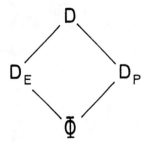

Fig. 5 Two-dimensional dynamic Boolean sublattice

reaction. But this sublattice is trivial, since its meaning is simple being or not being and we already know that the mesomery reaction takes place. To avoid the dilemma that we cannot make any non-trivial statement about the reaction as a whole one has to take recourse to the statistical interpretation. Every experiment should be executed in such a way that one can decide which is the correlated structure of the edge sets of the graphs. Therefore we have to normalize the Boolean sublattice L_D following the principle of valency conservation. In particular we choose the 1-norm $\|x\|_1 = 1$, labelling the weights p to the graphs $a \subset L_D$ (Plath, 1988; Simmons, 1963, pp. 217 – 242).

$p(a) = r$

$p(a^\perp) = 1 - r$

$p(a) + p(a^\perp) = 1$

$0 \leq r \leq 1$

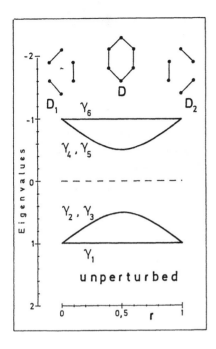

Fig. 6 The Eigenvalue-functions $\gamma_i = \gamma(r)$ of benzene for the dynamic Boolean lattice with the $\|x\|_1$- norm

For any value r there exists a weighted graph $D(r)$ with the corresponding adjacency matrix $\underline{D}(r) = r\underline{D}_E + (1-r)\underline{D}_P$. The eigenvalues $\gamma(r)$ vary with the variation of the weights. If the eigenvalue functions "cross" each other – exactly: if they are not differentiable in their points of contact – the reaction is thermally forbidden, otherwise the reaction is thermally allowed as shown for benzene in Fig. 6. This statement is well known to chemists since it is nothing else than the mathematical formulation of the Woodward-Hoffmann rule (Plath et al., 1992; Hass & Plath, 1983).

The concept of the logical analysis of chemical formulae is not restricted to the mesomeric reaction of benzene, but can also be applied to various other reactions, as for example the pericyclic reaction, the keto-enol-tautomerism, the Diels-Alder-reaction, etc., which are more important classes of reactions in chemistry, biology and medicine.

But if we assume that each weighted structure can be realized, one can ask for the information which is to gain if we carry out the structure for example via an additional reaction. The information H one can obtain at maximum from a normalized two-atomic ($p = 2$) Boolean lattice is given by the Hartley-formula (Plath, 1988; Rényi, 1982, p. 29).

$$H = ld2^P = ld2^2 = 2(\text{bit})$$

This maximum value stands for the number of questions one has to ask in order to identify all four elements of the two- atomic Boolean lattice. Assuming that we deal only with a mixture of the structures D_E and D_P, one question is already answered, resulting in information 1. Furthermore, if we specify a certain ratio of the mixture of structures D_E and D_P, the information is given by the Shannon entropy.

$$H(r) = -\sum_{i=1}^{2} r_i ld r_i \quad \text{with} \quad i = 1, 2 \quad \text{and} \quad r_1 + r_2 = 1$$

Instead of the simple eigenvalue problem which is similar to the Hückel eigenvalue problem $\underline{D}\,\underline{C} = \underline{I}\,\underline{C}\,\underline{\Gamma}$, in which each structure is taken as an individual classical structure, we now have to deal with an extended eigenvalue problem (Plath, 1988)

$$\underline{D}_r\,(1 + H(r))\,\underline{C}(H) = \underline{I}\,\underline{C}(H)\,\underline{\Gamma}(H)$$

$C(H)$:   matrix of eigenvectors
$\Gamma(H)$:   vector of eigenstates
I: unit matrix

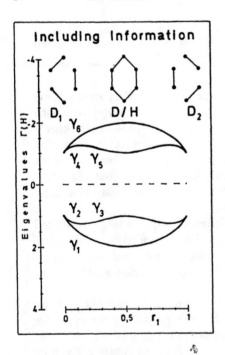

Fig. 7 The eigenvalue functions $\gamma_i = \gamma_i(r)$ of benzene for the dynamic Boolean lattice with the information norm

for which the eigenvalues $\gamma(H) \epsilon \Gamma(H)$ and the eigenvectors $C(H)$ are functions of the information H which one gains if one would fix the uncertain systems in the state in question (Fig. 7). Whereas one gets a minimum of the sum of the three lowest eigenvalues (each of them is weighted at maximum by the value two) for $r = 0.5$ in the case of the classical description of the molecular structure, a double minimum with a flat potential barrier comes up

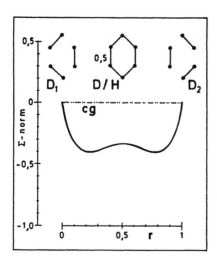

Fig. 8 The sum of weighted eigenfunctions of benzene as a function of r with respect to the information norm

in the case of taking into consideration the uncertainty of the superimposed chemical structures (Kekulé structures) of benzene (Fig. 8).

This is indeed a quantum mechanical description based on chemical structure theory, which results in a chemically realistic picture of the mesomeric reaction. This double minimum means that the system can stay in both of the two minima and one can find the system in each of them. There is no classical topological gestalt in this molecule since we cannot completely diminish the holistic character of the molecular system. It is exactly this non-vanishing holistic feature which lets the system exhibit bistability.

3 Bistability in the kinetic description of chemical reactions

Let us now switch from the molecular level of chemistry to its macroscopic level, i.e. to chemical kinetics. On this level we only know the continuous description of matter as for example the concentration of substances and temperature. Therefore we regard the previously described (Haberditzel, Jaeger & Plath, 1984) heterogeneous catalytic oxidation of methanol, using a Pd/Al_2O_3 support catalyst (0.5 weight % Pd, amount of the catalyst: about 20 mg) in a flow-reactor. The complete oxidation is a strong exothermic reaction,

$$CH_3OH + 3/2O_2 \rightarrow CO_2 + 2H_2O \qquad \Delta H^{353} = -603.3 \text{kJmol}^{-1} \qquad (1)$$

whereas the uncomplete oxidation is much less exothermic:

$$2CH_3OH + O_2 \rightarrow HCOOCH_3 + 2H_2O \qquad \Delta H^{353} = -373.2 \text{kJmol}^{-1} \qquad (2)$$

So we can follow the reaction by its heat production or by its temperature production respectively. Under certain conditions (flow rate of the gaseous reactants: 40 ml/min, CH_3OH content in the inlet: 4 vol %, temperature of the reactor: $T_{RR} = 353K$) it is possible to run this reaction under a very strange regime. The reaction then exhibits sudden changes between two states which are stable for a certain time; we would like to describe this kind of behaviour by the dutch adjective "wispelturig", which expresses very well what happens in this reaction (Fig. 9). Both states differ very much in their product distribution; in the upper state almost only CO_2 and H_2O are produced as shown in equation (1), whereas in the lower state of the temperature production a considerable amount of the ester $HCOOCH_3$ (about 18 % of the conversion) is produced but almost no CO_2 as shown in equation (2).

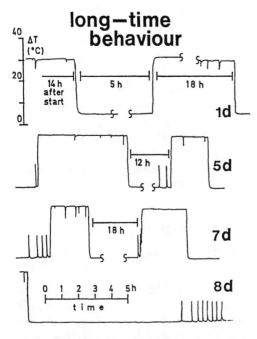

Fig. 9 Dynamical bistability of the temperature production ΔT in the catalytic oxidation of methanol. This behaviour could be observed for 8 days (Haberditzel et al., 1984)

So there are two different channels of reaction, which are correlated to the two states of the temperature production of the reaction. If we neglect for the moment the autonomous change of the reaction between its two states, we can call this reaction a bistable system. The sigmoid arctan-function which

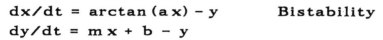

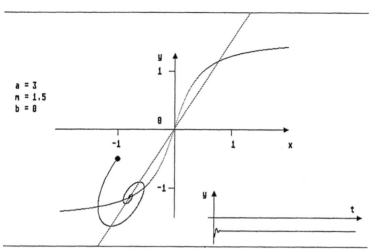

Fig. 10a Ordinary bistability. The trajectory turns to the lower stable intersection point of the two zero isoclines

intersects a linear function could be seen as a very simple mathematical description of the essential features of the system.

Different aspects of this reaction have been modelled a long time ago using mathematical tools (Dress, Jaeger & Plath, 1982; Plath & Prüfer, 1987) and also taking into account special chemical knowledge (Schuster, 1987; Plath, 1991). Thus we have developed the general idea of the strange behaviour of the catalyst. Here we are looking for the most simple description of the bistability and later on at the sudden change between the two states.

In order to catch the main ideas of bistability and "wispelturig" behaviour let us elaborate the very simple mathematical model.

If the linear function intersects the sigmoid function three times, the outer two intersection points are stable, whereas the intersection point in the middle becomes an unstable fix point (Fig. 10a). If one shifts the linear function from the original position to the right, the lower stable point and the unstable one come closer to each other and the basis of attraction of the stable fix point diminishes. If both fix points meet, they annihilate each other and only one, the upper stable fix point, remains. However, the system somehow "feels" that there was a stable fix point in the vicinity of the annihilation point, wherefore the trajectory of the system is directed towards the annihilation point coming from the side of the former stable fix point, whereas the trajectory flows away from this point if it steps into the region of the former unstable fix point (Fig. 10b).

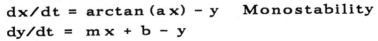

$$dx/dt = \arctan(ax) - y \quad \text{Monostability}$$
$$dy/dt = mx + b - y$$

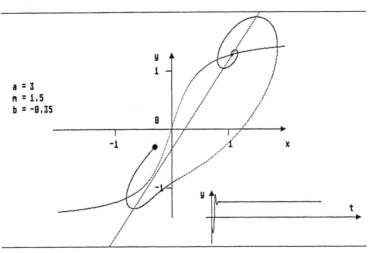

Fig. 10b Monostable situation. The trajectory seems to reach finally the former lower stable point (see Fig. 10a) before it ends up in the stable fix point.

Now we can make use of these properties in order to find a system of differential equations which constantly shifts the linear function away from the actual state of the system, i.e. the actual value of the variable $y = f(x)$. Depending upon the chosen coefficient bb the system holds the upper or the lower level as long as we like. But finally the system autonomously changes its level of production and we end up in a limited cycle with a very long period (Fig. 11). But the dynamic behaviour of such systems contains much more than simply a two level dynamic bistability. Figure 12a shows that the system can also switch between almost periodic oscillations, which show large amplitudes and persist in their lower level.

The main idea in this simple model is that one can divide the whole system into two parts: the fast dynamics in the variables x and y and the slow dynamic in the parameters m or b of the temporal behaviour of y. For example the linear function is shifted slowly by a parallel motion (Fig. 11). But of course one could also rotate the linear function around its origin getting similar results (Fig. 12a), which for some sets of parameters are in very good agreement with the experiments executed under other conditions (Fig. 12b).

It is rather difficult to interpret chemically the simple dynamic system which we have described above. In order to get closer to chemical reality we have to take into account the coupling between the rate of the heat production $d(T)/dt$, which is proportional to the rate of the combustion of the alcohol,

```
dx/dt = arctan (a x) - y        Dynamical
dy/dt = mx + b - y              Bistability
db/dt = bb y
```

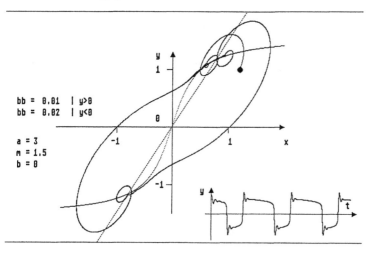

Fig. 11 Dynamical bistability occurs in the simple model. The trajectory finally reachs a limit cycle, with two regions of very slow movement.

and the rate of the cooling $d(\Delta T)/dt$ of the system by its loss of heat to its surroundings. Because of the dependency of the reaction rate on the temperature which follows the Arrhenius law, one gets a non-linear differential equation, the zero isocline of which is a sigmoid function

$$dT/dt = p \cdot \Theta_{CO} \cdot \frac{b \cdot \exp \cdot (-a/T)}{1 - b \cdot \exp(-a/T)} - \Delta T$$

whereas the cooling rate is a linear function

$$d(\Delta T)/dt = T - T_{rr} - \Delta T$$

p: summarizes the enthalpy and the volume-space velocity
Θ_{CO}: degree of coverage of the catalyst surface with CO molecules
a: E_a/R, ratio of the activation energy and the gas constant
b: product of the maximal reaction rate and the ratio of the reactor volume and the flow rate of the gaseous reactants
T_{rr}: temperature of the heat bath

Instead of shifting the linear function we can also shift the sigmoid function by varying the factor b or the exponent a. During one run of the simulation the position of the linear function is kept constant with respect to the normal experimental conditions. Again one can observe bistability for a fixed set of parameters. In order to get dynamic bistability the sigmoid function

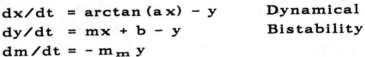

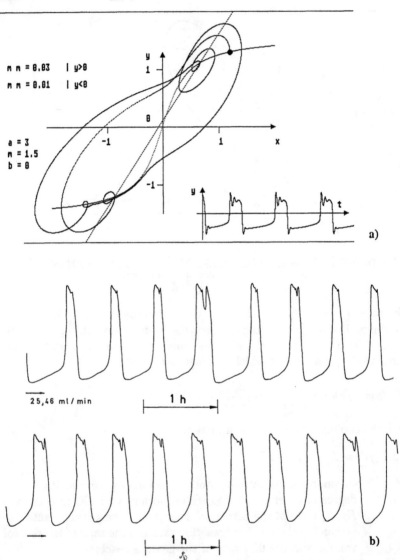

Fig. 12 Dynamical bistability: a) with rotated linear function in the simple model and b) in the chemical experiment

is shifted to lower temperatures if the system stays at its lower level or to higher temperatures if the system stays at its upper level.

$db/dt = -kb(\Delta T - 1/2(\Delta T)_\infty)$

Depending upon the chosen parameters one can observe: a limit cycle behaviour which exhibits the pattern of dynamic bistability, ordinary limit cycles, period doubling and chaotic behaviour. We observed all these types of behaviour in our chemical system (Figure 13a,b).

$$dT/dt = p\, t_{co}\, b\, e^{-a/T}/(1 - b\, e^{-a/T}) - \Delta T$$
$$d\Delta T/dt = (T - T_{RR}) - \Delta T$$
$$db/dt = -k_b(\Delta T - \Delta T_{\lim}/2)$$

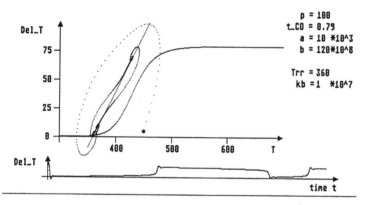

Fig. 13a Dynamical bistability in the thermodynamical model. The trajectory reaches a limit cycle remaining extraordinarily long at the two levels

Since there is a second reaction – the incomplete oxidation – besides the complete oxidation of methanol, one can extend this model introducing a second sigmoid function with only a small heat production. Using conditions under which the complete oxidation alone runs in the dynamic bistable scenario, the coupled reaction exhibits spike-like ignitions (Figure 14a) which have also been observed experimentally (Figure 14b).

The inverse function $T = g(\Delta T)$ of the double sigmoid function $\Delta T = f(T)$ resembles a polynomial of degree five in the variable T. On the other hand we derived a coupled system of two non-linear kinetic differential equations from the analysis of the set of bifurcations of the catalytic oxidation of ethanol (Engel-Herbert et al., 1990). There obviously exists a transformation of the coordinates which changes the function $\Delta T = f(T)$ into a system we have discussed in detail. The function $T = g(\Delta T)$ may allow an thermo-

$$dT/dt = p\, t_{CO}\, b\, e^{-a/T}/(1 - b\, e^{-a/T}) - \Delta T$$
$$d\Delta T/dt = (T - T_{RR}) - \Delta T$$
$$db/dt = -k_b(\Delta T - \Delta T_{lim}/2)$$

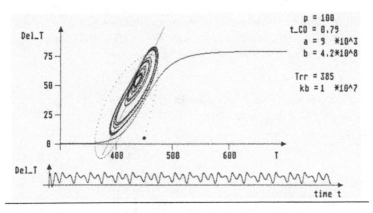

Fig. 13b Occurrence of chaotic motions in the thermodynamical model

chemical interpretation of the variables of the pure mathematical description of the ethanol oxidation.

But besides the concrete chemical system we have discussed above, there are more general questions correlated to the problem of bistability. The stable states are not necessarily fix points. It is well known that in a bistable two variable system

$$dx_1/dt = x_1^5 - ax_1^3 + bx_1 - x_2$$
$$dx_2/dt = x_1^2 - cx_1 + d - x_2$$

there exists a stable fix point within a stable limit cycle (Fig. 15a,b).

But it seems very unlikely to change temporarily between these two states in the bistable regime. The system is attracted either by one or the other stable attractor.

Let us emphasize the question if we can find a simple system which exhibits dynamic multistability (bistability) – as discussed above – where the different attractors could be fix points, limit cycles or strange attractors. The construction of such systems is based on the idea to couple a simple two or three variable system, which allows oscillation or even chaos, with a dynamic bistable system. The coupling can be chosen in such a way that the bistable system changes only one parameter of the oscillating system without being influenced itself by this partial system. In this way one can observe bistability between limit cycles or between limit cycles and fix points. If we use the two differential equations of order 5 and 2 for the oscillating system and the set of functions for the bistable partial system, in the change of the –

$$dT/dt = p\ t_{co}\ b\ e^{-a/T}/(1-b\ e^{-a/T}) - \Delta T$$
$$d\Delta T/dt = m(T - T_{RR}) - \Delta T$$
$$db/dt = -k_b(\Delta T - \Delta T_{lim}/2)$$

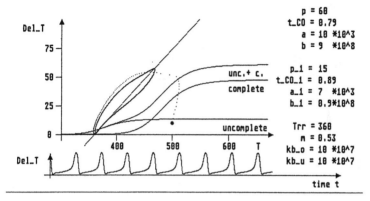

Fig. 14a Spike-like ignitions in the extended thermodynamical model. This model takes into account the complete and the uncomplete catalytic oxidation as observed experimentally

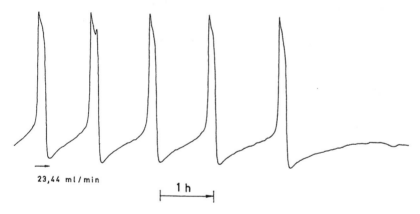

Fig. 14b Temperature production of the catalytic methanol oxidation as a function of time

for example oscillating – states a "wispelturig" behaviour becomes evident (Fig. 16). Instead of ending in a limit cycle a slightly chaotic variation of the changes seems to occur. This special way of coupling two systems is also of importance for studying the question of the new qualitative pattern beyond chaos which may arise if one increases the number of variables in the set of differential equations.

$$\varepsilon\, dx_1/dt = -p_{15}\, x_1^5 + p_{13}\, x_1^3 - p_{11}\, x_1 - x_2$$
$$dx_2/dt = p_{22}\, x_1^2 + p_{21}\, x_1 - p_{20} - x_2$$

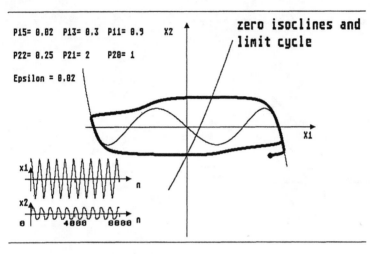

Fig. 15a Limit cycle behaviour

$$\varepsilon\, dx_1/dt = -p_{15}\, x_1^5 + p_{13}\, x_1^3 - p_{11}\, x_1 - x_2$$
$$dx_2/dt = p_{22}\, x_1^2 + p_{21}\, x_1 - p_{20} - x_2$$

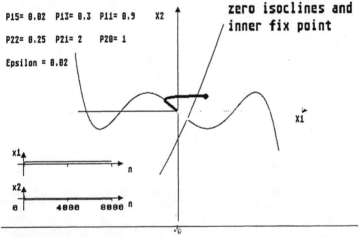

Fig. 15b Fix-point behaviour in the model of the ethanol oxidation as proposed by H. Engel-Herbert (1990)

$$\varepsilon\, dx_1/dt = -p_{15}\, x_1^5 + p_{13}\, x_1^3 - p_{11}\, x_1 - x_2$$
$$dx_2/dt = p_{22}\, x_1^2 + p_{21}\, x_1 - p_{20} - x_2$$

$$du/dt = \arctan(a_b\, u) - v$$
$$dv/dt = m_b\, u + b - v$$
if $v \geq 0$: $db/dt = b_p\, v$; else: $db/dt = b_m\, v$

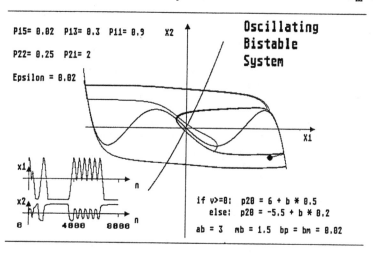

Fig. 16 Dynamical bistability between an upper oscillatory state and a lower fix point based on the model proposed by H. Engel-Herbert (1990) and the dynamical bistable system of Fig. 11. The trajectory does not end up in a limit cycle

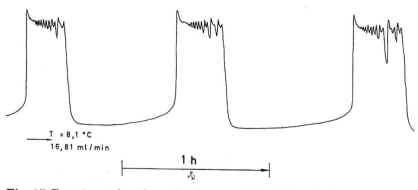

Fig. 17 Experimental evidence for dynamical bistability of the temperature production ΔT between an upper "oscillatory" state and a lower fix point behaviour. This was observed for the oxidation of methanol

Acknowledgement

We are very much indebted to E.C. Hass who works out the details of the graph theoretical description of the molecular rearrangements and to A. Haberditzel, E. van Raaij, J. Schweckendiek and R. Ottensmeyer who have done the experimental work concerning the methanol and ethanol oxidation.

References

Amman, A.: Das Gestaltproblem in der Chemie: Die Entstehung molekularer Form unter dem Einfluß der Umgebung, Gestalt-Theory 14, 4 (1992), 228–65

Dress, A., Jaeger, N.I., Plath, P.J.: Zur Dynamik idealer Speicher. Ein einfaches mathematisches Modell, Theoret. Chim. Acta (Berlin), 61 (1982), 437– 60

Engel-Herbert, H., Plath, P.J., Ottensmeyer, R., Schnelle, T., Kaldasch, J.: Dynamics of the Heterogeneous Catalytic Oxidation of Ethanol-II. Qualitative Modelling of Dynamic Features, Chemical Engineering Science 45 (1990), 955–64

Haberditzel, A., Jaeger, N.I., Plath, P.J.: Langzeitverhalten der bistabilen Oxidation des Methanols an Pd-Trägerkatalysatoren, Z. Phys. Chemie, Leipzig, 265, 3 (1984), 449–63

Hass, E.C., Plath, P.J.: Der λ-Formalismus – Eine kontinuierliche Beschreibung chemischer Reaktionen auf der Grundlage diskreter Reaktionsmodelle, Z. Chemie 22, 1 (1982), 14–23

Hass, E.C., Plath, P.J.: The Multi-Dimensional λ-Model – A Graph Theoretical / Algebraic Approach to Describe Mechanistics Aspects of Complex Chemical Reactions, in: Chemical Applications of Topology and Graph Theory (ed.: King, R.B.), Studies in Physical and Theoretical Chemistry Vol. 28, Elsevier, Amsterdam, Oxford, New York, Tokyo (1983), p.p. 405–19

Herzel, H., Plath, P.J., Svensson, P.: Experimental evidence of homoclinic chaos and type-II-intermittency during the oxidation of methanol, Physica D48 (1991), 340–352

Plath, P.J., Prüfer, H.: Ein stochastischer zellulärer Automat als Modell einer heterogen katalysierten Reaktion, Z. Phys. Chemie (Leipzig), 268, 2 (1987), 235–49

Plath, P.J.: Diskrete Physik molekularer Umlagerungen, Teubner Texte zur Physik, Bd. 19, BSB B. G. Teubner Verlagsgesellschaft, Leipzig (1988)

Plath, P.J., Hass, E.H., Kramer, M.: Graph Theory and the Mechanistic Description of Chemical Processes, in: Chemical Graph Theory, Reactivity and Kinetics (ed.: Bencher, D., Rouvrey, D.H.), Abacus Press / Gordon & Breach Science Publishers, Philadelphia, Reading, Paris, Montreux, Tokyo, Melbourne (1992), p.p. 99–153

Plath, P.J., Hass, E.H.: Logic of Chemical Ideas, in: Chemical Applications of Topology and Graph Theory (ed.: King, R.B.), Studies in Physical and Theoretical Chemistry Vol.28, Elsevier, Amsterdam, Oxford, New York, Tokyo (1983), p.p. 392–404

Primas, H.: Chemistry, Quantum Mechanics and Reductionism, Lecture Notes in Chemistry Vol. 24, Springer-Verlag Berlin, Heidelberg, New York (1981)

Rényi, A.: Tagebuch über Informationstheorie, VEB Deutscher Verlag der Wissenschaften, Berlin (1982), p. 29

Schuster, H.: Mathematische Modellbildung der Dynamik der heterogen katalysierten Oxidation des Methanols. Numerische Behandlung eines diskreten mathematischen Modells von über Diffusion miteinander gekoppelten chemischen Speichern, Dissertation, Universität Bremen (1987)

Simmons, G.F.: Introduction to Topology and Modern Analysis, Mc Graw-Hill Book Company, New York, San Francisco, Toronto, London (1963), p.p. 217–42

Ugi, I., Marquarding, D., Klusacek, H., Gokel, G., Gillespie, P.: Chemie und logische Strukturen, Angewandte Chemie 82, 18 (1970), 741-71

Perception of Ambiguous Figures: A Qualitative Model Based on Synergetics and Quantum Mechanics

G. Caglioti

Dipartimento Ingegneria Nucleare, Politecnico di Milano, Via Ponzio, 34/3, I-20133 Milano, Italy

Abstract: A qualitative model is presented for visual perception, based on synergetics and quantum mechanies. It is proposed that the logic underlying the process of perception, in particular of multistable ambiguous figures, is the same as the non-Aristotelian logic underlying quantum mechanics.

1 Introduction

Perception leads from a disordered state of the functionally engaged part of the mind to an ordered state via a dynamic instability involving a symmetry breaking (Haken, 1979).

This process can be modelled in the frame of synergetics (Haken, 1983), focusing on the interiorized figure and its time evolution. The observer explores the proposed figure and correlates the sensory stimuli promoted by the figure itself, so producing the interiorized figure. The attention and curiosity of the observer act as a control parameter on the evolution of the interiorized figure. This evolution can be schematized in terms of the overdamped motion of a low mass particle in a deformable landscape representative of the attention potential. The shape of this landscape depends on the proposed figure, as well as on the several archetypes impressed in the observer's brain by the genetics and/or the previous sensory experience. As the control exerted by the mind on the figure increases, these archetypes escavate attractors in the landscape of the attention potential, driving the process of perception toward a dynamic instability. A critical state is eventually reached where the interiorized figure falls into coincidence with the deepest attractor, matching more closely the proposed figure. *Visual thinking* (Arnheim, 1969) is then formed, occupying the mind as an order parameter.

As the critical state of this dynamic instability is reached, the disorder of a yet meaningless interiorized figure and the order of an already organized visual thinking merge together. While accomplishing the process of perception, the mind reaches a critical state. And this state, where two mutually incompatible aspects of the same reality coexist, is characterized by ambiguity (Caglioti, 1983).

In this contribution I'll focus on the process of perception and, in particular, on the perception of bistable (and multistable) ambiguous figures (Caglioti, 1992a).

A characteristic feature of the perception of bistable ambiguous figures is that the landscape of the attention potential presents two equivalent alternate attractors (Caglioti, 1992a). Once the dynamic instability has been reached thanks to the synergetic action of neurons responsible for the evolution of the interiorized figure, visual thinking doesn't keep still in the attractor that was reached first. To a difference with respect to the perception of nonambiguous figures, the visual thinking of bistable ambiguous figures undergoes a series of reversions between the two attractors. These reversions bear striking analogies with the resonant behaviour of a valence electron in chemical bond during a process of measurement of molecular spectroscopy according to quantum mechanics (Caglioti, 1992a). This dynamic behaviour of the order parameter encourages to model its qualitative features in terms of quantum mechanics.

In order to show how appropriate the analogies are, we'll devote section 6 to the quantum mechanics description of the dynamic behaviour of such valence electron in the simplest diatomic molecule: The hydrogen molecule-ion.

The objectives of this paper are:
- to emphasize the role of synergetics and symmetry breaking in perception;
- to propose an analogy between the above series of reversions, on the one hand, and the quantum resonance of the chemical bond, on the other hand;
- to suggest that the logic underlying quantum mechanics seems more helpful than the current Aristotle's logic of the *tertium non datur* for an interpretation of the dynamics of perception of the ambiguous figures. And, viceversa, that the dynamics of perception of ambiguous figures lends itself to a direct interpretation of the fundamentals of quantum mechanics (Caglioti, 1990)
- to stress that ambiguity, the coexistence of two mutually incompatible aspects of the same reality in a single structure, is a permanent cultural value (Caglioti, 1992a, b).

2 The roles of symmetry and symmetry breaking in quantum mechanics and in the concept of information

Symmetry is *invariance against transformation*. *Symmetry* is *indiscernibility of a transformation*. In other words, *symmetry* is a *no-change as the outcome of a change*.

Take a glass and rotate it around its geometrical axis. The glass remains identical to itself. No changes can be perceived as a consequence of the rotation. That axis is then a symmetric element of the glass.

Take the two twin candelabras on the altar of the church and interchange their positions. The configuration remains identical to itself. No changes can be perceived as a consequence of the permutation. We are dealing here with an example of exchange symmetry.

The exchange symmetry is perhaps the most important form of symmetry.

The presence of exchange symmetry introduces a sort of interaction between identical particles, even if these particles are separated by a large distance. For instance, in condensed matter, the magnetic interaction between two neighbour magnetic dipoles in magnetite is by orders of magnitude weaker that the "interaction" between magnetic dipoles far apart in the same magnet, originated by the exchange symmetry.

The exchange symmetry is realized in an ideally perfect way at the atomic level. In an atom the indistinguishability of two electrons is at the root of the Pauli exclusion principle, responsible for the ordering of the elements in the Mendeleev periodic table.

Furthermore, the indistinguishability of the photons or the helium atoms or the Cooper pairs of electrons is at the root of the behaviour of the condensed bosons in a macroscopic quantum structure such as the laser or the superfluid helium or the superconductors respectively.

Should the exchange symmetry be switched off, matter and radiation would change their properties and behaviour radically, and in a unconceivable way. To extrapolate and visualize the consequences of these changes could constitute an instructive task for scholars engaged in the development of virtual realities.

Symmetry is synonymous with conservation and invariance. As such, it is associated with the permanent, characteristic properties of a structure. The group of the symmetry elements of a structure contributes to define the meaning of the structure.

We all know that energy is conserved. Therefore it is not surprising that the most representative operator of a structure is the energy operator or the Hamiltonian. The Hamiltonian, **H**, sums up all symmetry transformations of the structure. The hamiltonian is a permanent reference of the structure.

In quantum mechanics the eigenstates ψ_n of **H** define symmetric, time independent ways of being of the structure, and the eigenvalues E_n associated to these stationary configurations are the possible energies of the structure itself. Here the integers **n** indicate that natura facit saltus: There are quantum jumps between the discrete energy levels associated to qualitatively different ways of being of the structure.

If the structure is isolated and no energy flows into it or away from it, its energy remains constant. The effective form of H, its ψ_n's and E_n's are the essence of a (quantum) structure: Any description of a quantum system is given be done in terms of H, ψ_n and E_n.

However, in order to proceed from the description to the knowledge, measurements must be done of the observables of the quantum structure. And yet no experimental method is available allowing to measure e.g. the energy E_n of a system in a stationary state ψ_n without removing the system away from ψ_n: Even the most representative observable, the Hamiltonian, cannot be measured without producing radical changes of the actual value of the energy at the moment of the observation.

The act of measure is irreversible – before the measurement I didn't know, after the measurement I know – and is incompatible with the symmetric and stationary nature of the eigenstates.

The concept of symmetry is somehow paradoxical. To perceive the symmetry implies a before and an after. And the invariance inherent in symmetry is incompatible with a before and an after. Therefore, at least in principle, to catch symmetry implies to break it. We'll expand this argument in Sect. 6.

There is a cost for breaking a symmetry, in terms of both energy and time.

Since symmetry implies time invariance, a certain amount of time is needed to break symmetry, so as to abandon a timeless dimension. Furthermore, since the Hamiltonian is representative of all the symmetry transformations of the structure, a certain amount of energy is needed to pull a system away from a stationary, symmetric eigenstate of the Hamiltonian toward a different final state.

The action needed to break the symmetry of an initially balanced configuration so as to produce an information, can be envisaged, at least dimensionally, as the product of the above time by the above energy.

In physics the action is indeed the product of an energy by a time.

The elementary action is the Planck's constant, h. Perhaps, at an elementary level, the act of symmetry breaking leading to the bit – the unit of information – might require indeed a minimum action of the order h.

In this perspective information could be defined as the outcome of an action leading to a form.

3 Bistability in perception and the dynamic structure of the bit of information

Consider a bistable ambiguous figure (Fig. 1). This figure enables us to visualize a representation of the dynamic structure of the bit of information.

By construction, this two-dimentional figure possesses a centre of inversion symmetry: By inverting any element of it with respect to its geometrical center, no changes can be perceived: The image remains identical to itself.

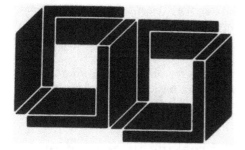

Fig. 1 Franco Grignani, Psychoplastic structure, cm^2 70 × 70 (1968), detail

Initially we look at this figure as a well balanced structure. But after few seconds, from the heap of sensory stimuli visual thinking rises and a transition from flat to spacial suddenly occurs. The common wall shared between the cubic moduli (or, mutatis mutandis, the covalent electron shared between two protons in a hydrogen molecule-ion (see Sect. 6)) is hooked by the visual thinking (Arnheim, 1969) and is then seen to become part of the rhs (or the lhs) cube. The center of symmetry disappears and the wall centered on it falls e.g. toward the rhs. However, immediately after, this wall starts oscillating from right to left and again to right (like the electron exhibiting the charge transfer spectrum under the action of an electromagnetic field in a resonant chemical bond (see Sect. 6)).

Where is then the central wall? Initially it appeared to be equally shared between left and right. Two are the possible ways to reach such equipartition, and these ways are mutually incompatible:
- the wall belongs 50 % to the lhs and 50 % to the rhs modulus

or
- the wall belongs neither to the lhs nor to the rhs modulus.

However, in the critical stage of the process of formation of an information, the central wall belongs 50 % to the lhs modulus and 50 % to the rhs modulus and, at the same time, it doesn't belong neither to the lhs nor to the rhs modulus. This situation might appear fastidious to some naive observers, or intriguing and stimulating to others. All of them feel urged to remove the ambiguity by taking a decision. A fluctuation, and it's catastrophe! Once the decision is taken, the central wall appears to fall toward the right hand side cube or, according to the sign of the by now climbing fluctuation, toward the left hand one: At this stage the center of symmetry has disappeared, the configuration is no more balanced and one bit of information – right or left – has been produced. As anticipated, the perception of a bistable ambiguous figure helps thus to visualize the dynamic structure of the bit of information.

In Sects. 4–6 we'll analyze more in detail the dynamics of ambiguity characteristic of the process of perception of this class of figures.

Concluding the present section we stress that the measurement in quantum mechanics and the (visual) perception in everyday life both allow to obtain an information. But one must be prepared to pay a price for this information: To remove symmetry.

Symmetry breaking is then a prerequisite for obtaining an information. Symmetry removal and symmetry breaking are basic features of the process of perception.

4 Synergetics and symmetry breaking in perception

Synergetics is the cooperation of many subsystems interacting non-linearly within a structure. Under the control of external resources of e.g. energy, information or matter, the structure selforganizes itself leading to spontaneous formation of spatial, temporal or functional order (Haken, 1983).

As indicated in the introduction, perception can be envisaged as a process of selforganization of the interiorized figure, leading to the formation of visual thinking.

Notwithstanding its chaotic features, nothing is perhaps more ordered than the thought. Indeed, as H. Haken states, "the order parameters are ultimately the thoughts" (Haken, 1983). But order as a synonym of organization and correlation, on the one hand, and symmetry as a synonym of indiscernibility and invariance, on the other hand, are in opposite relation. Therefore a reduction of symmetry and a removal of indiscernibility are prerequisites for the formation of visual thinking.

During perception the reduction of symmetry occurs in two steps.

Initially the eyes scan the figure, trying to identify in it its "interesting" features. According to the synergetic model of pattern recognition, a potential landscape $V(\mathbf{q})$ is introduced in the order parameter space, whose gradient acts as the main force driving the time evolution $d\mathbf{q}/dt$ of the interiorized figure. This procedure suggests that a criterion – perhaps *the* criterion – regulating this evolution at the neurophysiological level is to discard all mutually indistinguishable regions of the proposed figures as not deserving to be stored in the memory. In other words if A and B = A are two regions related one to the other one by exchange symmetry, at least at first they will be relegated automatically to the background. Thereafter, as a rule, they will play a marginal role in the process of evolution of the interiorized figure.

While this continuous removal of exchange symmetry proceeds, the regions of the proposed figure characterized by high gradient, e.g. its contours, emerge more and more vividly from the background and selforganize themselves aiming to the formation of a coherent pattern.

As indicated in the introduction, the second step of perception is accomplished suddenly, as soon as a critical state is reached where the interiorized figure falls into coincidence with one of the archetype attractors. At this point visual thinking is formed, and the order parameter develops embracing the interiorized figure as a whole unit. But contextually a hierarchy is established among the elements of the pattern: Among elements that, at the very beginning, were contributing with equivalent weights to the formation of the interiorized figure. The onset of such differentiation is indeed a symmetry breaking.

It becomes now clear that in this critical state correlation, order and visual thinking coexist with and take the place of homogeneity, symmetry and unawareness. This critical state is thus characterized by its ambiguity.

5 Ambiguity as a unavoidable ingredient of the process of perception

Tertium non datur is a rule, attributed to Aristotle in the Middle Ages, according to which "any substance – e.g. a man, a horse etc. – cannot host opposite qualities simultaneously" (Aristotle, 1988).

In other words, according to Aristotle's logic, at any instant one can assign either a quality or its opposite to a substance.

Beside these two alternatives, a third possibility seems not granted: Tertium non datur. Although "the essential character of the substance seems in being constituted for hosting the opposites", the substance can thus host opposite qualities only at different times, and never simultaneously. The tertium non datur has influenced the western culture through centuries. The need of certainty has induced us to rely upon it as a reference, so dispelling ambiguities. Ambiguity is annoying. We feel inclined to believe that "the truth is unique". We insist that clear yes or no answers must be given to our questions.

Indeed many persons exposed for the first time to some bistable ambiguous figures, find them fastidious.

However our claim for reassurance has been undermined with the development of the Gestalt psychology and, after the Great World War, in the Twenties, with the advent of quantum mechanics and with the agnostic position adopted by e.g. L. Wittgenstein and L. Pirandello.

Ambiguity is indeed an intrinsic and intriguing feature of the process of perception. Ambiguity is continuously experienced by our mind: Every act of perception culminates into a dynamical instability of the interiorized image, where the incoherent heap of sensory stimuli and the coherent visual thinking merge into coincidence. Since, in turn, perception is a basic ingredient of life and a prerequisite of survival, we should look at ambiguity not so much as to a fastidious feature of the perceptive catastrophe, but rather as to a fixed course towards thoughts and emotions (e.g. Caglioti, 1987).

6 Dynamics of perception of bistable ambiguous figures and resonance of the chemical bond

In the introduction we have proposed an analogy between the dynamics of reversions during the perception of a bistable ambiguous figure and the resonance of the chemical bond in molecular spectroscopy.

The aim of this section is to show that the mind operates during perception according to the logic underlying quantum mechanics. Therefore we'll now deep into the above analogy, focusing on the spectroscopy of the hydrogen molecule-ion (Herzberg, 1950).

This molecule is made of two positive nuclear unit charges – two protons – and one negative unit charge – the valence electron.

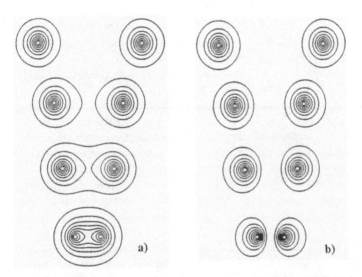

Fig. 2 (a) The gerade, centersymmetric orbitals and (b) the ungerade, centerantisymmetric orbitals of the hydrogen molecule-ion. This molecule is obtained by aggregation of two protons bonded by one valence electron. Like in a military map, along the contour lines the probability of presence of the electron is constant. Note that the center of symmetry of the field of the presence probability of the electron described by the ungerade orbital is not accessible to the electron. These molecular orbitals are computed for six values of the interprotonic distance, decreasing from top to bottom toward the equilibrium value. Although the center of symmetry is the most important characteristic of all the corresponding structures, it is not marked in the drawings

Like for any quantum system, the position of the electron binding the two protons can be defined only probabilistically in terms of a wavefunction whose square modulus provides the probability that the electron is found in the unit volume in the neighborhood of the protons.

There are in practice two wavefunctions suitable to describe the state of this electron.
- the gerade wavefunction, g, representative of a state where an important fraction of the electronic charge is localized around the center of symmetry half a way between the two protons (Figure 2a)
- the ungerade wavefunction, u, representative of a state where no electronic charge is localized around the center of symmetry half a way between the two protons (Figure 2b).

The above two states are centersymmetric and centerantisymmetric respectively: By inverting the electron coordinate with respect to the center of symmetry g remains unaltered, while u changes its sign. (Figure 3a)

Two constant energy levels, E_g and E_u, are associated to g and u respectively. (Figure 3b)

The electronic charge described by g is distributed predominantly in the central region of the molecule. Such charge distribution can be envisaged as a shield between the two positive nuclear charges. This shield screens the two nuclei more efficiently than the electronic charge described by u. Correspondingly g describes the bonded molecule in its fundamental state, and E_g turns out to be smaller than the energy E_0 of the unbounded system made of a proton and a single hydrogen atom in its fundamental state. Both E_g and E_0 are in turn smaller that the energy E_u associated to the antibonding state u of the hydrogen molecule-ion.

So much for the description of the two symmetric possible states.

But in order to remove the uncertainty or to acquire information and *knowledge* about the electron distribution, a spectroscopic *measurement* becomes necessary.

The measurement must be performed by an electromagnetic radiation whose frequency is tuned to the exchange frequency corresponding to the difference between the energy levels, E_u and E_g, associated to the above states. The electric component of this resonant radiation field hooks the electron and drives it from left to right and viceversa. But at the critical state, marking the transition from the *stationarity* of the *description* to the *irreversibility* accompanying any process of *measurement*, the electronic charge around the center of symmetry belongs equally both to the lhs and rhs atom *and, simultaneously*, it belongs neither to the lhs nor to the rhs atom.

Therefore, during the spectroscopic observation of the bistable diatomic hydrogen molecule-ion, the valence electron, hooked by the electric component of the electromagnetic resonant radiation, is exchanged between the two bonded atoms in the so called *charge transfer spectrum* (Herzberg, 1950).

The frequency of this periodic exchange is determined by the difference:

(1) $\quad 2K = E_u - E_g.$

In turn, this difference is a measure of the coupling between the two nuclear positive charges, as provided by the bonding electron. Naturally this

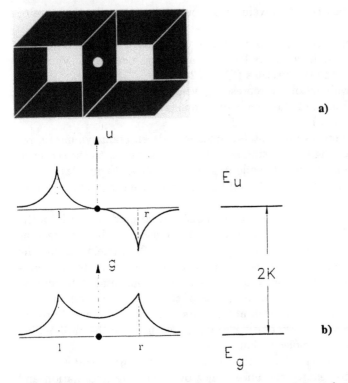

Fig. 3 The graphic counterpart of the centersymmetric and centerantisymmetric "eigenstates", **g** and **u** respectively, of the hydrogen molecule-ion. The molecular orbitals of Figure 2 are obtained from **g** and **u** by computing their square moduli. On the right the "energy levels" E_u and E_g associated to the **u** and **g** eigenstates are presented. These levels differ by the exchange energy **2K** (see eq. (1) in the text). In turn **2K/C**, where **C** is a psychoneurographic parameter covering in perception a role analogous to that covered by the Planck constant in quantum mechanics, measures the average frequency of perceptual reversion of the central wall in the bistable ambiguous figure (see also Figure 5)

coupling would become smaller in an idealized situation where the proton charges would be frozen further apart than at their actual equilibrium distance (Figure 4): In the limit of very large internuclear distance the time taken by a round trip reversion would become infinitely long (Fig. 5).

The reader will visualize this situation by comparing the reversion frequencies of e.g. the three or four binary structures at the bottom of Figure 5: These frequencies manifestly decrease as the "internuclear" distance between the two cubic "atomic" moduli increases. These frequencies decrease also if the whole size of the structure increases (Borsellino, 1982). These remarks outline one of the objective reasons for which the aesthetic emotions stimulated by an original painting (I am thinking here, e.g. to *l'Eglise d'Auvers* by

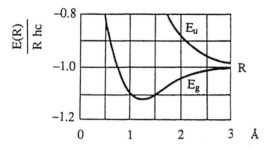

Fig. 4 Dependence of the energy levels E_g and E_u of the hydrogen molecule-ion on the internuclear distance **R** between the two protons. When the distance **R** is very great, the system's energy becomes that of a hydrogen atom in the groundstate energy level $E_0 = -13.6$ eV. E_0 is the reference level and corresponds to -1.0 Rydberg. As **R** decreases, this reference level splits in two. The lower level (E_g) refers to the bonding orbital described in first approximation by the eigenstate **g**, and the upper one (E_u) refers to the antibonding orbital described by the eigenstate **u**.

V. van Gogh) are usually far more violent than those offered by a smaller-sized reproduction of it (Paris vaut bien une Messe!).

The analogy between the dynamic perception of a bistable ambiguous figure and the spectroscopic measurement of the charge transfer spectrum of the hydrogen molecule-ion seems so stringent that we feel induced to push it further. Consider the sequence of reversions starting at the critical state of the perception process: This sequence could be formalized mathematically adopting the same formalism utilized in molecular spectroscopy.

In practice a linear combination of the **g** and **u** states is initially computed, the weights of these states being chosen as equal for a balanced bistable structure and as different for an unbalanced figure such as the "Plastic permutation in the field" by Franco Grignani (1959).

Thereafter the modulus square of this linear combination is calculated in order to obtain the time dependence of the apparent position P(t) of the central wall in Fig. 1.

In explicit form, the sequence of reversion is described in terms of the eigenstates left (**l**) and the right (**r**) of the hydrogen atoms centered on the proton positions l and r in Figure 3 as follows:

(2) $\quad P(t) = \left| \frac{1}{\sqrt{2}} \left(g \exp -i\frac{E_g}{C}t + u \exp -i\frac{E_u}{C}t \right) \right|^2.$

C is a parameter which should cover, in perception, a role similar to that of the Planck constant $h/2\pi$ in quantum mechanics, and the gerade and ungerade states for the condensed graphic diatomic molecule at the bottom of Figure 1 are (see e.g. Caglioti, 1974):

(3) $\quad u = \frac{l+r}{\sqrt{2}} \exp -i\frac{E_g}{C}t$

Fig. 5 Franco Grignani, Graphic condensation, cm² 70 × 70 (1968). Each of the six binary structures in this figure is composed of a pair of modules. By construction, each structure possesses a center of symmetry: By inverting any element of a binary structure with respect to its center, a structure identical to the original one is obtained. The process of perception of the "condensed" structure at the bottom of this figure is analysed in the text and described synthetically below. During the observation of this structure the mind collects in the memory and interiorizes the sensory stimuli coming from the image. While controlling the initially scattered and disordered stimuli, the mind processes and correlates them: The aim is to build synergetically an ordered pattern around the center of symmetry of the structure. There are two possible patterns that the stimuli can form while the control exerted on them by the mind increases. Both patterns assign the central wall 50 % to the left hand side (lhs) module, and 50 % to the rhs module. But while for one of the two patterns (pattern **g**, Figure 3a, the center of symmetry is equally shared by the lhs and the rhs modules, for the other pattern (pattern **u**, Figure 3b, the center of symmetry belongs neither to the lhs nor to the rhs module. These patterns are thus mutually in conflict, and yet each one separately is fully compatible and consistent with the logic and operative system of the mind. As the control parameter increases further, a critical stage is reached finally where one could state: The center of symmetry is equally shared by the lhs and rhs modules **and, simultaneously**, the center of symmetry belongs neither to the lhs nor to the rhs module. The mind is thus trapped in an ambiguous state where two conflicting, irreconcilable patterns coexist. This critical state of the mind is uneasy, the mind is anxious to escape through any possible way out. A small fluctuation, and it's catastrophe! A dynamic instability in the pattern of the stimuli occurs leading perception from flat to spacial; the center of symmetry disappears and, as symmetry breaking occurs, the order parameter – the visual thinking – starts to rise driving the initially common wall between the two cubes toward the lhs or the rhs module. From an originally symmetric or balanced situation the "center of symmetry" falls either left or right: In order to perceive symmetry we must destroy it: But, at the same time, a bit of information is created. Actually, visual thinking cannot remain fixed in any of these two positions: Soon a dynamics of perpetual reversions sets in, resembling the resonant behavior of the chemical bond during the spectroscopic observation of a diatomic molecule by the charge transfer spectrum

$$(4) \quad g = \frac{l-r}{\sqrt{2}} \exp{-i\frac{E_u}{C}t}.$$

In terms of the exchange frequency, **2K/C**, eq. (2) can be written in the form:

$$(2') \quad P(t) = \exp\left(-i\frac{E_g + E_u}{2C}t\right)\left(l\cos\frac{E_g}{2C}t - ir\sin\frac{E_u}{2C}t\right).$$

The mind of any layperson lacking scientific education is thus capable to perform algebraic operations involving e.g. the imaginary unit **i**, without specific instructions.

Fig. 6 Franco Grignani, Psychoplastic structure, cm² 70 × 70 (1968), details merging into a graphic condensation, like n atoms. Perceptually, one can control this multistable figure and organize it in a way reminescent of a conduction electron under an alternate electric field, surfing on a wave packet and bouncing on the top and bottom walls

This human mathematical skill is not surprising: Every time we cross the street we take only few tenths of a second in order to estimate the trajectory of a car in our reference frame. Somehow the funambulist is an excellent mathematician. As an additional example, let's spend a minute to perceive Figure 6. This resonant modular structure offers an example of perceptual multistability. Once the critical state has been reached, the common walls between contiguous moduli move upward and/or downward, depending on the chaotic will of the perceiver. However there is a special way to perceive this figure: One could form a dynamic pattern where, starting e.g. from the top wall of the figure, like in an arpeggio the walls come downward one after the other in succession, and bounce against the bottom wall upward, again in succession.

This behaviour corresponds to the description of the motion of a wave packet representing a conduction electron in a metal according to quantum mechanics. The examples proposed above suggest that through the introspective analysis of the process of perception our mind can attain scientific forms of knowledge without structuring them in their appropriate explicit mathematical formulation: Art itself is a form of knowledge, and the artist's intuition often preceeds scientific formalizations by centuries or even millennia.

Let's come back to the analogy presented before between the spectroscopic resonance of the valence electron in the chemical bond, on the one hand, and the bistability in perception of an ambiguous figure presented in Sect. 3, on the other hand. The analogy becomes now evident: The equipartition of the central wall, equally shared by two cubic moduli according to the two mutually incompatible ways postulated in that section, resembles closely the balanced distribution of the electron charge in the covalent bond between the two nuclei of the hydrogen molecule-ion according to the gerade and ungerade states defined at the beginning this section.

This analogy between the process of perception of undecidible, unconcretable, illusory, paradoxical and ambiguous figures, on the one hand, and the spectroscopic observation of quantum structures on which physicists use to base their self-confidence and their certainties of the "exact" sciences, on the other hand, fits appropriately in a conceptual scheme elaborated several years ago by I. Prigogine and I. Stengers (Prigogine & Stengers, 1977). In order to characterize the incompatibility between the stationariness of the states in whose terms the quantum structures must be described (cfr. the first and second pattern in Figure 2) and the irreversibility of the processes of measurement, they write (Prigogine & Stengers, 1977):

> Since stationary states are states where nothing happens we could state nothing about them if they were effectively stationary; it is only when a system makes a transition from a stationary state to another stationary state that a number becomes accessible, thanks to quantum mechanics. .... The formalism of the quantum mechanics is based on the simultaneous assertion of two mutually incompatible terms, each one being necessary in order to allow the other to have a physical meaning. The irreversible process is necessary in order to reach a knowledge of the stationary state: A state that contextually ceases being stationary; quantum mechanics can describe the process exclusively in terms of transition between stationary states, and considers the process as totally incomprehensible without this reference to concepts in terms of which the process itself is totally meaningless.

The results obtained so far could be extended also to the critical states of macroscopic physical system undergoing equilibrium phase transformations or nonequilibrium dynamic instabilities.

For instance a binary alloy undergoing a disorder-order transformation in its critical state is ordered, is disordered and, simultaneously, is both ordered and disordered. Furthermore, a liquid heated from below undergoing the Rayleigh-Bénard diffusive-convective instability, when the Rayleigh number reaches the critical value, is still, is in motion, and, simultaneously, is both still and in motion.

One could argue that the above critical states occur, in nature, only in pathological and exceptional conditions. However, in the depth of our mind, critical states are the rule rather than the exception. As living systems, we continuously process information. Our life proceeds along a jagged watershed, made of options and choices. Whenever we find ourselves on the watershed of an option we must come to a decision. Thus our mind is bound to come across critical states. And in these critical states the rule of the tertium non datur becomes questionable.

In conclusion, the critical state of any physical transformation, including the perceptive catastrophes occurring continuously in our mind, is characterized by symmetry breaking, and exhibits ambiguous features that play a crucial role in the onset of the transformation itself. However these features seem not compatible with the tertium non datur.

7 Conclusions

Ambiguity can be defined as the coevolution or coexistence of two mutually incompatible aspects of a same reality inside a single structure.

The role of ambiguity in art and in quantum physics, in the critical state of equilibrium phase transformations and in the dynamical instabilities of non equilibrium systems has been discussed in the frame of synergetics. Among these dynamic instabilities we have focused our attention on perception. The act of perception leads to the acquisition of an information and triggers the process of thinking. Perception is accomplished by a comparatively slow synergetic process of symmetry removal, followed by a sudden symmetry breaking accompanied by the removal of ambiguity: From an initially balanced and symmetric situation, one reaches thus an unbalanced, informed or ordered situation.

Our thinking – a continuous process of elaboration of either semantic or aesthetic information – proceeds thus on the lanceolate watershed of ambiguity.

An analogy has been proposed between the perception of bistable ambiguous figures and the spectroscopic observation of the hydrogen molecule-ion according to quantum mechanics. The analogy is stringent. One feels thus induced to infer from it that the logic underlying the process of perception is the same as the non Aristotlean logic underlying quantum mechanics.

This logic assigns a central role to ambiguity. Therefore ambiguity, as a fixed course towards perception, thoughts and emotions, rises to the role of a permanent cultural value (e.g. Caglioti, 1987).

Acknowledgements

I am grateful to Franco Grignani, master of visual communication and design, whose work is for me an ever active source of inspiration since twenty years. Figures 1, 5 and 6 constitute his specially devised graphic solutions to the problems of perceptual multistability that have arisen during enlightening discussions on the dynamics of ambiguity and visual thinking.

Thanks are due to Professor Carlo E. Bottani, who has computed and prepared Figure 2. Heartfelt thanks to Carla Cattaneo who edited the manuscript with intelligence and precision.

References

Aristotle (1988): Opere 1, Organon. Laterza: Biblioteca Universale.
Arnheim, R. (1969): Visual Thinking. Berkeley, C.A.: Regents of the University of California.
Borsellino, A., Carlini, F., Riani, M., Tuccio, M.T., DeMarco, A., Penego, P. & Trabucco, A. (1982): Effects of visual angle on perspectives reversal for ambiguous pattern. Perception 11, 263–273.
Caglioti, G. (1974): Introduzione alla Fisica dei Materiali. Bologna: Zanichelli.
Caglioti, G. (1983): Simmetrie Infrante nella Scienza e nell'Arte. Milano: CLUP.
Caglioti, G. (1986): The world of Escher and physics. In: H.S.M. Coxeter, M. Emmer, R. Penrose & M.L. Teuber (Eds.), M.C. Escher: Art and Science, pp. 287. Amsterdam: North Holland.
Caglioti, G. (1987): Synergetics and Weltanschauung. In: R. Graham & A. Wunderlin (Eds.), Lasers and Synergetics; A Colloquium on Coherence and Self-organization in Nature. Berlin: Springer.
Caglioti, G. (1990): The Tertium non Datur in Aristotle's Logic and in Physics, Proceedings of the International Conference on Mechanics, Physics and Structure of Materials; A Celebration of Aristotle's 23 Centuries, Thessaloniki, Aug.19–24, 1990. Trieste: Manuscript.
Caglioti, G. (1992a): The Dynamics of Ambiguity. Berlin: Springer.
Caglioti, G. (1992b): Ambiguity in Art and Science, Proceedings of the Workshop on Art and Science, ENEA/WAAS, Vinci, 11–13 December, 1992. Trieste: Manuscript.
Ditzinger T. & Haken, H. (1994): A synergetic model of multistability in perception. (in this volume).
Haken, H. (Ed.) (1979): Pattern Formation by Dynamic Systems and Pattern Recognition. Berlin: Springer.
Haken, H. (1983): Synergetics – An Introduction. Berlin: Springer. (2nd ed.).
Haken, H. (1991): Synergetic Computers and Cognition: A Top-down Approach to Neural Nets. Berlin: Springer.
Haken, H. & Stadler, M. (Eds.) (1989): Synergetics of Cognition. Berlin: Springer.
Herzberg, G. (1950): Molecular Spectra and Molecular Structure I - Spectra of Diatomic Molecules. New York: Van Nostrand Reinhold.
Prigogine, I. & Stengers, I. (1977): La nouvelle alliance. Scientia 112, 287–332, 617–653.

Name Index

Abelard 107
Acuna, C. 386
Adams, P. 272
Adrian, E.D. 424, 436
Aertsen, A. 436
Alberti, L.B. 116–118
Albertz, J. 138
Aleksanov, S.N. 376, 383
Allazetta, A. 183, 272, 418
Allison, T. 385, 388
Altafullah, I. 374, 383, 385
Ames, A. 102
Amit, D.J. 13, 18
Amman, A. 442, 443, 460
Ammons, C.H. 272
Ammons, R.B. 272
An der Heiden, U. 203–205, 216
Andersen, R.A. 386
Anderson, J.A. 6, 19, 173, 184, 256, 273, 295, 308, 328, 390, 405
Anderson, J.R. 86, 96
Anstis, S. 275, 280, 281, 288, 308, 309
Anstis, S.M. 49, 52, 53, 68
Antebi, D.L. 82
Aristotle 106, 109, 135, 243, 247, 464, 469, 477, 478
Arnheim, R. 463, 466, 478
Attneave, F. 7, 18, 111, 137, 173, 183, 280, 308, 390, 404
Aufseß, A. 383, 404
Augustinus, A. 241, 243, 252

Babich, S. 272
Babiker, I.E. 82
Babloyantz, A. 184, 427, 436
Bacon, R. 107
Bahnsen, P. 57, 67
Bakker, L. 86, 96

Balhorn, H. 354
Ballmer, T.T. 240
Barlow, J.S. 411, 418
Bartolini, B. 183, 272, 418
Basalou, L.W. 239
Başar, E. 24, 43, 82, 357, 358, 361–364, 368, 370–373, 375–377, 379–388, 393, 404, 405, 436
Başar-Eroglu, C. 5, 18, 71, 82, 357, 358, 374, 375, 377, 379, 383, 384, 389, 391, 393, 403, 404, 416, 418
Bashore, T. 404
Bauden, P. 388
Bauer, R. 436
Baust, R.F. 295, 309
Beck, J. 134, 137
Beeler, G.W. 411, 419
Behne, K.E. 20
Belcher, S.J. 412, 418
Bencher, D. 460
Bendat, J.S. 362, 384
Berg, P. 83
Berger, H.J.C. 81–83
Bergum, B.O. 79, 82, 272
Bergum, J.E. 79, 82, 272
Biederman, I. 57, 59, 67
Biernisch, M. 240
Birbaumer, N. 358, 384, 403–405
Bischof, N. 187, 201, 203, 209–213, 215, 216
Bodenstein, G. 411, 418
Bodis-Wollner, I. 385
Boff, K.R. 308, 309
Bonaparte, M. 436
Bongartz, W. 83
Borelli 422
Boring, E.G. 390

Borsellino, A.   175, 183, 256, 257, 259, 261, 262, 268, 272, 308, 328, 412, 418, 472, 478
Borst, A.   109, 137
Botwinick, J.   272
Bourget, M.   385
Boyd, E.H.   374, 384
Boyd, E.S.   384
Braasch, M.   404
Bradley, D.R.   100, 101, 137
Bragdon, H.R.   378, 385
Brandeis, D.   413, 416, 418
Brandimonte, M.A.   100, 137
Bressler, S.L.   184, 432, 436
Brinley, J.F.   272
Broggi, G.   309
Brooks, V.   114, 137
Brosch, M.   436
Brown, K.T.   272, 295, 303, 308
Brown, L.E.   384
Brügelmann, H.   354
Bruner, J.   272
Brunner, E.J.   20
Bryan, W.L.   16, 18
Buchanan, J.   139, 155
Buchanan, S.   184
Buchwald, J.S.   374, 384, 385
Buffart, H.   86, 89, 96
Bullinger, M.   217
Bullock, T.H.   376, 383–388, 404, 405, 436
Bunz, H.   23, 43, 71, 83, 167, 175, 183
Burke, B.   429, 436, 437
Burt, P.   279, 288, 308
Buzsaki, G.   379, 384

Caelli, T.   290, 305, 308
Caglioti, G.   6, 18, 38, 43, 463, 464, 469, 473, 477, 478
Caianiello, E.R.   338, 353
Cain, W.S.   432, 437
Calvert, J.E.   80, 82
Campenhausen, C. von   346, 348, 353
Canavan, A.G.M.   384, 404
Carello, C.   99, 138, 140, 155
Carlini, F.   272, 308, 478
Carlsson, A.   81, 83
Carmesin, H.-O.   81, 82, 84, 337, 339, 350, 352–354
Casasent   331
Case, P.   43, 83, 159, 163, 165, 167, 169, 184
Cattaneo, C.   478

Cavanagh   331
Cederberg, J.N.   350, 354
Chapman, R.M.   378, 384, 385
Chauvel, P.   388
Christiaens, D.   89, 96
Chung, S.   239
Clarke, F.J.J.   412, 418
Cohen, J.   404
Cohen, R.   83
Coles, M.G.H.   386
Collet, P.   275, 308
Commons, M.L.   201
Cools, A.R.   81–83
Cooper, L.A.   224, 239
Cooper, L.N.   335
Cornelis, H.M.B.   404
Cornwell, H.   272
Costa, L.D.   387
Courson, R.   272
Courtois, G.   410, 419
Cowan, W.M.   386
Cox, P.W.   94, 97
Cracco, R.Q.   385
Crandall, P.   383
Crawford, H.J.   80, 83
Creutzfeldt, O.D.   411, 418
Crick, F.   175, 183, 337, 339, 354
Cross, D.V.   195, 201
Curchesne, E.   374, 385
Cutting, J.E.   139, 141, 154

Dali, S.   6
Davis, J.   17, 18
Dawson, M.R.W.   279, 286, 308
de Graef, P.   89, 96
Debecker, J.   374, 385
Deecke, L.   387, 388, 396, 401, 402, 404, 405
DeGuzman, G.C.   175, 182, 184
DelleDonne, M.   6, 18
DeMarco, A.   183, 272, 295, 296, 301, 308, 418, 478
Demiralp, T.   357, 358, 361, 364, 370–373, 379, 380, 382, 384, 385, 401, 404
Descartes   422
Desmedt, J.E.   371, 374, 385
Destexhe, A.   427, 436
Dickerson, L.W.   385
Diekmann, V.   386, 388
Ding, M.   43, 83, 159, 165, 167, 169, 184
Dirlich, G.   80, 83

Ditzinger, T.   6, 11, 18, 31, 37, 38, 41, 43, 71, 83, 169, 173, 183, 255–257, 272, 295, 299, 301, 308, 335, 390, 404, 412, 418, 478
Dodwell, P.L.   275, 308
Domany, E.   354
Donahue, W.T.   272
Donchin, E.   378, 385, 388
Dorst, R.   14, 19
Drake, D.   138
Dress, A.   451, 460
Dürer, A.   116–119
Duffy, C.J.   386
Dugger, J.   272
Duncker, A.   251, 252
Dvorak, I.   419

Eckhorn, R.   11, 19, 430, 436
Eckmann, J.P.   275, 308
Edelman, G.M.   386, 388
Egatz, R.   272
Egelhaaf, M.   109, 137
Ehrenfels, Chr.von   14, 19
Ehrlich, D.   273
Elbert, T.   81, 83, 384, 402–405
Elffers, J.   99, 137
Ellis, W.D.   51, 67
Elsner, N.   353, 354
Émile   241, 252
Emmerton, J.   201
Emrich, H.M.   80, 83
Engel, A.K.   19, 385
Engel, F.   6, 20
Engel, K.A.   436
Engel, P.   411, 418
Engel-Herbert, H.   455, 458–460
Epstein, W.   61, 64, 67
Erke, H.   52, 67, 202, 411, 412, 419
Ertel, S.   14, 19, 21, 235, 239
Escher, M.C.   6, 20, 124, 127, 478
Etevenon, P.   388
Euclid   119, 350
Eysenk, H.J.   83

Fantz, R.L.   133, 137
Faude, E.   354
Feingold, E.   80, 83
Fender, D.H.   411, 419
Fernald, R.D.   109, 137
Feuerlein, W.   207, 216
Finlay, D.   290, 305, 308
Fischgold, H.   387

Fisher, G.H.   6, 19, 162, 183, 295, 308, 390
Fishman, J.   239
Flohr, H.   24, 43
Flores d'Arcais, G.B.   67, 137
Fodor, J.A.   53, 67, 93, 96, 107, 137
Fonseca e Castro, S.L.   307
Foote, S.L.   387
Ford, M.F.   82
Forestell, P.   6, 20
Fourier, J.   144, 145, 147, 153
Fox, P.T.   387
Foxley, B.   242, 252
Freeman, J.A.   335
Freeman, W.J.   6, 11, 15, 19, 20, 92, 96, 379, 385, 400, 404, 421, 425, 427–430, 432, 436, 437
Freud, S.   422, 423, 436
Freude, G.   401, 404
Frykholm, G.   139, 155
Fuchs, A.   43, 139, 145, 149, 154, 155, 159, 183, 184
Fuchs, T.   335
Fuster, J.M.   372, 380, 385

Gaillard, A.W.K.   386, 405
Galambos, R.   376, 385
Gale, A.G.   96
Gall, W.E.   386
Gardiner, C.W.   296, 297, 308
Garrett, J.   273
Gastaut, H.   387
Gendlin, E.T.   249, 252
Georgopoulos, A.   386
Gerbino, W.   57, 67, 100, 134, 137
Gerstner, W.   341, 354
Gevins, A.S.   419, 436
Gheorghiu, V.A.   20, 80, 82, 83
Gibson, B.S.   96
Gibson, J.J.   47, 67, 86, 91, 96, 106–108, 137, 278, 308
Giese, F.   243, 252
Gillespie, P.   461
Gillis, W.M.   13, 21
Girgus, J.   272
Glass, L.   150, 155, 160, 162, 183
Gökhan, N.   404
Gönder, A.   384
Goethe, J.W. von   134, 137, 243, 252
Goff, E.R.   374, 385
Goff, W.R.   388
Gokel, G.   461
Goldberg, M.E.   387

Name Index

Goldman-Rakic, P.S.   370, 385
Goldmeier, E.   104, 137
Gonzales-Estrada, T.M.   436
Goodenough, D.R.   94, 97
Gordon, K.   273
Grady, D.   330, 335
Gräser, H.   52, 67, 80, 83, 390, 404
Graham, R.   478
Graves, A.   379, 386
Gray, C.M.   11, 19, 376, 379, 385, 427, 430, 436
Gray, J.A.   84
Greer, C.A.   437
Gregory, R.L.   134, 137, 239
Greist-Bousquet, S.   18
Greitschus, F.   18, 383, 389, 418
Griffitts, C.H.   272
Grignani, F.   466, 473, 475, 478
Grözinger, B.   388
Grossberg, S.   335, 337, 354, 388, 423, 436
Grünewald, G.   404
Grünewald-Zuberbier, E.   404
Grzywacz, N.M.   286, 310
Guckenheimer, J.   275, 306, 308
Guillou, S.   388
Gutowitz, H.   137

Haas, R.   41, 43, 139, 443
Haberditzel, A.   449, 450, 460
Haken, H.   6, 11, 15, 18–21, 23, 24, 27, 28, 31, 32, 38, 41, 43, 67, 68, 70, 71, 83, 84, 96, 105, 137, 139–141, 145, 149, 154, 155, 159, 167, 169, 171, 173, 175, 183–187, 191, 201, 203, 204, 217, 255–257, 272, 273, 275, 293, 295, 299, 301, 307–309, 311–313, 315, 317, 321, 323, 328, 330, 334, 335, 389, 390, 404, 405, 407, 409, 412, 417–419, 463, 468, 478
Halgren, E.   358, 374, 383, 385, 388
Harnad, S.   165, 183
Harris, J.P.   81, 82, 84
Harrison, J.B.   358, 374, 385
Harter, N.   16, 18
Harvey, E.M.   63, 68
Harvey, L.O.   339, 354
Hass, E.C.   443, 444, 446, 460
Hassler, F.   165, 183
Hearst, E.   188, 202
Heath, H.A.   80, 83, 273
Hebb, D.O.   313, 315, 330, 337, 339, 353, 354, 377, 382, 429, 437

Heffley, E.   385
Heider, F.   247, 252
Heinze, H.J.   357, 385
Held, R.   183, 308
Helmholtz, H. v.   108
Helson, H.   169, 183
Hemmen, J.L. van   338, 341, 354
Hendler, T.   405, 416, 419
Henggeler, B.   414, 419
Henle, M.   67
Herrnstein, R.J.   201
Herzberg, G.   470, 471, 478
Herzel, H.   460
Heslenfeld, D.   94, 96
Hildebrandt, G.   411, 418
Hillyard, S.A.   357, 358, 378, 385–387, 395, 402, 404, 405
Hilton, H.J.   57, 67
Hinton, G.E.   19
Hochberg, J.   64, 67, 85, 89–91, 93, 96, 114, 124, 137
Hock, H.S.   19, 71, 84, 275, 277, 280, 281, 288, 291, 292, 295, 297, 299, 301–304, 308
Höffding, H.   17, 19, 61, 62, 64, 67, 76
Hoenkamp, E.   141, 155
Hörmann, H.   243, 252
Hoeth, F.   9, 19, 52, 67, 281, 308, 391, 405
Hogeboom, M.   86, 91, 96
Hollandsworth, J.G.   437
Holmes, P.   275, 306, 308
Holroyd, T.   43, 83, 159, 174, 175, 182, 184
Holst, E. von   182, 184
Hommel, J.   402, 404
Honig, W.K.   187, 201
Hopf   315
Horsthemke, W.   280, 307, 308
Horstink, M.W.I.   82, 83
Horvath, E.I.   43, 83, 139, 150, 155, 159, 160
Hüppe, A.   137
Hull, C.L.   421, 423, 437
Hume   108
Hummel, J.E.   57, 67
Huy, N.T.   385

Iberall, A.S.   165, 183
Inanaga, K.   386
Intrator, N.   335
Isozaki, H.   386
Ivins, W.M. jr.   117, 119, 137

Ivry, R. 137

Jaeger, N.I. 449, 451, 460
Jaspers, R. 82
Jastrow, J. 77, 78, 99
Jellinek, E.M. 207, 208, 217
Jiang, W.J. 305, 309
Johanson, A.M. 273
Johansson, G. 41, 43, 139, 141, 143, 153, 155
John, E.R. 388
Johnson Jr., R. 358, 384–386, 388
Johnson, F. 96
Jordan, W. 436
Julesz, B. 160, 184, 223, 237, 239
Jung, R. 423

Kaldasch, J. 460
Kanizsa, G. 5, 10, 17, 19, 47, 57, 61, 67, 76, 78, 83, 92, 93, 96, 100, 102, 124, 137, 138, 185, 201, 269, 273, 416, 418
Kant, E. 108
Karakullukcu, E. 404
Karamürsel, S. 404
Karrer, R. 404
Katayama, S. 411, 419
Katayama, T. 335
Katayama, Y. 374, 385
Kauer, J.S. 437
Kaufman, L. 308, 309
Kawamoto, A.H. 6, 19, 173, 184, 256, 273, 295, 308, 328, 390, 405
Keesey, U.T. 411, 419
Kekulé, A. 442, 443, 448
Kelso, J.A.S. 19, 23, 27, 38, 43, 71, 83, 84, 139, 140, 145, 155, 159, 160, 163, 165, 167, 169, 172, 174, 175, 182–184, 187, 201, 275, 277, 280, 281, 287, 291–293, 295–297, 299, 301, 303–305, 307–310
Kemmler, L. 21, 239
Kenemans, L. 94, 96
Kennedy, J.M. 93, 96
Kent, P.F. 424, 437
Kienker, P.K. 6, 19
Kimura, T. 419
King, R.B. 460
Klatt, D.H. 305, 309
Kleinen, G. 20
Klemme, M. 273
Klimesch, W. 376, 379, 387
Klintman, H. 9, 19, 80, 83, 273
Klusacek, H. 461

Knauss, T.A. 411, 419
Knight, R.T. 273, 358, 373, 374, 386
Kobs, M. 9, 20
Koch, C. 175, 183, 337, 339, 354
Köhler, W. 2, 6, 11, 14, 19, 28, 43, 51, 64, 66, 67, 69, 83, 108, 110, 135, 137, 185, 201, 242, 252, 255, 257, 273, 295, 309, 328, 337, 354, 390, 400, 405, 423, 425, 437
Koenderink, J.J. 110, 137
König, P. 19, 385, 436
Koepchen, H.P. 20, 21, 84, 184, 309, 404, 405, 419
Koffka, K. 51, 91, 96, 110, 135, 137, 275, 281, 308
Kohl, K. 94, 96
Kohonen, T. 13, 19
Kolers, P.A. 52, 67, 303, 309
Kolev, V. 379, 386
Kornhuber, H.H. 386–388, 396, 402, 404, 405
Korte, A. 288, 309
Koslow, S.H. 385
Kosslyn, S.M. 201
Koukkou, M. 408, 416, 417, 419
Kramer, M. 443, 460
Kremen, I. 60, 68
Kreyscher, M. 337, 354
Kris, E. 436
Kristeva, R. 404
Kristiansen, K 410, 419
Krohn, W. 20
Kruse, P. 5–11, 16–18, 20, 21, 28, 43, 49, 62–73, 76, 80–84, 187, 191, 202, 256, 269, 270, 273, 281, 301, 303, 307, 309, 337, 353, 354, 389–391, 400, 404, 405, 412, 418, 419
Kruse, W. 436
Kubovy, M. 275, 309
Kühn, R. 338, 341, 354
Künkel, H. 378, 386
Künnapas, T. 263, 265, 268, 273, 328
Küppers, G. 20
Kuhn, T.S. 6, 20
Kunzendorf, R.G. 419
Kutas, M. 386

Labov, W. 228, 229, 239
Lacroix, D. 405
LaGournerie, J. de 101, 102, 119, 122, 137
Lancet, D. 424, 437
Land, M.F. 109, 137

Name Index

Landa, J.S. 71, 84, 311
Landis, T. 417, 419
Lane, H.L. 195, 201
Lang, E. 230, 240
Lang, M. 376, 377, 386
Lang, W. 386
Langton, C.G. 109, 137
Lass, N.J. 309
Lausberg, H. 244, 252
Leeman, F. 99, 137
Lefever, R. 280, 307, 308
Leftshetz, S. 160, 184
Lehmann, D. 6, 20, 407, 408, 411–414, 416–420
Leibing, U. 388
Leibniz 421
Leibowitz, H.W. 308
Lesèvre, N. 376, 387
Levit, H. 126, 131, 133, 137
Li, C.N. 239
Li, W. 109, 137
Libet, B. 415, 419
Lichtenberg, G.C. 107
Liégeois-Chauvel, C. 388
Lienert, G.A. 393, 405
Lindauer, M.S. 295, 309
Lindner, K. 405
Linke, P.F. 49, 50, 67
Llinas, R.R. 379, 386
Locke, J. 108, 173
Lockhead, G.R. 67
Lopes da Silva, F.H. 379, 386
Lorenz, K. 7, 20
Lovejoy, A.O. 138
Loveless, N. 385
Lu, F. 71, 84
Lu, T. 311, 335
Luccio, R. 10, 17, 19, 47, 61, 67, 102, 137, 138, 185, 201, 269, 273
Luchins, D.J. 423, 437
Luck, S.J. 385, 395, 402, 405
Luer, G. 138
Lugiato, L.A. 293, 309
Lutzenberger, W. 83, 402, 404, 405
Lynch, J.C. 373, 386

MacClelland, J.L 135, 138, 333, 335
Mach, E. 7, 390
Mallot, H.A. 307
Malmo, R.B. 273
Maltsman, I. 385
Mandell, A.J. 24, 43, 173, 183
Mangun, G.R. 385

Manneville, P. 182, 184
Mantegna, A. 118, 119
Marquarding, D. 461
Marr, D. 108, 133, 134, 137, 240, 278, 303, 309
Martin, R. 184
Marx, K. 6, 20
Masotoshi, T. 386
Maturana, H.R. 108, 110, 137, 185, 201
Maurer, K. 430, 436
McCarthy, G. 374, 388
McClintock, P.V.E. 309
McCulloch, W.S. 311, 313, 337, 354
McGuigan, F.J. 384
McLulich, D.A. 6, 20
Mefferd, R.B. 274
Mendeleev, D.I. 465
Merri, M. 309
Metelli, F. 134, 137
Metzger, W. 7, 9, 14, 20, 57, 67, 101, 134, 138, 187, 202, 236, 240, 252, 280, 281, 291, 309
Michaels, C.F. 99, 138
Michel, C.M. 420
Michotte, A. 234, 240
Miezin, F.M. 387
Migler, B. 191–193, 195, 196, 200, 202
Mikhailov, A.S. 20, 83
Millenson, J.R. 191–193, 195, 196, 200, 202
Miller, R. 359, 375–377, 386
Mita, T. 419
Mizuki, Y. 372, 376, 377, 386
Moiré 150, 155, 160, 183
Moles, A. 207, 217
Mondrian, P. 133, 134, 138
Moore, C.A. 94, 97
Morinaga, S. 57, 67
Moss, F. 309
Mosteller, F. 272
Mostofsky, D.I. 187, 202
Motte-Haber, H. de la 20
Motter, B.C. 386
Mountcastle, V.B. 358, 373, 386, 388, 400, 405
Mozell, M.M. 424, 437
Münte, T.F. 378, 386
Mumford, D.B. 201
Mundy-Castle, A.C. 376, 386
Munk, M. 436
Musolino, A. 388

Name Index

Näätänen, R.   357, 358, 371, 380, 386, 387
Nam-Gyoon, K.   140, 155
Nattkemper, D.   89, 96
Navon, D.   89, 96
Nawrot, M.   275, 309
Nebelytsyn, V.D.   79, 84
Necker, L.A.   6, 7, 20, 31, 47, 69, 83–85, 89–91, 94–96, 99–101, 114, 125, 135–137, 173–175, 177, 221, 237, 255, 261, 268, 272–274, 328, 390, 403–405, 412, 419
Neisser, U.   239
Neuhaus, W.   288, 309
Neville, H.J.   386, 387
Newell, A.   248, 252, 408, 419
Newton, I.   421
Nichols, D.J.   411, 419
Nicki, R.M.   6, 20
Nishihaba, H.L.   240
Nishijima, H.   386

O'Connor, T.A.   374, 387
O'Donnell, B.F.   403, 405, 416, 419
O'Shaughness, M.   273, 412, 419
Ockham   107
Özesmi, Ç.   371, 384
Ogden, C.K.   12, 20
Ohman, A.   385
Oiwa, K.   335
Orbach, J.   80, 83, 273
Ostwald, W.   252
Ottensmeyer, R.   460
Oyama, T.   263, 265, 266, 268, 273
Ozaki, H.   413, 419

Packard, N.H.   109, 137
Pal, I.   413, 419
Paller, K.A.   358, 374, 387
Palm, G.   436
Palmer, S.E.   89, 96
Paluzzi, S.   65
Pandya, A.S.   43, 139, 145, 155
Parasuraman, R.   384, 386, 388
Parnefjord, R.   381, 383, 387
Pavlekovic, B.   20
Pavlov, I.P.   79, 80, 84, 423
Peirce, C.S.   14, 39, 241
Pelton, L.   273
Penengo, P.   272, 308, 478
Penrose, R.   101, 124, 478
Perez, R.   160, 183
Perkell, J.B.   305, 309

Perkins, D.N.   113, 138
Pernigo, M.A.   309
Petersen, S.E.   370, 373, 387
Peterson, M.A.   62–64, 66–68, 85, 89–91, 93, 96
Petry, H.M.   100, 101, 137
Petsche, H.   376, 379, 387, 400, 405
Pfaff, S.   21
Pfurtscheller, G.   376, 379, 387
Phillips, G.   150, 155, 275, 280, 310
Phillipson, O.T.   81, 82, 84
Piaget, J.   337, 338, 350, 352–354
Pick, H.L.   154
Picton, T.W.   357, 358, 371, 378, 385, 387
Pidoux, B.   388
Piersol, A.G.   362, 384
Pineda, J.A.   374, 387
Pirandello, L.   469
Piranesi, G.B.   125, 126, 130, 131, 133
Pirenne, M.H.   122, 138
Pitts, W.H.   311, 313, 337, 354
Planck, M.   466, 471, 473
Plath, P.J.   6, 20, 441, 443–449, 451, 460
Plato   106
Pöppel, E.   217
Pomeau, Y.   182, 184
Pomerantz, J.R.   67, 275, 309
Poole, E.W.   411, 419
Poppelreuter, W.   236, 240
Portal, A.   93, 96
Porter, E.   273
Posner, M.I.   370, 387
Postman, L.   272
Poston, T.   237, 240, 302, 309
Pozzo   124, 128
Pratt, J.B.   138
Prazdny, K.   137
Preobrashenskaya, L.A.   383
Prigogine, I.   185, 202, 476, 478
Primas, H.   442, 460
Primi, C.   65
Prinz, W.   89, 96
Pritchard, W.S.   374, 387
Proffitt, D.R.   139, 154
Prüfer, H.   451, 460
Psaltis, D.   331, 335
Pylyshyn, Z.   107, 108, 138

Quintilian, M.F.   243, 252

Rabin, M.D.   432, 437

Name Index

Raczaszek, J.   43, 83, 159, 163
Rahn, E.   383, 404
Raichle, M.E.   387
Ramachandran, V.S.   49, 52, 53, 68, 275, 280, 281, 309
Raphael   119, 122, 124
Rappelsberger, P.   376, 379, 387, 405
Rausch, E.   185, 188, 202
Rechenberg, I.   109, 138
Reeß, J.   388
Regan, D.   358, 378, 387, 402, 405
Reisberg, D.   273, 412, 419
Reitboeck, H.J.   11, 19, 436
Rémond, A.   376, 387, 419, 436
Renault, B.   386
Rentschler, I.   339, 354
Rényi, A.   447, 460
Repp, B.H.   305, 309
Requin, J.   184
Reuter, H.   9, 20
Reynolds, G.S.   196, 198, 200, 202
Riani, M.   272, 308, 478
Richards, I.A.   12, 20
Richards, W.   183
Richter, D.W.   353, 354
Richter, P.H.   6, 18, 21
Rinesi, S.   183, 272, 418
Ritter, M.   239
Ritter, W.   374, 378, 387, 388, 405
Robbin, J.S.   272
Roberg, D.   335
Robert, H.   126
Robinson, D.L.   373, 387
Rock, I.   60, 61, 64, 67, 68, 272, 275, 309
Rockstroh, B.   83, 384, 403–405
Röschke, J.   383
Rösler, F.   358, 385, 387, 405
Rogers, A.K.   138
Rohrbaugh, J.W.   384, 386, 388
Roscelin   108
Rosen, B.   383
Rosenthal, R.   83
Ross, J.   6, 18
Roth, G.   1, 69, 84, 138
Roth, W.T.   378, 387
Rousseau, J.J.   241, 242, 244, 252
Rouvrey, D.H.   460
Rubin, E.   31, 47, 57, 60, 61, 66, 68, 76, 222, 390
Ruchkin, D.   385
Rumelhart, D.E.   135, 138, 333, 335
Runeson, S.   139, 155

Rustin, J.   133, 134, 138

Sachs, L.   363, 387
Saint-Exupéry, A. de   204, 217
Sakai, K.   328, 335
Sakata, H.   386
Santayana, G.   108, 138
Santisi, A.   65
Scheid, P.   404
Scherb, W.   388
Schiepek, G.   20
Schiffman, H.R.   17, 18
Schiller, P. von   8, 20, 49–52, 54, 68, 281, 282, 310
Schleidt, W.M.   7, 20
Schlosberg, H.   411, 420
Schmidt, S.J.   138, 185, 202
Schmidt-Heinrich, E.   405
Schmielau, F.   384
Schnelle, T.   460
Schnyt, M.   99
Schöner, G.   6, 11, 19, 20, 71, 84, 140, 155, 275, 277, 280, 281, 287, 291–293, 295–297, 299, 301, 303–305, 308–310
Scholz, J.P.   155
Schoonover, R.A.   384
Schramm, U.   384
Schröder   7, 80, 135, 390
Schürmann, M.   357, 379, 383, 384, 386, 404
Schütt, A.   383
Schult, J.   384
Schulten, K.   354
Schumacher, L.E.   19
Schumaker, J.F.   83
Schuster, H.   451, 460
Schuyt, M.   137
Schweckendiek, J.   460
Schwegler, H.   14, 20, 203–205, 216, 337, 339, 353, 354
Scotus   107
Searle, J.R.   426, 437
Sejnowski, T.J.   19
Sekuler, A.B.   89, 96
Sekuler, R.   150, 155, 275, 280, 309, 310
Sellars, R.W.   138
Serlio   125, 129
Shannon, C.E.   447
Shaw, J.C.   248, 252
Shaw, R.   427, 437
Sheikh, A.A.   419
Sheng, Y.   311, 331, 335

Name Index

Shepard, R.N.   62, 68, 109, 124, 127, 136, 138, 224, 239, 240
Shepherd, G.M.   372, 387, 437
Sherif, M.   75, 84
Shimizu, H.   307
Shlesinger, M.F.   183
Short, P.   6, 20
Siddle, D.   385
Siegel, R.K.   201, 202
Simmons, G.F.   445, 461
Simon, H.A.   248, 252
Simson, R.   374, 387
Singer, W.   19, 379, 385, 436
Skapura, D.M.   335
Skarda, C.A.   11, 15, 20
Skinner, B.F.   189
Skinner, J.E.   423, 427, 436, 437
Skrandies, W.   413, 419
Smit, D.   91, 94, 96
Smith, M.E.   358, 374, 385, 388
Smitsman, A.   89, 96
Sokolik, M.   388
Solley, C.   273
Solodovnikov, V.V.   388
Song, S.   385
Soodak, H.   165, 183
Spencer, D.D.   388
Sperling, G.   279, 288, 308
Sporns, O.   388
Squire, L.R.   387
Squires, K.   388
Squires, N.K.   405, 416, 419
Squires, N.S.   374, 378, 384, 388
Stadler, C.   6, 20, 441
Stadler, M.   5, 7, 9–11, 14, 16–21, 28, 43, 49, 67–73, 76, 81–84, 96, 137, 140, 155, 171, 183, 187, 191, 201, 202, 236, 239, 240, 269, 273, 281, 301, 303, 307, 309, 311, 337, 353, 354, 389–391, 400, 404, 405, 411, 412, 418, 419, 478
Stampfer, H.G.   357, 358, 368, 371, 373, 384, 388
Standing, L.   272
Stapleton, J.M.   374, 383, 385, 388
Starr, A.A.   374, 387
Staunton, G.B.   387
Stefanelli, M.   39, 44
Steinmetz, M.A.   386
Stelmach, G.E.   184
Stengers, I.   476, 478
Stewart, I.   302, 309
Stier, B.   383
Stins, J.   91, 96

Strasburger, H.   339, 354
Strik, W.K.   413, 414, 419, 420
Strong, S.A.   138
Strüber, D.   5, 11, 16, 18, 20, 21, 69, 73, 82, 84, 269, 273, 301, 303, 309, 389, 390, 404, 405, 412, 418, 419
Stukat, K.G.   80, 84
Stuss, D.T.   357, 371, 387
Subokawa, T.   385
Sukiyama, T.   385
Sutton, S.   358, 388
Svensson, P.   460
Swinney, D.A.   85, 86, 88, 96
Szu, H.H.   6, 21, 71, 84, 311, 312, 317, 331, 335

Takii, O.   386
Talbot, W.H.   386
Talmy   230
Tefler, B.   312, 317, 335
Teuber, H.-L.   308
Teuber, M.L.   478
Thetford, P.   273
Thiery   135
Thomas, J.P.   308, 309
Tononi, G.   376, 388
Torii, S.   266, 273
Tortrat, S.   388
Trabucco, A.   272, 308, 478
Treisman   402
Tretter, F.   203–205, 216, 217
Tschacher, W.   20
Tsuda, I.   430, 437
Tsunashima, Y.   419
Tuccio, M.T.   272, 308, 478
Tueting, P.   385, 388, 404
Tuller, B.   43, 83, 159, 163, 165, 167, 169, 184
Turner, W.   134
Turvey, M.T.   140, 155

Ugi, I.   441, 461
Ullman, S.   278, 279, 286, 303, 310
Ullsperger, P.   401, 404
Ungan, P.   363, 377, 384
Urcueioli, P.J.   187, 201
Uttal, W.R.   128

Vainstein, I.I.   383
van den Bercken, J.H.L.   82, 83
van der Twell, L.H.   362, 388
van Dijk, B.   430, 436
van Gogh, V.   472

van Hoof, J.J.M. 82
van Leeuwen, C. 7, 21, 85, 86, 89, 91, 94, 96
van Leeuwen, L. 89, 96
van Raaij, E. 460
van Staendonck, K.P.M. 82, 83
Vaughan, H.G. 374, 385, 387, 388
Vegt, J. 86, 89, 96
Vetter, G.H. 185, 188, 202
Viana Di Prisco, G. 427, 436
Vickers, D. 256, 263, 273
Villablanca, J. 409, 410, 419
Vogt, S. 10, 21
Voss, A. 301, 308
Vukovich, A. 6, 9, 21, 241, 246, 252

Wackermann, J. 413, 416, 419, 420
Wada, S. 335
Wagenaar, W.A. 41, 44
Wagner, K. 335
Walk, R.D. 154
Wallace, B. 273
Wallace, J.G. 134, 137
Wallace, S. 139, 155
Wallach, H. 295, 309
Walter, D.O. 376, 388
Webster 226
Wehner, T. 72, 84, 269, 273, 281, 301, 309, 337, 354, 391, 405
Weidenbacher, H.J. 63, 68
Weitenberg, N. 91, 94, 96
Werbos, P. 312, 335
Wertheimer, M. 6, 13, 14, 21, 49, 50, 64, 66, 68, 102, 138, 187, 196, 202, 251, 252
Westphal, K.P. 376, 388
Widrow, B. 311, 331, 335

Wieland, B.A. 274
Wildgen, W. 6, 9, 14, 21, 31, 44, 221, 236, 240
Wiley, J. 384, 386
Williams, D. 150, 155, 275, 280, 310
Williamson, P.D. 388
Winter, R. 311, 331, 335
Witkin, H.A. 94, 97
Witte, W. 14, 21, 252
Wittgenstein, L. 6, 21, 241, 249, 252, 469
Wood, C.C. 374, 388
Wood, C.D. 384
Woods, D.L. 358, 378, 379, 386, 388
Woodworth, R.S. 411, 420
Wulf, F. 92, 97, 110, 138
Wunderlich, D. 227, 240
Wunderlin, A. 478
Wundt, W. 7, 124, 138, 411
Wyse, L. 335

Yamaguchi, Y. 307
Yang, Q. 437
Yao, Y. 429, 436, 437
d'Ydewalle, G. 89, 96
Yin, T.C.T. 373, 386, 388
Yokoyama, S. 419
Yuille, A. 286, 310

Zanone, P.G. 303, 310
Zaus, M. 14, 21
Zeno 421
Zimmer, A.C. 7, 21, 70, 84, 99, 102, 104, 109–111, 138
Zola-Morgan, S. 387
Zubin, J. 388

Subject Index

actuality 135, 432
ADALINE network 332
adaptation 269, 302, 303, 390
aesthetic information 477
alcoholism 204
anchor stimuli 244
animal cognition 188
anticipation 93, 381
apparent motion 8, 48, 54, 65, 72, 201, 278, 281, 290, 291, 391
appetence 215
artificial neural network 311–313, 331, 335, 435
associative memory 33
asymmetry 232
attention 18, 36, 37, 41, 60, 90, 112, 124, 168, 169, 247, 258, 358, 379, 380, 401, 410, 412, 417, 426, 428, 431, 434, 463
attention parameter 255, 269, 299, 311, 313, 315, 317, 328, 330, 334
attraction 203, 213, 215, 279, 302, 306
attractor 5, 10, 13, 14, 16, 17, 42, 71, 73, 79, 81, 135, 160, 167, 168, 171, 182, 228, 278–280, 285, 288, 291, 293, 299, 306, 313, 315, 316, 318, 334, 389, 390, 430, 456, 463, 464, 468
attractor neural network 13, 14
auditory perception 9
auto-inhibition 113
autokinetic phenomenon 75
autonomous order formation 185, 187
aversion 215

back error propagation 313
back error propagation network 312, 316, 330

basin of attraction 293, 296, 297, 341, 345, 421, 429
Bénard instability 6
benzene molecule 442
bifurcation 5, 140, 168, 187, 203, 205, 206, 285, 295, 306, 315, 353, 389, 430
biological rhythms 255
bistability 221, 237, 291, 293, 442, 449, 453, 455, 456, 459, 466
bistable molecules 6
blindsight 13
brain 24, 25
brain computing network 312
brain dynamics 421, 428, 433, 435
brain dysfunctions 1
brain electric field 409, 410, 413, 414, 417, 418
brain electric microstate 414–416, 418
brain functional macrostate 410
brain functional microstate 407, 413
brain functional state 408
brain models 423, 428, 435
brain-mind relationship 2
brain-state-in-a-box 328
Buridan point 216

camouflage 223, 251
catastrophe 238, 467
catastrophe theory 207, 237
cell assembly 429
cellular automata 109
central perspective 124
chaos 429, 456, 457
chaos generation 11
chaos theory 1
chaotic activity 427
chaotic attractor 10, 42, 429
chaotic behaviour 203, 455

Subject Index

chaotic dynamic model 430
chaotic dynamics 426
chaotic itinerancy 430
chaotic motion 456
chaotic processes 431, 432
chaotic systems 431
circular apparent motion (CAM) 8, 72, 74, 75, 77
circular causality 39
closeness 51
cognitive activities 375
cognitive components 377
cognitive dysfunctions 1
cognitive flexibility 9, 80, 81
cognitive functions 408
cognitive order formation 5, 400
cognitive psychology 337, 383, 435
cognitive science 433
cognitive self-organization 70, 72
cognitive style 94
coherent motion 276
coherent oscillations 11
coincidence neurons 343
collective behavior 15
collective processes 18
collective state 338, 341, 346, 352, 353
collective variable 140
collinearity 62
complexity 339, 345, 346, 349–351, 401
complexity – velocity dilemma 340
connectionist 311
conscious 418
consciousness 414, 415, 435
constancy of form 123
constancy of meaning 243
constructivistic 185, 187
context influence 91
control parameter 15, 25, 27, 28, 105, 140, 163, 171, 186, 187, 463
cooperativity 275–277
cortical activity patterns 426
creativity 80
critical fluctuations 27, 71, 185, 186, 191
critical realism 108
critical slowing down 71, 140, 185, 195
cultural relativism 107
cusp 206, 237

decision making 380, 381
decisions 5
depth effect 116
deterministic chaotic systems 427

direction bistability 349
direction of apparent motion 70
disambiguation 31, 85, 86, 90, 101, 136, 229, 390
disorder 223
distributed processes 136
dopaminergic system 81
dynamic brain models 422, 423
dynamic modelling 203
dynamical systems 203

Ebenbreite 57, 60
ecological perception 99
ecological realism 91
ecological validity 48, 49, 81, 100
EEG 1, 71, 82, 94, 182, 255, 358, 360–363, 376, 379, 381, 382, 389–394, 400, 401, 403, 410–413, 416, 421, 423–427, 432, 434, 435
eigenvalue 447, 448
Einstellung 57
emergence of meaning 241, 251
emergence of new qualities 15, 24
emergent temporal structures 409
equilibrium 92, 205–207, 209, 213, 215
equilibrium states 203
event related potential (ERP) 357, 358, 363, 368, 370–372, 374, 378, 380, 382, 402, 410, 413
evoked potential (EP) 1, 357, 361, 363
evolutionary algorithms 109
experimental phenomenology 5, 47

factor of proximity 56
factor of similarity 50
field theory 108
field-like processes in the brain 12
figural aftereffect 60
figural goodness 114
figural organization 10
figure and ground 47, 57, 60, 61, 63, 65, 70, 77, 101, 221, 224, 400, 417
figure and ground stimuli 64
figure and ground tristability 7
fix-point attractors 10
fix-point behaviour 458, 459
flexibility 176
fluctuations 69, 79, 81, 140, 168, 170, 182, 186, 192, 194, 196, 257–259, 278, 280, 290, 295, 296, 301, 334, 390, 397, 409, 411, 412, 467
fluctuations of attention 411
fluctuations of complex patterns 7

focused attention   380
form constancy   117, 118, 124
frame neurons   344
frames of reference   15

gamma activity   426
gamma range   434
geometric projections   7
geometry networks   339
gestalt   111, 134, 237, 281, 441, 443, 449
gestalt factors   186
gestalt organization   10
gestalt psychologist   64, 91, 108, 135, 249, 303, 400
gestalt psychology   91, 236, 295, 469
gestalt qualities   14, 15
gestalt theory   28, 70, 102, 185, 187, 243, 291
global preference   86, 88
goal-directed behavior   423, 435
goal-oriented behavior   425
good continuation   51, 114
grandmother cells   25

habituation   271, 429
higher order invariant   86
higher-order cognitive processes   64
higher-order information   87, 89
holistic perception   443
homogeneity   223
human gait   144, 150
human motion   141
hypercube   115
hypnosis   82
hysteresis   11, 28, 71, 80, 159, 160, 162, 165, 167, 170, 182, 185, 196, 199, 200, 271, 275, 277, 281, 291–293, 295, 302, 304

illusory contours   124, 416
illusory triangle   417
imagination   80, 390
imaging system   231
individual differences   80
infinite regression   421
inhomogeneity   223
instability   11, 79, 80, 82, 140, 160, 171, 182, 187, 191, 194, 295, 344, 463, 469, 476, 477
intentional movement   401
intentionality   18, 426
intentions   60

interactive realism   106, 109, 110
interpersonal differences   72
intrinsic dynamics   279, 287, 288
isomorphistic relation   14

language   224, 225, 233, 234, 242, 243, 249, 252, 433
language perception   9
lapse of meaning   13
learning   42, 69, 188, 301, 426, 430
learning by insight   14
learning curve   16
limit cycle   203, 209, 457, 459
limit cycle attractor   389
limit cycle behaviour   455
linguistic problem solving   241
local precedence   86, 88

meaning   1, 5–7, 9–17, 19, 21, 23, 29, 42, 60, 69, 72, 85, 86, 109, 163, 164, 224–227, 243, 245, 246, 257, 432
meaning creation   250
meaning inconsistency   244
mental imagination   223
mental rotations   224
metastability   173
methodological aspect of multistability   49
methodological window   71
mind-brain-relation   18
minimum principle   111
modularity of mind   93
movement patterns   41
multistability of meaning   9
multistability of symmetry axes   7
music   9, 14

naive realism   107
natural neuronal networks   2
network ecology theory (NET)   85, 86, 92
neural network   1, 5, 12–14, 16, 17, 42, 69, 70, 81, 85, 296, 328, 337–339
neuronic equation   338
neurons   342, 343, 346–349, 352, 422–425, 428, 432, 433, 464
nominalism   107
non-equilibrium phase transitions   28, 141, 185, 186, 389, 400
non-linear phase transitions   69, 71, 187
non-summativity   15

nonlinear dynamics 140, 159, 182, 276, 302
nonlinear neural dynamics 182
nonlinear oscillators 177

oddball paradigm 361, 364, 380
omitted stimulus paradigm 361, 364, 368, 371
optical bistability 6
order 223
order parameter 15, 16, 24, 26, 27, 30, 31, 33, 39, 42, 72, 76, 140, 141, 154, 186, 187, 257, 259, 315, 317, 463, 468
orientation multistability 349
orienting response 412
originality in thinking 80
oscillation 256, 258, 259, 299, 319, 328, 330, 334, 361, 362, 424, 426, 428, 452, 456

P300 358, 359, 374, 376–378, 389, 395, 403
paradigm shifts 6
parallel analysis 63
parallel distributed networks 2
Parkinson's disease 81
past experience 60
pattern formation 276
pattern recognition 16, 17, 32, 33, 41, 42, 76, 141, 154, 255, 257, 313, 335, 341
perceptual organization 47, 275–282, 286, 292, 295, 302, 303, 305, 306
perceptual switching 389, 391, 394, 397, 400–403
perceptual switching related positivity 389, 395, 396, 403
perceptual switching related potentials 398, 399, 402
phase relation 177
phase transition 140, 200, 295, 314, 323, 334
phenomenal causality 234, 235
phenomenal world 108
phenomenological approach 24, 204
phenomenological synergetics 71
phonetic symbolism 14
physical world 108–110, 135
point attractor 168, 306
point repeller 168
postsynaptic potential 422

potential landscape 35, 36, 42, 43, 71, 73, 74, 81, 105, 106, 167, 168, 186, 214, 257, 315, 468
Prägnanz 102, 111, 185, 186, 198
Prägnanzbereich 189
Prägnanzstufen 187
prior experience 64
probability of first occurrence 271
problem solving 249, 251
proximity 51
psychoneural identism 2
psychophysical experiments 255, 256, 260
psychosomatics 82
pure cognitive paradigm 370

quantum mechanical system 443
quantum mechanics 465, 469, 470, 472, 473, 476, 477

radical constructivism 7, 12, 109, 110
random dot kinematogram 139, 140, 150, 151, 154, 160
random fluctuations 27
rate of apparent change (RAC) 269–271
reaction lattice 445
readiness potential 397, 401, 402
realism 106–108
reality 7, 15, 99, 109, 134, 135, 432, 442
recognition network 345
recognition of the orientation 347
relative phase 140, 141, 288, 291
relative timing 282–284, 286
repelling 215, 307
repellor 287
representational realism 108
residence time 297, 302
response-locked positivity 395
restructuring of the figure 189
return map 297
reversible figures 6, 7, 48
reversion time 37, 255, 256, 259, 261–263, 268, 272
rhetoric 244, 247, 252
rhetorical figures 244

satiation 113, 296, 301, 390
saturation 258, 328, 329
saturation of attention 296
schizophrenia 80–82
selective adaptation 277, 303
selective attention 359, 379

selective visual attention   373
self-evaluation   2
self-organization   1, 5, 23, 69, 71, 73, 79, 108, 140, 182, 185–188, 190, 194, 390, 409, 411, 431, 433, 435, 468
self-satiation   11, 12
semantic scale   2, 13, 18, 233, 235, 477
semiotic relation theory   14
sensory suggestibility   80
serial reproduction   11
shape recognition   63
signal detection   380
simplicity   114
simulation   262, 264, 270, 271, 288, 289, 295, 297, 298, 300, 318
singularity   102, 103, 111, 125, 136
size constancy   118
slaving principle   26, 31, 33, 39, 313
slaving-process   15
solipsism   106
space perception   136
spatial coherence   427
spatial frequency   89, 136
spatial rotation   223
spontaneous reversals   9, 297, 304
spontaneous switching   301, 330
SQUID   172, 173, 182
stability   11, 102, 136, 176, 186, 275, 276, 280, 295, 301, 304, 344
stages of multistable perception   10
stimulus frequency   270
stimulus generalization   188, 191
stimulus-response theories   5
stochastic forces   307
strange attractor   456
stream of consciousness   415, 418
stroboscopic alternative motion (SAM)   8, 47, 50, 72–75, 78, 80, 201, 255, 269, 272, 282, 296, 389, 391, 392
stroboscopic motion   144
suggestibility   80
suggestive influences   80
switching between attractors   10
switching point   293
switching probability   299
switching process   174
switching time   175, 275, 298, 300, 301
symmetric   286, 299, 341

symmetry   57, 64, 111, 113, 114, 116, 131, 136, 175, 283, 305, 307, 319, 464–467, 469, 471
symmetry breaking   28, 80, 113, 117, 186, 187, 191, 194, 463, 464, 466, 468, 477
synapses   338–340, 351, 377, 425, 429
synaptic modification   296
synchronism   290
synchronization   391
synchrony   286
synergetic computer   149, 255, 312–316, 328
synergetics   1, 6, 23–28, 32, 33, 39, 42, 71, 73, 76, 105, 139, 140, 159, 169, 171, 174, 182, 185, 187, 191, 200, 204, 256, 257, 272, 407, 413, 417, 463, 464, 468, 477
syntactic ambiguity   231–233, 236, 239

temporal stability   295, 302
tendency towards stability   103, 125
textual ambiguity   236, 237
textures   223
theory of colour   134
thermodynamical model   456, 457
thinking   390
time scale   303
top-down approach   17, 49, 59, 123, 124, 133, 313, 315
top-down influence   15, 16, 47, 66, 76, 277
transparency   135
transphenomenal world   108, 135
transposibility   15

unstable equilibrium   51

vector-field   285, 287
verbal suggestions   75
virtual contours   101
visual imagery   415
visual thinking   463, 464, 467, 468, 474, 478

weak causal effects   15

Zürich model of dynamical distance regulation   209

Springer Series in Synergetics
Editor: Hermann Haken

Synergetics, an interdisciplinary field of research, is concerned with the cooperation of individual parts of a system that produces macroscopic spatial, temporal or functional structures. It deals with deterministic as well as stochastic processes.

1 **Synergetics** An Introduction
   3rd Edition   By H. Haken
2 **Synergetics** A Workshop
   Editor: H. Haken
3 **Synergetics** Far from Equilibrium
   Editors: A. Pacault, C. Vidal
4 **Structural Stability in Physics**
   Editors: W. Güttinger, H. Eikemeier
5 **Pattern Formation by Dynamic Systems and Pattern Recognition**
   Editor: H. Haken
6 **Dynamics of Synergetic Systems**
   Editor: H. Haken
7 **Problems of Biological Physics**
   By L. A. Blumenfeld
8 **Stochastic Nonlinear Systems**
   in Physics, Chemistry, and Biology
   Editors: L. Arnold, R. Lefever
9 **Numerical Methods in the Study of Critical Phenomena**
   Editors: J. Della Dora, J. Demongeot, B. Lacolle
10 **The Kinetic Theory of Electromagnetic Processes**   By Yu. L. Klimontovich
11 **Chaos and Order in Nature**
   Editor: H. Haken
12 **Nonlinear Phenomena in Chemical Dynamics**   Editors: C. Vidal, A. Pacault
13 **Handbook of Stochastic Methods**
   for Physics, Chemistry, and the Natural Sciences   2nd Edition
   By C. W. Gardiner
14 **Concepts and Models of a Quantitative Sociology** The Dynamics of Interacting Populations   By W. Weidlich, G. Haag
15 **Noise-Induced Transitions** Theory and Applications in Physics, Chemistry, and Biology   By W. Horsthemke, R. Lefever
16 **Physics of Bioenergetic Processes**
   By L. A. Blumenfeld
17 **Evolution of Order and Chaos**
   in Physics, Chemistry, and Biology
   Editor: H. Haken
18 **The Fokker-Planck Equation**
   2nd Edition   By H. Risken
19 **Chemical Oscillations, Waves, and Turbulence**   By Y. Kuramoto
20 **Advanced Synergetics**
   2nd Edition   By H. Haken
21 **Stochastic Phenomena and Chaotic Behaviour in Complex Systems**
   Editor: P. Schuster
22 **Synergetics – From Microscopic to Macroscopic Order**   Editor: E. Frehland
23 **Synergetics of the Brain**
   Editors: E. Başar, H. Flohr, H. Haken, A. J. Mandell
24 **Chaos and Statistical Methods**
   Editor: Y. Kuramoto
25 **Dynamics of Hierarchical Systems**
   An Evolutionary Approach
   By J. S. Nicolis
26 **Self-Organization and Management of Social Systems**   Editors: H. Ulrich, G. J. B. Probst
27 **Non-Equilibrium Dynamics in Chemical Systems**
   Editors: C. Vidal, A. Pacault
28 **Self-Organization** Autowaves and Structures Far from Equilibrium
   Editor: V. I. Krinsky
29 **Temporal Order**   Editors: L. Rensing, N. I. Jaeger
30 **Dynamical Problems in Soliton Systems**
   Editor: S. Takeno
31 **Complex Systems – Operational Approaches** in Neurobiology, Physics, and Computers   Editor: H. Haken
32 **Dimensions and Entropies in Chaotic Systems** Quantification of Complex Behavior   2nd Corr. Printing
   Editor: G. Mayer-Kress
33 **Selforganization by Nonlinear Irreversible Processes**
   Editors: W. Ebeling, H. Ulbricht
34 **Instabilities and Chaos in Quantum Optics**   Editors: F.T. Arecchi, R.G. Harrison
35 **Nonequilibrium Phase Transitions in Semiconductors** Self-Organization Induced by Generation and Recombination Processes   By E. Schöll

Springer-Verlag and the Environment

We at Springer-Verlag firmly believe that an international science publisher has a special obligation to the environment, and our corporate policies consistently reflect this conviction.

We also expect our business partners – paper mills, printers, packaging manufacturers, etc. – to commit themselves to using environmentally friendly materials and production processes.

The paper in this book is made from low- or no-chlorine pulp and is acid free, in conformance with international standards for paper permanency.

Printed in the USA
CPSIA information can be obtained
at www.ICGtesting.com
LVHW011131150324
774517LV00040B/1578